Introduction
to Analysis

A Series of Books in Mathematics

Editors: R. A. Rosenbaum, G. Philip Johnson

Introduction to Analysis

Bernard Kripke

University of California, Berkeley

W. H. Freeman and Company San Francisco and London

Preface

This book is the outgrowth of a course I taught at the University of California, Berkeley, during the Spring semester of 1966. It was the fourth semester of an honors sequence in calculus, at which point the course undertook a transition from calculus to modern abstract analysis. It is to that transition that this book is addressed.

I am one of those teachers who is rarely satisfied with any other teacher's way of doing things. It has been my custom to write Dittoed lecture notes according to my taste, which I pass out to my students. In this case, the notes were so well received that I was encouraged to offer them to a wider audience. To make them accessible to students in other schools, I have expanded the introductory section to include material on sets, counting, and functions, which my students had studied the previous semester. This section contains a few definitions and conventions of notation that should be looked over even by someone who is already familiar with its contents. I have also included material on integration and complex functions, which was part of my original plan for the course but which I did not have time to complete. I had intended to put a section on Fourier series in Chapter IX, but changed my mind when I realized that Chapter IX threatened to overwhelm the rest of the text. However, the Bibliographical remarks suggest several references on Fourier series. In addition, among the problems in the Appendix for Chapter IX, Section B, are several that work out the rudiments of the theory of Fourier series.

As it stands, there is too much here to cover in one semester. It would fit into two quarters or, at a pace slower than the forced march I maintained that Spring, it could fill two semesters. My advice to a teacher who planned to use the text would be to *omit the material on determinants and eigenvalues* and some of the applications, as I did. There is nothing to prevent a student from reading the rest on his own.

My exposition is frankly opinionated. Were it not, I could hardly see any justification for adding one more to the heap of textbooks on calculus that are available. I have tried to avoid writing an encyclopedia but to include a body of related matter attractive enough to sustain the interest of a potential mathematician by its substance, without the support of a body of artificial "applications to real life." On the other hand, I have tried to be honest and explicit about the reasons why *mathematicians* study these things, as I see it.

A few remarks about this and that: (1) *Exercises.* Most of the exercises are hard, but hints are given. Most exercises are intended to clear up potential misunderstandings, to offer counterexamples, to present important new ideas, and otherwise to enlarge the reader's appreciation of the main body of material. There are a few computational exercises in cases where the theorems present important algorithms. All exercises are solved, either in the Answers at the back of the book or in the Teachers Manual. Those starred exercises whose solutions are contained in the Answers are (a) especially difficult exercises, with the exception of a few which anticipate material to be presented later in the text, (b) especially important exercises whose answers are referred to in the text proper, or which present theorems omitted, for one reason or another, from the body of the text, and (c) all computational exercises. The Appendix contains additional routine problems. They are included for the benefit of teachers who do more drill work, but it is my feeling that the exercises in the body of the text are sufficient to ensure that a student commands the material. Answers (not solutions) to these problems are in the Teachers Manual.

(2) ϵ *and* δ. The use of ϵ and δ in this book needs no apology—there is no other way to do the work. For example, it is clear that the proof of the existence of continuous, nowhere differentiable functions in Chapter IV actually needs epsilonics; ϵ and δ are not introduced merely for the sake of pedantry. Mathematicians have known all along why they use ϵ and δ. It is precisely when they first prove theorems in "advanced calculus" that students are admitted to the secret.

(3) *Illustrations.* The main content of this book is analysis. In many cases, there is no better way to understand an argument than to master the analytical proof. In others, pictures aid understanding. Mathematicians use pictures to discover some proofs. I have not tried to hide this fact from students.

(4) *References to the text.* The full name of a Theorem is something like "Theorem IV C 1," meaning Theorem 1, Section C, Chapter IV. In Chapter IV, this is shortened to "Theorem C 1," and in Section IV C it is shortened to "Theorem 1."

(5) The symbol ● is a *stop sign*. See the end of Section I C.

I have profited immensely from the assistance of Mr. Eric Anderson, who, as a freshman, was one of my students in the Spring of 1966. He has read the manuscript text (except Chapter IX) with great care, has made numerous helpful suggestions and comments, and has generally assisted me in transforming a rough set of notes, written in haste, into something more like a book. My gratitude is also due to the other students in that class for their perceptive criticism and for turning me on thrice weekly by their enthusiasm for mathematics.

I am also glad to express my appreciation to Mrs. Lynn Schell for her fine job of typing.

BERNARD KRIPKE

January 15, 1968

Contents

Introduction
to Analysis

CHAPTER I

Introduction

This book is devoted to a reasonably careful re-examination, criticism, and extension of the guts of calculus—the notion of limit. By "careful," I mean that a serious attempt will be made to state hypotheses and prove theorems, with a minimal use of the phrases "it can be shown that" or "it is shown in more advanced texts that." By "reasonably," I mean that I shall not use all the care of which contemporary mathematicians are capable, and in particular that a detailed criticism of the arguments to be presented would point out that certain questions about the foundations, having to do with sets, non-constructive arguments in favor of existence, the degree to which a given collection of axioms characterizes the real number system, and so on, are worthy of more investigation than I shall undertake. For the sake of clarity I shall shortly elaborate these ideas in a discussion of my private conception of mathematical rigor.

The large plan of the book is to examine a familiar notion (for example, convergent sequences of real numbers), study it in depth, then single out from this study certain key ideas, examine them in their own right, and finally apply them to new problems. Thus from a study of how to solve nonlinear equations involving a single real variable, we shall be led to the notion of a contraction of a metric space. Contractions can then be used to prove the existence and uniqueness of solutions to ordinary differential equations. This pattern will be repeated several times, on ever-higher levels of abstraction.

A. Rigor

Rigor means stiffness or rigidity, as in *rigor mortis*. In mathematics it is opposed to carelessness, sloppiness, fuzzy thinking, handwaving, and frequently to understanding. There is a hoary school of mathematical exposition that holds that one test of rigor is the absence of pictures. It arose in reaction to the habit of making proofs by the argument, "the picture makes it look that way." When the reaction reached the point of banishing even those illustrations that aid our understanding—even those that were responsible for the discovery of arithmetical proofs—it had gone too far. Others seem to feel that rigor is best achieved by exposition whose structural principle is theorem ———— proof ———— theorem ———— proof ————, and so on. My own opinion is that the nature of rigor changes from generation to generation, being always an outgrowth of criticism and suspicion.

The crucial fact is that mathematicians, from time to time, come to feel that the work of their forebears was careless, incomplete, ugly, or simply wrong. Gradually they learn to avoid certain errors, and to evolve practices, canons, and rituals designed to keep the dragon from the door. There is no reason to believe that this process has terminated in our era. On the contrary, we ought to assume that what we find rigorous today will seem woefully flabby by the time we are middle-aged. Or it may happen that the rigor of today will be the nit-picking of tomorrow.

What are mathematicians worried about? Let us list some bugaboos.

1.   MISTAKES. Most textbooks and most research papers contain misprints or mistaken calculations. I would make some deliberately in my lectures just to keep my students on their toes if experience had not proved that a conscious effort on my part to make mistakes is entirely unnecessary.

At a recent meeting of the American Mathematical Society, a paper announcing the solution to a famous problem was withdrawn at the last minute. Rumor had it that the authors had made a mistake in the evaluation of an elementary definite integral, which spoiled their proof. It had escaped the scrutiny of a number of their colleagues who had examined the manuscript, no doubt taking for granted that such elementary calculations would be carried out correctly.

2.   HANDWAVING. When one uses the words "it is clear that," he should have verified privately in advance that the assertion is indeed clear, and that it is clear *why* it is clear. There is in our textbooks and journals an immense store of examples of assertions (actually false) whose truth was claimed to be clear. We know from much experience that the words "it is clear" often mask a refusal to do hard work. For example, it was once thought to be clear that every continuous function is differentiable, except possibly at a few bad points. This is now known to be false.

There are also cases where true statements, claimed to be evident, turned out to be very difficult to prove. The Jordan Curve Theorem, which asserts that a simple closed curve divides the plane into exactly two disjoint connected parts, is an example. Jordan, who recognized that a proof was needed, published one that was false. Correct proofs only came later.

Finally, an assertion may be clear once the reader sees a trick, but the trick may be hard to find. I recall that when I was a graduate student, I was reading a book by one member of my faculty, Professor X, under the direction of Professor Y. I came to a point where X wrote, "It is easy to see that." I couldn't see it. After trying for a month, I asked Y how to do it. He didn't know, but referred me to the author, X. X had forgotten. However, as a result of these conversations, I had attained a new insight into the problem. I tried this new idea. With it in hand it *was* easy to see that the assertion was true. It just wasn't easy to see why it was easy to see.

3. UNSTATED ASSUMPTIONS. Much of nineteenth-century mathematical criticism was devoted to making explicit the unstated assumptions in earlier work. For example, it was discovered that Euclid's *Elements* made use of unstated assumptions about order and completeness, and that a principle in set theory called the "Axiom of Choice" had been used without formal recognition for some time.

4. PHILOSOPHICAL DIFFICULTIES. We assume ordinarily that it always makes sense to state of a real number x that $x = 0$ or $x \neq 0$. However, we often cannot actually determine which is the case. In numerical analysis, for example, the inevitable presence of round-off errors usually makes it impossible to do more than assert that a number is *close to zero*, or is zero *within rounding error*. This fact vitiates the usefulness of certain methods for (exemplia gratia) finding roots of polynomials, since these methods depend on the determination whether or not a certain real number is *exactly zero*. Here is a more subtle example. Fermat made the claim, unproved during the years since 1637 when it was proposed, that there are no quadruples $\langle a, b, c, n \rangle$ of positive integers such that $n > 2$ and $a^n + b^n = c^n$. Suppose we enumerate all the quadruples of positive integers—$q_1, q_2, q_3, q_4, \ldots$. For example, first count off the quadruples $\langle a, b, c, n \rangle$ for which $a + b + c + n = 4$, then those for which $a + b + c + n = 5$, and so on. In this way, every quadruple is counted just once. The beginning of such an enumeration might look like that in Table 1.

Let us define a real number s as follows: s shall have the decimal expansion $0.s_1 s_2 s_3 s_4 \ldots$, where the mth digit s_m is 0 if q_m is *not* a counterexample to Fermat's assertion (that is, if $q_m = \langle a, b, c, n \rangle$, where $a^n + b^n \neq c^n$ or $n \leq 2$) and $s_m = 1$ if q_m is a counterexample ($a^n + b^n = c^n$ and $n > 2$). Thus $s = 0$ if, and only if, there is no counterexample to Fermat's conjecture— that is, if and only if his claim is true. Although we can calculate as many decimal digits of s as we please in a simple way, it is just as hard to tell whether $s = 0$ as to solve the Fermat problem. For example, it is easy to see that if we enumerate the quadruples of positive integers as above, the

Table 1

q_1	$\langle 1, 1, 1, 1 \rangle$	$(a + b + c + n = 4)$
q_2	$\langle 2, 1, 1, 1 \rangle$	$(a + b + c + n = 5)$
q_3	$\langle 1, 2, 1, 1 \rangle$	
q_4	$\langle 1, 1, 2, 1 \rangle$	
q_5	$\langle 1, 1, 1, 2 \rangle$	
q_6	$\langle 2, 2, 1, 1 \rangle$	$(a + b + c + n = 6)$
q_7	$\langle 2, 1, 2, 1 \rangle$	
q_8	$\langle 2, 1, 1, 2 \rangle$	
q_9	$\langle 1, 2, 2, 1 \rangle$	
q_{10}	$\langle 1, 2, 1, 2 \rangle$	
q_{11}	$\langle 1, 1, 2, 2 \rangle$	
q_{12}	$\langle 3, 1, 1, 1 \rangle$	
q_{13}	$\langle 1, 3, 1, 1 \rangle$	
q_{14}	$\langle 1, 1, 3, 1 \rangle$	
q_{15}	$\langle 1, 1, 1, 3 \rangle$	
q_{16}	$\langle 2, 2, 2, 1 \rangle$	$(a + b + c + n = 7)$
q_{17}	$\langle 2, 2, 1, 2 \rangle$	
q_{18}	$\langle 2, 1, 2, 2 \rangle$	
q_{19}	$\langle 1, 2, 2, 2 \rangle$	
q_{20}	$\langle 3, 2, 1, 1 \rangle$	

first twenty digits of s are all 0. Indeed, all but the fifteenth of the first twenty quadruples fail to be counterexamples to Fermat's conjecture because n is not greater than two, and the fifteenth is not a counterexample because $1^3 + 1^3 \neq 1^3$.

Some mathematicians have argued on philosophical grounds that it doesn't make sense to affirm that either $x = 0$ or $x \neq 0$ when we have no hope of deciding which is the case. They argue similarly that it is meaningless to assert the existence of a number that cannot be calculated. For example, we could get such a number by defining the number x to be 0 or 1 according to the truth or falsity of Fermat's conjecture. If the reader thinks Fermat's problem is too easy, let him replace it by some other hard problem.

There are many degrees of mathematical rigor, and our standards continually change. It is futile to try to be absolutely or perfectly rigorous. In this book, I shall attempt to maintain a standard comparable to that of contemporary research journals, and substantially higher than that of most introductory texts on calculus, but I make no claim that it is impossible to be more careful. The level of rigor at which I choose to operate will be determined by my taste, my desire to be brief, and my estimate of what college students will find congenial.

B. Remarks on sets and counting

In mathematics, as in everyday life, we frequently have to talk about sets, classes, families, collections, groups, bunches, bevies, herds, and assemblages of things. One of the great accomplishments of nineteenth century mathematics was Georg Cantor's development of the theory of sets, which is a body of learning about sets, classes, collections, and so forth. Most mathematicians use set theory just because it provides them with a convenient and general language in which to formulate their work. Sometimes, however, they use some of the serious theorems of set theory, almost all of which are controversial in one way or another.

Set *language* is so useful that it has gained universal acceptance. Set *theory* is interesting in large part because in its early form it was found to contain contradictions that were difficult to cure, and some of the proposed cures were so radical that they threatened to kill the patient. Moreover, philosophers seized upon set theory as a means of answering their old questions about why mathematical truths seem so much truer than other kinds of truth by showing that mathematics is a part of philosophy. Naturally, the discovery of contradictions in set theory caused quite a stir among these philosophers, and has kept them busy for many years. Nowadays set theory has so thoroughly penetrated mathematical thinking that it is becoming fashionable to teach it to school children long before they have any use for set language or any understanding of the technical and philosophical questions raised by set theory.

For our purposes, the most elementary terminology and formulas will be sufficient, together with a few facts about counting. More abstract treatments of algebra and analysis frequently employ the *Axiom of Choice* or *Zorn's Lemma*, which are still somewhat controversial, or the even more controversial *Continuum Hypothesis*. We shall have to use only the *Countable Axiom of Choice*, which seems so plausible as to pass almost universally as an unstated assumption.

We shall call a collection of things a **set,** a **class,** or a **family,** according to considerations of euphony. For our purposes, all three words mean the same thing. If S is a set, the things in it are called its **elements** or **members,** and the relation **x is in S** is written symbolically $x \in S$. $x \notin S$ stands for "x is *not* in S." Two sets are considered to be the same if they have the same members, even if they are described in quite different ways. For example, let S be the set consisting of those integers between 1 and 10 that are equal to the sum of their proper divisors (such as $6 = 1 + 2 + 3$), and let T be the set of those even integers that are greater than 1, less than 10, and not a power of 2. Then S and T each has a single member, the integer 6. Although the descriptions of S and T are different, we consider S and T to be the same set.

By convention, we allow the existence of an empty set, $\varnothing$, which contains nothing at all. It can also be described as the collection of all odd integers that are multiples of 2, the collection of all negative squares of real numbers, and the collection of all thirteen-legged pink Russian wolfhounds. A few crotchety mathematicians are reluctant to call the empty set a set, but they run into difficulty when they try, for example, to talk about the set of counterexamples to Fermat's conjecture. No one knows whether it is empty or not. Therefore, if one were reluctant to call the empty set a set, he wouldn't know whether the collection of counterexamples to Fermat's conjecture is a set or not. Even so, it is easy enough to tell whether or not any given quadruple of positive integers is in it.

If S is a set and T is a set, each of whose members is in S, then we call T a **subset** of S. Conventional notations for this relationship are $T \subseteq S$ and $S \supseteq T$. Notice that if $T \subseteq S$ and $S \subseteq T$, then S and T contain the same things, so they are equal.

If S is a collection of sets (a set of sets), then $\bigcup S$, the **union** of S, is the set of all things that are in at least one of the sets in S. For example, let S be the collection of all sets that contain exactly two integers and nothing else. Then $\bigcup S$ is the collection of all integers, since each integer n is contained in a set of two integers—for example, $\{n, n + 1\}$.

Likewise, if S is a collection of sets, $\bigcap S$, the **intersection** of S is the set of all things that are in all of the sets in S. If S is, as above, the collection of all sets of two integers, then $\bigcap S = \varnothing$.

Figure 1 is a schematic representation of union and intersection. In this figure, S is a set containing two disks in the plane, say the disks of radius 3 about the points $(-2, 0)$ and $(2, 0)$ in the x-y-plane. Call these disks D_1 and D_2, so that D_1, for example, consists of the points in the plane whose distance from $(-2, 0)$ is less than or equal to 3 (Figure 1,A). Then $\bigcup S$ consists of the points which are in at least one of D_1 and D_2 (the shaded area in Figure 1,C). $\bigcap S$ consists of the points that are in both D_1 and D_2 (the shaded area in Figure 1,E).

Figure 1,E shows the **difference** of the sets D_1 and D_2, the set $D_1 - D_2$ of points that are in D_1 but *not* D_2.

If S contains just two things, D_1 and D_2, it is convenient to name S by the symbol $\{D_1, D_2\}$, which shows explicitly the contents of S. Similarly $\{1, 2, 5, 7\}$ is the set containing just the integers 1, 2, 5, and 7, and $\{1, 2, 3, \ldots, 10\}$ is the set containing the first 10 positive integers. The three dots in this latter symbol indicate that the reader is supposed to figure out how the sequence 1, 2, 3, 4, 5, 6, 7, 8, 9, 10 proceeds by examining its first three terms, and then realize that the progression stops at 10. Rather than use the clumsy notation $\bigcup \{D_1, D_2\}$ to represent the collection of things that

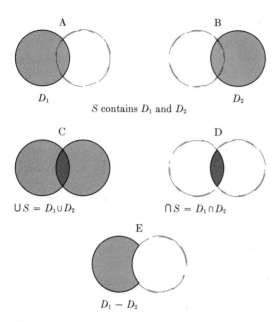

Figure 1

are in at least one of the sets in $\{D_1, D_2\}$, it is more convenient to write $D_1 \cup D_2$, as shown in Figure 1,C. Similarly, $D_1 \cap D_2$ is shorthand for the set $\cap \{D_1, D_2\}$ of things that are in both D_1 and D_2. $D_1 \cup D_2 \cup \cdots \cup D_6$ means the set of things which are in at least one of $D_1, D_2, D_3, D_4, D_5,$ and D_6. $\cap_{n=1}^{\infty} D_n$ means the set of things that are in every one of the sequence of sets $D_1, D_2, D_3, \ldots$.

A convenient way to name a set when we cannot simply list its elements is

$\{x \in S : x \text{ has the property } P\}$.

This symbol describes the subset of S consisting of things in S that have the property P. For example, if **Z** is the set of **integers** ("**Z**" stands for the German word *Zahl*, which means *number*),

$\{x \in \mathbf{Z} : 1 < x < 10, x \text{ is even, and } x \text{ is not a power of 2}\}$

is the set containing only the integer **6**. When it is clear what S is, this symbol may be abbreviated to

$\{x : x \text{ has the property } P\}$.

However, experience has shown that this latter convention can be dangerous when the over-set S is not clearly described.

Troubles seem to arise when S is too large. Here is a famous example. Suppose there were a set of all sets, A. Then we could define the Russell set R to be $\{x \in A : x \notin x\}$. Now it is impossible that $R \in R$, for then R would have to be an object having the characteristic property of things in R—namely, that $R \notin R$, which is a contradiction. On the other hand, it is impossible that $R \notin R$, for then R would be a set, having the characteristic property of things in R, so that R would have to be in $R: R \in R$. This is another contradiction. There seem to be two ways out of this impasse, both of which have been tried by various mathematicians and philosophers. One way is to change the rule that, for any set S and any well-defined property P, there is a subset of S consisting of just those things in S that have the property P. The other way is to deny the existence of a set of all sets, so that there is no possibility of defining the Russell set to be a subset of something that doesn't exist. I find this latter approach to be intuitively appealing. I like being able to define at will subsets of any set I have already managed to construct. I am willing to conclude that if there were a set of all sets, then we would be in trouble because of Russell's paradox. Therefore, there can be no such set. This is like the following argument. Suppose there were a male barber in Seville who shaved every man in town who did not shave himself, but no one else. Then this barber would not be allowed to shave himself, but if he didn't shave himself, he would have to. Therefore, there can be no such barber.

We might be willing to admit that the phrase "x is interesting" is too controversial to allow us to state with confidence that there is a subset of the set of integers consisting precisely of those that are interesting. The condition is too imprecise. We might interpret Russell's paradox as a demonstration that the sentence "x is a set" is too controversial to allow us to state with confidence that there is a set consisting of precisely those things that are sets. This does not mean that we can never confidently call something a set. Just as men generally will agree that 0, 1, and 2 are interesting integers, so mathematicians generally agree that there is a set of integers. People will not agree, however, that 1,253,428,791,233,502 is an interesting integer. In the same way, they may not agree that certain sentences define sets.

The purpose of these remarks is not to prepare the way for elaborate precautions in my use of set theory, but to be frank about the fact that difficulties do arise in set theory and that they deserve attention and reflection. Readers who would like to pursue these questions can find further discussion in *Naive Set Theory* by P. R. Halmos (Van Nostrand, 1960), in *Mathematical Logic* by W. van O. Quine (Harvard University Press, 1951), and in the first part of *Introduction to Metamathematics* by S. C. Kleene (Van Nostrand, 1952). The latter two books have extensive bibliographies. There has also been much recent interest in set theory stimulated by the discoveries of Paul Cohen, which have not yet been given a popular exposition so far as I know.

The set $\{a, b\}$ is equal to the set $\{b, a\}$. Suppose I want to give an order of precedence to the pair $\{a, b\}$ of things—for example, to make a the first of them and b the second. No order of precedence is indicated by the symbol $\{a, b\}$. I use the symbol $\langle a, b \rangle$ to indicate an **ordered pair**—that is, a set $\{a, b\}$

of two things with an order: a is first, b is second. To describe an ordered pair, $\langle a, b \rangle$, I need to do two things: give the set $\{a, b\}$ containing the elements of the pair, and the set $\{a\}$ containing the first element of the pair. I must say which two things are involved and which is first. The description of an ordered pair $\langle a, b \rangle$ involves the specification of two sets: one, $\{a, b\}$, containing the elements of the pair, and another, $\{a\}$, containing the first element. To specify an ordered pair $\langle a, b \rangle$, I must specify a pair of sets $\{\{a, b\}, \{a\}\}$. This ingenious technique for describing an ordered pair in terms of a pair of unordered sets is due to Norbert Wiener. Note that if $\{\{a, b\}, \{a\}\}$ is the set corresponding to an ordered pair $\langle a, b \rangle$, then

$$\{a\} = \{a, b\} \cap \{a\} = \cap \{\{a, b\}, \{a\}\},$$

and

$$\{a, b\} = \{a, b\} \cup \{a\} = \cup \{\{a, b\}, \{a\}\}.$$

If $\{a\} = \{a, b\}$, then $b = a$. If $\{a\} \neq \{a, b\}$, then $\{b\} = \{a, b\} - \{a\}$. That is, the first and second elements of the pair $\langle a, b \rangle$ can be recovered from the corresponding pair of sets $\{\{a, b\}, \{a\}\}$ by the operations of union, intersection, and difference. In technical set theory, it is common to define the ordered pair $\langle a, b \rangle$ to be the set $\{\{a, b\}, \{a\}\}$. For our purposes, it is more natural to think of ordered pairs as familiar objects, which can be specified by giving sets of the form $\{\{a, b\}, \{a\}\}$. In technical set theory, however, it is advantageous to have to deal with things of only one kind, sets, rather than having to deal with ordered pairs as a separate species.

The relation $y = \sin x$ has the special property that when x is known, y is determined. It is called a *functional relation*. Since such relations are characteristic of analytic geometry and applied mathematics, they have been subjected to intensive study for several centuries. Originally, "function" seemed to refer to any relation given by a simple formula, but with a large measure of confusion about formulas such as $x^2 + y^2 = 1$, where for each real number x, there may be 0, 1, or 2 real values of y that satisfy the equation. Sometimes it was said that this relation defines two functions, $y = \sqrt{1 - x^2}$ and $y = -\sqrt{1 - x^2}$, and sometimes $x^2 + y^2 = 1$ was not considered to be a functional relation at all.

Gradually the notion of function changed, in large part because of developments of the nineteenth century. Mathematicians discovered more and more complicated ways of constructing functions, so that the notion of a function as a relation defined by a simple formula became inadequate for their investigations. They undertook careful and sophisticated investigations of implicitly defined functions, analytic continuation, and other matters in which it was crucial to specify just which things were related by a functional relation. In the twentieth century, they began to study relations between functions in much the same way as they had earlier studied relations between numbers.

The result is that today the notion of function has taken on many meanings. In set theory and modern analysis and topology, functions are commonly characterized in terms of sets, with attention focused on the property that in a functional relation such as $y = \sin x$, when x is known, y is determined. In applied mathematics, the notion of a function as a relation defined by a fairly simple formula is still popular. In the study of algebraic function fields, the focus is on the formula that defines a function, rather than on the fact that a function relates things.

In this book, the functions to be considered will not generally be defined by simple formulas. Because we shall consider topological questions and implicitly defined functions, it will be essential to specify just which things are related by a functional relation. The modern set-theoretic characterization of a function is most suited to our purposes.

We shall think of a function as an apparatus that pairs off elements of some set D with the elements of a set R. That is, *a function f from D to R* assigns to each element x of D an element $f(x)$ in R. To describe a function f it is enough to list the ordered pairs $\langle x, f(x) \rangle$, that is, to give a *set* of ordered pairs. Once given an x, we can look in this list for the ordered pair $\langle x, f(x) \rangle$ whose first element is x, and read off $f(x)$, the second member of the ordered pair. Describing a function by a set of ordered pairs is like giving a table of the function. For example, the set $\{\langle 1, 1 \rangle, \langle 2, 4 \rangle, \langle 3, 9 \rangle, \langle 4, 16 \rangle, \ldots\}$ corresponds to Table 2.

Table 2	
n	n^2
1	1
2	4
3	9
4	16
.	.
.	.
.	.

More generally, any relation can be described by a set of ordered pairs, the pairs of things that are so related. If R is a relation, it is described by

$$\{\langle x, y \rangle \colon x \text{ bears the relation } R \text{ to } y\}.$$

Conversely, suppose S is a set of ordered pairs. Then we can define a relation as follows: x bears the relation R to y if, and only if, $\langle x, y \rangle \in S$. The relation is that of being the first and second members of an ordered pair in S.

Not all relations are functional relations, and so not all sets of ordered pairs describe functions. The characteristic property of a functional relation $y = f(x)$ is that when x is known, y is uniquely determined. If this relation is described by a set S of ordered pairs, then the fact that when x is known, y is determined, corresponds to an assertion about S: *for each x, there is at most one y such that $\langle x, y \rangle \in S$.* Otherwise, y would not be uniquely determined by the condition that $\langle x, y \rangle$ be in S.

If S is a set of ordered pairs, the **domain** of S is

$\{x : x$ is the first member of a pair in $S\}$.

The **range** of S is

$\{x : x$ is the second member of a pair in $S\}$.

The fact that f is a function from D to R can be described as follows in terms of the set of ordered pairs $S = \{\langle x, f(x) \rangle\}$: *the domain of S is D and the range of S is contained in R.*

In technical set theory, it is convenient to define a function to be a set f of ordered pairs such that for each x, there is at most one y such that $\langle x, y \rangle \in f$. Then *the value $f(x)$ of the function f at the argument x in its domain is the unique y such that $\langle x, y \rangle \in f$.* As in the case of ordered pairs, this has the technical advantage of allowing set theory to deal with only the one species, sets, rather than two, sets and functions. Unlike the case of ordered pairs, however, we shall find it convenient to use this device of technical set theory in this book. We therefore make the following formal definition.

1. DEFINITION. In this book, a **function** is a set f of ordered pairs such that for each x, there is at most one y with the property that $\langle x, y \rangle \in f$. The **domain** of f is

$\{x : x$ is the first element of a pair in $f\}$.

The **range** of f is

$\{y : y$ is the second member of a pair in $f\}$.

If x is in the domain of f, **f(x)** is the unique y such that $\langle x, y \rangle \in f$. f is **a function from A to B** if A is the domain of f and B includes the range of f. ●

When we have occasion to describe a function by a formula such as $y = \cos x$, we have three natural courses of action. There may already be an accepted name for the function. In this case, we can call the function "cos." Another course of action is to make up a name for the function: "the function f such that $f(x) = \cos x$." This can be abbreviated by the notation "the function $f : x \mapsto \cos x$," or still more briefly "the function $x \mapsto \cos x$." Using a modification of this arrow notation, we can write "$f : A \to B$" or "$A \xrightarrow{f} B$" to mean "f is a function from A to B." Note the distinction: an arrow without a tail, $\rightarrow$, goes between the domain and range, an arrow with a tail, $\mapsto$, goes between elements of the domain and range. In all these cases it is essential to specify the domain of f, but it is permissible to allow this specification to be implicit in the context.

We also say that a function $f: A \rightarrow B$ **maps** A into B. ("Map" is used by analogy to the function that assigns to each place in Nevada the corresponding place on a map of Nevada.) f is said to map A **onto** B if the range of f is equal to all of B. f is **one-to-one** if for each $y \in B$, there is *at most one* $x \in A$ such that $f(x) = y$. If $f: A \rightarrow B$ is both one-to-one and onto, then there is a function $f^{-1}: B \rightarrow A$ defined by $f^{-1} = \{\langle y, x \rangle : \langle x, y \rangle \in f\}$. That is, for each $y \in B$, $f^{-1}(y)$ is the unique $x \in A$ such that $f(x) = y$. The fact that f is one-to-one guarantees that f^{-1} is a function; the fact that f is onto guarantees that the domain of f^{-1} is B.

If $f: A \rightarrow B$ and $S \subseteq A$, then we define the **image f(S) of S by f** to be $\{f(x) : x \in S\}$. In particular, $f(A)$ is the range of f. If f is one-to-one and onto, and $T \subseteq B$, then

$$f^{-1}(T) = \{f^{-1}(y) : y \in B\} = \{x \in A : f(x) \in T\}.$$

This last expression makes sense even if f is neither one-to-one nor onto. For any function $f: A \rightarrow B$, and any subset T of B, we define the **inverse image f^{-1}(T) of T by f** to be $\{x \in A : f(x) \in T\}$. If f is not onto, $f^{-1}(T)$ may be empty even when T is not. If f is not one-to-one, $f^{-1}(T)$ may consist of more than one point even when T consists of only one point.

2.   PROPOSITION. If $f: A \rightarrow B$ and S and T are subsets of B, then $f^{-1}(S \cup T) = f^{-1}(S) \cup f^{-1}(T), f^{-1}(S \cap T) = f^{-1}(S) \cap f^{-1}(T),$ and $f^{-1}(S - T) = f^{-1}(S) - f^{-1}(T)$. If U and V are subsets of A, $f(U \cup V) = f(U) \cup f(V)$.

Proof. This is an immediate consequence of the definitions.   ●

It is *not* generally true that $f(S \cap T) = f(S) \cap f(T)$, nor that $f(S - T) = f(S) - f(T)$. For example, let A and B be two nonempty sets whose intersection is empty. Let $f = \{\langle x, 1 \rangle : x \in A \cup B\}$. Then $f(A) = f(B) = f(A \cup B) = f(A) \cap f(B) = \{1\}$, and $f(A \cap B)$ is empty, since $A \cap B$ is empty. Likewise, $f(A) - f(B) = \{1\} - \{1\} = \varnothing$, but $f(A - B) = f(A) = \{1\}$.

If $A \xrightarrow{f} B \xrightarrow{g} C$ ($f: A \rightarrow B$ and $g: B \rightarrow C$), then we can define the composite function $g \circ f: A \rightarrow C$ by the formula $g \circ f: x \mapsto g(f(x))$. Then $(g \circ f)^{-1}(S) = f^{-1}(g^{-1}(S))$ for any $S \subseteq C$.

To say that a set A has n elements is to say that its elements can be counted: $1 \mapsto a_1, 2 \mapsto a_2, \ldots, n \mapsto a_n$, where $A = \{a_1, \ldots, a_n\}$. To put it differently, there is a function $f: i \mapsto a_i$ that maps $\{1, 2, 3, \ldots, n\}$ one-to-one onto A. We can say that A and $\{1, \ldots, n\}$ are equally numerous. It was Cantor's idea to use this method to compare the sizes of infinite sets as well.

3. DEFINITIONS. The sets A and B are **equally numerous** if, and only if, there is a function f mapping A one-to-one onto B. By definition, A is **at least as numerous** as each of its subsets, and A is **at least as numerous** as B if B is equally numerous with a subset of A. Let N be the set of positive integers, $N = \{1, 2, 3, 4, \ldots\}$. A set A is **countable** if N is at least as numerous as A. A is **countably infinite** if it is infinite and countable. ●

One of the surprising facts about these definitions is that a subset of A can be equally numerous with A. For example, $f: n \mapsto 2n$ is a one-to-one map from N onto the set of positive even integers, so the set of positive even integers is equally numerous with the set of integers. The interesting thing is that not all infinite sets are equally numerous.

The next theorem will never be used in what follows, but it is included with its proof because it helps to show that Definition 2 is reasonable.

4. SCHROEDER-BERNSTEIN THEOREM. *If A is at least as numerous as B and B is at least as numerous as A, then A and B are equally numerous.*

Proof. A moment's reflection will show that the theorem is *not* an obvious consequence of the definitions. The surprising thing is that it is so easy to prove. What we know is that there is a one-to-one function $f: A \to B$, which need not map A onto all of B, and that there is a one-to-one function $g: B \to A$, which need not map B onto all of A. We have to construct, using f and g, a one-to-one function h from A onto all of B.

Let us say that the element x of A is an **ancestor** of each of the elements $x, f(x), g(f(x)), f(g(f(x))), g(f(g(f(x)))), \ldots$, and likewise that an element y of B is an **ancestor** of each of the elements $y, g(y), f(g(y)), g(f(g(y))), \ldots$. Not every element in A or B need have an ancestor different from itself, since f and g need not be onto. Let us say that x is an **initial ancestor** of y if x is an ancestor of y, but x has no ancestor other than itself. Not every element of A or B need have an initial ancestor, but no element of A or B can have more than one, because f and g are one-to-one.

Now let

$A' = \{x \in A : x \text{ has no initial ancestor}\},$

$A'' = \{x \in A - A': \text{the initial ancestor of } x \text{ is in } A\},$

$A''' = \{x \in A - A': \text{the initial ancestor of } x \text{ is in } B\}.$

Note that every element of A is in exactly one of the sets A', A'', A'''. Moreover, if $x \in A'''$, then x must be in $g(B)$. Therefore we can define a function $h = A \to B$ as follows:

$$h = \{\langle x, y \rangle : \text{either } x \in A' \cup A'' \text{ and } y = f(x) \text{ or } x \in A''' \text{ and } y = g^{-1}(x)\}.$$

Let us show that h is one-to-one and onto.

Suppose that $h(x) = h(x')$. If x and x' are both in $A' \cup A''$, then $f(x) = h(x) = h(x') = f(x')$. Then $x = x'$ because f is one-to-one. Likewise, if x and x' are both in A''', then $x = g(g^{-1}(x)) = g(h(x)) = g(h(x')) = g(g^{-1}(x')) = x'$. In the remaining case, one of x and x', say x, would be in $A' \cup A''$ and the other one, x', would be in A'''. Then we should have $x' = g(g^{-1}(x')) = g(h(x')) = g(h(x)) = g(f(x))$. It would follow that x is an ancestor of x', so that x would also have to be in A'''. Since this contradicts the assumption that $x \in A' \cup A''$, this last case cannot occur. Thus h is one-to-one.

Now let $y \in B$. If $g(y) \in A' \cup A''$, then y cannot be an initial ancestor, so $y = f(x)$ for some x. But then the initial ancestor of x is likewise the initial ancestor of $g(y) = g(f(x))$, so $x \in A' \cup A''$. That is, $y = f(x) = h(x)$. Otherwise, $g(y) \in A'''$, so that $h(g(y)) = g^{-1}(g(y)) = y$. In either case, y is in the range of h. Thus h is onto.

5.   THEOREM. If A is at least as numerous as B and B is at least as numerous as C, then A is at least as numerous as C. In particular, any subset of a countable set is countable.

Proof. By assumption, there are a one-to-one function g from C onto a subset of B and a one-to-one function f from B onto a subset of A. Then $f \circ g$ is a one-to-one function from C onto a subset of A.

6.   THEOREM. The union of two countable sets is countable.

Proof. Let A and B be countable sets. By assumption, there are a one-to-one function f from A into $\mathbf{N}$ and a one-to-one function g from B into $\mathbf{N}$. Define a function h from $A \cup B$ into $\mathbf{N}$ as follows:

$$h = \{\langle x, y\rangle \colon \text{either } x \in A \text{ and } y = 2f(x) \text{ or } x \in B - A \text{ and } y = 2g(x) + 1\}.$$

Then h is a one-to-one function from $A \cup B$ onto a subset of $\mathbf{N}$. (h is one-to-one because no integer is both odd and even.)

7.   DEFINITION. The **product** $A \times B$ of two sets A and B is $\{\langle a, b\rangle \colon a \in A, b \in B\}$.

8.   THEOREM. The product of two countable sets is countable.

Proof. If A and B are countable sets, then A and B are equally numerous with subsets S and T of $\mathbf{N}$. Then it is easy to see that $A \times B$ is equally numerous with $S \times T$; that is, if $f\colon A \to S$ and $g\colon B \to T$ are one-to-one and onto, then $h\colon \langle a, b\rangle \mapsto \langle f(a), g(b)\rangle$ is a one-to-one function from $A \times B$ onto $S \times T$. Thus it is enough to prove that the product of two subsets of $\mathbf{N}$ is countable. If $S \subseteq \mathbf{N}$, $T \subseteq \mathbf{N}$, then $S \times T \subseteq \mathbf{N} \times \mathbf{N}$. According to Theorem 5 it is then enough to prove that $\mathbf{N} \times \mathbf{N}$ is countable. Define a function $f\colon \mathbf{N} \times \mathbf{N} \to \mathbf{N}$ by the formula $f(x, y) = 2^x(2y + 1)$. If we can show that f is one-to-one, it will follow that $\mathbf{N} \times \mathbf{N}$ is equally numerous with the range of f, which is a subset of **N**. This will show that **N** $\times$ **N** is countable. Suppose $2^x(2y + 1) = 2^z(2w + 1)$, where $x \geq z$. Then $2^{x-z}(2y + 1) = (2w + 1)$. Now $2w + 1$ is odd, so $2^{x-z}(2y + 1)$ must be odd as well. But this can happen only if $x - z = 0$. Thus $x = z$. Then $(2y + 1) = (2w + 1)$, so $y = w$ as well. This shows that f is one-to-one.

9.   COROLLARY. The set **Q** of all *rational numbers* is countable.

Proof. For each positive rational number r, choose positive integers $n(r)$ and $d(r)$ such that $f = n(r)/d(r)$. Then the function $r \mapsto \langle n(r), d(r) \rangle$ maps the set **Q**⁺ of positive rational numbers one-to-one onto a subset of **N** $\times$ **N**. Thus **Q**⁺ is countable. It is clear that **Q**⁺ is equally numerous with the set **Q**⁻ of negative rational numbers. According to Theorem 6, **Q**⁺ $\cup$ **Q**⁻ is countable, and so is **Q** $=$ (**Q**⁺ $\cup$ **Q**⁻) $\cup$ $\{0\}$.   ●

The preceding proof provides a good example of an unstated assumption: that we could choose for each r integers $n(r)$ and $d(r)$ such that $r = n(r)/d(r)$. There is nothing untoward about making this assumption for any particular r—it is simply the definition of "rational number." We assumed more than that, however. We assumed that we could make infinitely many such choices all at once. This is an application of the **Axiom of Choice.** In the present case, the use of the Axiom of Choice seems so natural that it probably would have passed unnoticed had I not called attention to it. In fact, in this case it could have been avoided by expressing each positive fraction in lowest terms. From the full strength of the Axiom of Choice, however, some quite surprising conclusions can be reached, such as Zermelo's Well-Ordering Theorem. For this reason the Axiom of Choice is somewhat controversial (but not very much). It is still not my intention to digress into technical questions of set theory. Curious readers are referred to the references already cited.

Here is one consequence of the Axiom of Choice.

10.   THEOREM. For the set A to be at least as numerous as the set B, it is necessary and sufficient that there be a function f from A onto B.

Proof. Suppose there is a function f from A onto B. For each $b \in B$, choose an element $g(b)$ in $f^{-1}(\{b\})$, so that $f(g(b)) = b$. Then g is a one-to-one function from B into A, for if $g(b) = g(c)$, then $b = f(g(b)) = f(g(c)) = c$. Therefore, B is equally numerous with $g(B)$, which is a subset of A.

Conversely, suppose that A is at least as numerous as B, so that there is a one-to-one function g from B onto a subset of A. Choose any element b_0 of B (if B is empty, the empty set of ordered pairs is a function from A onto B). Let $f = \{\langle a, b \rangle$: either $a \in g(B)$ and $b = g^{-1}(a)$ or $a \in A - g(B)$ and $b = b_0\}$. Then f is a function from A onto B.

11.   COROLLARY. The union of countably many countable sets is countable.

Proof. Let $\mathfrak{F}$ be a countable family of sets, each of which is countable. Then there must be a function f from **N** onto $\mathfrak{F}$. By assumption, for each $n \in$ **N**, there is a function g_n from **N** onto the set $f(n) \in \mathfrak{F}$. Let $h(n, m) = g_n(m)$. Then h is a function from **N** $\times$ **N** onto $\mathbf{U} \,\mathfrak{F}$, so **N** $\times$ **N** is at least as numerous as $\mathbf{U} \,\mathfrak{F}$.

12. DEFINITION. A is **less numerous** than B if A is *not* at least as numerous as B. In particular, if **N** is less numerous than B, then B is **uncountable.**

Let us exhibit some uncountable sets.

13. DEFINITION. If A and B are sets, A^B is the set of all functions from B to A.

14. THEOREM (CANTOR'S DIAGONAL ARGUMENT). If A contains at least two elements and B is nonempty, then B is less numerous than A^B.

Proof. It must be shown that B cannot be at least as numerous as A^B. According to Theorem 10, it must be shown that there can be no function f from B onto A^B, or, what is the same thing, that if f is any function from B into A^B, then $f(B)$ is not all of A^B. Suppose then that $f \colon B \to A^B$. For each $b \in B$, $f(b)$ is an element of A^B; that is, $f(b)$ is a *function* from B to A. We can thus define a function F from $B \times B$ to A as follows: $F(\langle b, c \rangle)$ is the value that the function $f(b)$ takes at c. In other words, $F(\langle b, c \rangle) = (f(b))(c)$. Since A contains at least two elements, $A - \{F(\langle b, b \rangle)\}$ is nonempty for each $b \in B$. Thus for each $b \in B$ we can choose an element $g(b) \in A - \{F(\langle b, b \rangle)\}$. Then $g \colon B \to A$, so that g is an element of A^B. I shall now show that g cannot be in $f(B)$. Suppose, on the contrary, that g were equal to $f(b)$ for some $b \in B$. Then, in particular, $g(b)$ would equal $(f(b))(b) = F(\langle b, b \rangle)$. But $g(b)$ was chosen from $A - \{F(\langle b, b \rangle)\}$, so that $g(b) \neq F(\langle b, b \rangle)$. Thus g cannot equal $f(b)$.

15. COROLLARY. The set of all subsets of **N** is uncountable.

Proof. Let A be the set containing the two expressions "in" and "not in." To each subset S of **N** corresponds an element of $A^{\mathbf{N}}$, to wit, the function

$$f_S = \{\langle n, a \rangle \colon \text{either } n \in S \text{ and } a = \text{"in" or } n \notin S \text{ and } a = \text{"not in"}\}.$$

Indeed, the function $S \mapsto f_S$ is a one-to-one map from the set of all subsets of **N** onto $A^{\mathbf{N}}$. But $A^{\mathbf{N}}$ is uncountable.

16. COROLLARY. The set of real numbers is uncountable.

Proof. We shall show more: that the set of real numbers that have decimal expansions of the form $0.d_1 d_2 d_3 d_4 \ldots$, where the kth digit d_k is either 3 or **7**, is uncountable. In fact, this subset of the set of real numbers is in one-to-one correspondence with $\{3, 7\}^N$: to each function $f: N \rightarrow \{3, 7\}$, we make correspond the real number $\sum_{n=1}^{\infty} (f(n)/10^n)$, whose decimal expansion has $f(n)$ as its nth digit. ●

Finally, a few rules of set algebra.

17. PROPOSITION. Let M and U be sets, and let A and B be sets of sets. Then

(a) $M - (M - U) = M \cap U$,

(b) *distributive law:* $(\bigcup A) \cap (\bigcup B) = \bigcup \{S \cap T : S \in A \text{ and } T \in B\}$,

(c) *distributive law:* $(\bigcap A) \cup (\bigcap B) = \bigcap \{S \cup T : S \in A \text{ and } T \in B\}$,

(d) *De Morgan's law:* $M - \bigcup A = \bigcap \{M - S : S \in A\}$,

(e) *De Morgan's law:* $M - \bigcap A = \bigcup \{M - S : S \in A\}$.

Proof. Use the symbol "$\Leftrightarrow$" to abbreviate "if, and only if."

(a) If $x \notin M$, then x is in neither $M - (M - U)$ nor in $M \cap U$. If $x \in M$, then $x \in M \cap U \Leftrightarrow x \in U \Leftrightarrow x \notin M - U \Leftrightarrow x \in M - (M - U)$.

(b) $x \in (\bigcup A) \cap (\bigcup B) \Leftrightarrow x \in \bigcup A$ and $x \in \bigcup B \Leftrightarrow$ there are an $S \in A$ and a $T \in B$ such that $x \in S$ and $x \in T \Leftrightarrow$ there are an $S \in A$ and a $T \in B$ such that $x \in S \cap T \Leftrightarrow x \in \bigcup \{S \cap T : S \in A \text{ and } T \in B\}$.

(c) The proof of (c) is another trivial argument like the proof of (b).

(d) If $x \in M$, then $x \in M - \bigcup A \Leftrightarrow x \notin \bigcup A \Leftrightarrow$ there is no $S \in A$ such that $x \in S \Leftrightarrow$ for every $S \in A$, $x \notin S \Leftrightarrow$ for every $S \in A$, $x \in M - S \Leftrightarrow x \in \bigcap \{M - S : S \in A\}$.

(e) The proof of (e) is another trivial argument like the proof of (d).

Exercises

1. Show that if A is *any* set, A is less numerous than the set of all its subsets. (The empty set has *one* subset, $\varnothing$.)

2. Show that if A contains n elements and B contains m elements (n and m are integers), then A^B contains n^m elements. This is the motivation for the notation "A^B."

3. Once again show that the set of all sets causes trouble by proving **Cantor's Paradox:** If A is the set of all sets and S is the set of all subsets of A, then A is both less numerous than S and at least as numerous as S. *Hint:* $S \subseteq A$. Conclusion: whatever A may be, it is not the set of all sets.

4. Fill in the details of the following argument. Let $A = \{1, 2\}$. Then N^N is no more numerous than $(A^N)^N$, which is equally numerous with $A^{N \times N}$, which is equally numerous with A^N, which is no more numerous than N^N. Thus A^N and N^N are equally numerous.

5. Let A be the set of all real **algebraic** numbers—that is, the set of all real numbers x that satisfy an equation of the form $x^n + r_{n-1}x^{n-1} + \cdots + r_1 x + r_0 = 0$, where $r_0, r_1, \ldots, r_{n-1} \in Q$. Show that A is countable. *Hint:* Show that a polynomial $x^n + r_{n-1}x^{n-1} + \cdots + r_0$ has at most n real roots, and that there are at most countably many such polynomials.

6. Prove that there are uncountably many **transcendental** real numbers—that is, real numbers that are not algebraic.

C.   Remarks on proofs

My experience has shown that students of calculus frequently cannot distinguish a proof from an argument that merely shows that some assertion is plausible, do not understand what questions are at stake in a proof, and are uncertain which details to put in a proof and which to leave out. Traditionally students are expected to absorb this knowledge from the air of classrooms. I hope these few remarks will aid the process.

Originally a proof seems to have been simply an argument intended to convince, but now proofs have many other uses. I hope that my readers are convinced that there are countably many rational numbers but uncountably many reals, although these facts probably were not obvious, and may not even have seemed plausible. An argument makes an assertion **plausible** if it makes one *want* to believe the assertion. An argument is **convincing** if it makes one *believe*.

1.   DEFINITION. A student of mathematics who believes whatever he wants to believe is **gullible.** A student who is skeptical of what he wants to believe is **wise.**   ●

For example, it might seem *plausible* that there are more rational numbers than integers, for if we think of the rational numbers spread out in the usual way along a line, we see that between each two integers there are infinitely many rationals. Although it may be plausible that the rational numbers are more numerous than the integers, it is false.

2.   DEFINITION. An argument which proves a plausible assertion to be false, or an implausible assertion to be true, is **surprising.**   ●

When one reads a proof, it is a good idea to ask, "What is the author worried about? Why does he bother to address himself to this question? What details have to be verified to make this proof convincing?" If the answer to the last question uncovers details which are not mentioned explicitly in the proof, there are two likely explanations. The details may be so trivial that the author did not bother to mention them, or so easy to verify that they were deliberately left to the reader. On the other hand, the author may not have noticed that these details need attention. He may not know how to prove them, or they may be false.

For example, in the proof of the Schroeder-Bernstein Theorem, I made use of the fact that if x is the initial ancestor of y and z is any ancestor of y, then x is also the initial ancestor of z, but I never mentioned this fact explicitly. Instead, I assumed that my readers could figure this out for themselves without difficulty. By omitting an argument addressed to this point, I saved words and hoped to avoid obscuring the main ideas of the proof in a mass of detail.

For a second example, consider the following proof.

3.   THEOREM. $x^{1/x} \geq e^{1/e}$ for all $x > 0$.

Proof. The minimum of $f(x) = x^{1/x}$ occurs when $f'(x) = 0$. But

$$f'(x) = \frac{d}{dx}\left[e^{1/x \log x}\right] = \left[-\frac{1}{x^2}\log x + \frac{1}{x^2}\right]e^{1/x \log x}.$$

Setting $f'(x) = 0$, we find that $1/x^2 = (1/x^2)\log x$, or $\log x = 1$. Thus $x^{1/x}$ takes its minimum when $x = e^1 = e$. ●

Unfortunately, Theorem 3 is false. In fact, as $x \to \infty$, $x^{1/x} \to 1$, which is less than $e^{1/e}$. The difficulty arises from some details I neglected to check in my proof of Theorem 3. I neglected to show that $x^{1/x}$ does in fact attain its minimum at some $x > 0$, and I neglected to check that the critical point e is a relative minimum rather than a relative maximum. Alternatively, it could be said that my proof of Theorem 3 made use of two **unstated assumptions,** which in this case were **unwarranted.**

4.   DEFINITION. An expositor who makes use of unwarranted assumptions is **glib.**

5.   PRINCIPLE OF WISHFUL THINKING: If A is an assertion and X would be happy if A were true, X will want to believe A.

6.   PRINCIPLE OF REWARD AND PUNISHMENT: Mathematicians are rewarded for proving assertions and punished for failing to solve problems.

7.   THEOREM. A gullible student will believe any assertion he makes. A wise student will believe only true assertions.

Proof. According to the Principle of Reward and Punishment, a student will be made happy if he can prove one of his assertions. According to the Principle of Wishful Thinking, he will want to believe his assertions. By definition, a gullible student believes whatever he wants to believe. On the other hand, a wise student is skeptical of whatever he wants to believe. His skepticism will prevent him from being fooled by false assertions. ●

This last proof provides a good example of an unwarranted assumption: that a wise student always knows what he wants to believe. In fact, even wise students sometimes relax their skepticism when their will to believe is *unconscious*. That is why even wise students make mistakes.

8.   THEOREM. A glib author will fool a gullible student.

Proof. Gullible students want to believe assertions made by the authors of their texts. If the text is wrong, how can you be expected to learn anything?   ●

9.   THEOREM. Most authors are glib.

Proof. Being glib saves a lot of hard work in checking boring details. Moreover, authors of textbooks believe that they have a thorough understanding of their materials, good enough so that they know which things are true and which are false without having to do a lot of dreary busy-work.   ●

Of course it is glib to call the foregoing a proof. It really only makes assertion 9 plausible. To avoid being glib, it is a good idea to follow the following rule.

10.   FIRST LAW OF SCRUTINY. Never tell *yourself* that something is obvious. Before you omit a detail or say that something is obvious or trivial, check for yourself that it is both true and easy to verify. If you have trouble verifying it, write out the details of the proof.   ●

A proof leads us from the knowledge of one fact to the knowledge of another. It does not create facts out of thin air. Before we can prove anything, we must first accept some principles with which to start. According to the context, these first principles may be called "common knowledge," "what the Bible says," "prejudice," "the Constitution," "axioms," or "postulates." Likewise an attempt to define all our terms leads to circular definitions (assuming our vocabulary is finite). Accordingly, any mathematical argument must use some undefined terms or *primitives*.

It is considered good practice in mathematical writing to set down a list of the assumptions on which an argument is based. Such a list of axioms for a discussion about the real number system appears in Section II, A. The next section contains an argument intended to make the crucial *Axiom of Completeness* plausible. In the sense that it is intended to be convincing, the argument in the next section is a proof. However, our formal position will be that the Axiom of Completeness is one of the first principles that we do not attempt to prove. Officially, then, the argument in the next section is intended only to make it seem reasonable to you that we should accept the Axiom of Completeness as one of our first principles.

Careful scrutiny of the list of axioms in Section II, A reveals that the list is incomplete. It contains no list of axioms for set theory, although elementary set theory is used in the subsequent arguments about real numbers. There are other implicit assumptions that are not given formal recognition. This is typical of the way most mathematics is carried on. There is a social context of common knowledge for mathematicians. We usually attempt to set forth only those axioms that pertain to the subject immediately at hand. However, there have been attempts to axiomatize all the principles used in mathematical reasoning.

When the discovery of Cantor's Paradox and Russel's Paradox convinced mathematicians that something was wrong with their set theory, they feared that their methods of proof might be at fault or that they might be using unwarranted assumptions. One approach to this problem was Hilbert's program of Formalism. The Formalists proposed to mechanize the validation of proof, so that by simple checking, a mathematician or a computing machine could verify whether a mark on paper was an acceptable symbol in Formalist language, whether a series of marks on paper were a grammatically correct sentence, whether a sentence was an axiom, and whether a sequence of sentences were obtained by writing down axioms and applying acceptable rules of inference. To accomplish such a program, it was necessary to use language with more restricted symbols and grammatical rules than English— a special Formalist language had to be created. There could be no unstated assumptions. Everything taken for granted must meet the test of being an axiom. The acceptable methods of inference had to be restricted to a small number. Proofs had to be given in minutest detail with no steps omitted. Of course it was not required that the *discovery* of a proof be capable of being performed by a machine, but only that a machine could be programmed to verify proofs once they had been discovered.

The Formalists largely succeeded in carrying out this program, but with some surprising limitations. The famous Incompleteness Theorem of Gödel showed that any Formalist system powerful enough to include elementary arithmetic must either be inconsistent or **incomplete,** in the sense that there must be a grammatically correct assertion within the system that could neither be proved nor disproved.

The Formalists did not propose that mathematicians actually write out Formalist proofs in books and research journals. Formalist proofs are too long and tedious to be readable. They were interested, instead, in showing that their program could be carried out *in principle*. However, their criterion for proof has profoundly influenced our present standards. Nowadays an ideal proof should not merely be concise, literate, witty, and convincing. It should be possible to convert it into a Formalist proof without the exercise of any ingenuity, merely by filling in details in an obvious way and making obvious changes of notation. It is hard enough to achieve any one of these goals, not to say all at once, but we are expected to make an attempt. An ideal proof does *not* actually contain all the Formalist details—if it did it could not possibly be concise, witty, or literate, and it would probably be too enmired in trivia to be convincing. What one should strive to do, in accord with the First Law of Scrutiny, is to put in just enough detail that a reader can fill in as much more as he likes without having to be clever. Of course, the precise amount of detail required depends on the audience for which the proof is written. The proofs in this book, intended for undergraduates, would seem cluttered with trivia if they appeared in a research journal.

The activity of proving theorems has become ritualized, so that mathematicians commonly make their proofs in one of a small number of standard forms. Moreover, they prove assertions when there is no need to convince anyone of their truth. This ritualization of proving accounts for much of the activity of lecturers in calculus classes.

Often the proof of a theorem contains important information that is left out of its statement. This information may be omitted from the statement to make it more concise, witty, memorable, or surprising. It may also happen that the statement does give all the information that is needed for *most* applications. For example, the statements of Picard's and Peano's theorems on the existence of solutions to differential equations, which appear later in this book, omit the lengthy descriptions of the constructions of approximate solutions that are used in the proofs. However, a numerical analyst may use these constructions to calculate approximate numerical values for solutions. The proof of Picard's Theorem not only shows how to construct such numerical approximations, but also gives error bounds. Even if one was convinced by his knowledge of a physical situation described by a differential equation that the equation has a solution, he might still be interested in the method for calculating numerical values of the solution described in the proof of Picard's existence theorem.

A final word on the symbolism of proofs. If A and B are statements, "$A \Rightarrow B$" means "if A, then B." "$A \Leftrightarrow B$" means "A if, and only if, B" or "A is equivalent to B." Occasionally it will be easier to use these symbols than to write out "if . . . , then . . ." in words. We shall also continue to use the mark of punctuation, ●. This is a *stop sign*, which means that one discussion (a proof, definition, or the like) has come to an end, and that the next paragraph takes up a new matter. The stop sign will not be used when the stop is clear from the context. The *hand,* ☛, points out something of special importance.

D.   The real number system—introduction and motivation

The algebraic properties of the real number system have long been familiar to you. Their properties of order also should not be surprising. However, there is one characteristic property of completeness, the one that will be crucial to our subsequent studies, which may not seem evidently true. What follows is an attempt to make it seem desirable and *plausible*. It is *not* a proof of the completeness of the real number system. On the contrary, completeness will be one of our axioms, which will be used as the starting point for proofs.

Let us begin with some simple examples of limiting processes, most of which arise in geometry.

1.   The sequence .3, .33, .333, .3333, . . . increases, and its limit is $\frac{1}{3}$.

2.   Consider a square whose sides have length one unit. It seems plausible that the length of the diagonal should be represented by some number, d. What kind of a number can this be? Note that in Example 1 all the decimal fractions, .3, .33, . . . , as well as the limit are *rational* numbers. (A **rational** number is one that is the *ratio* of two integers.) According to Pythagoras' Theorem, the square of d must be 2. *No rational number has this property.*

Proof. If $\sqrt{2}$ were rational, it would be a ratio of two integers, $\sqrt{2} = p/q$. We may assume that this fraction is in lowest terms, and in particular that p and q are not both even integers. (In the contrary case, a common factor 2 could be canceled.) Then $p^2 = 2q^2$, so p^2 is even. Only an even integer can have an even square, so p must be even—$p = 2r$. It follows that $4r^2 = 2q^2$, or $2r^2 = q^2$; then, by repeating the preceding argument, q is even as well. This would contradict the hypothesis that p/q is in lowest terms. Thus there can be no rational number whose square is two.   ●

It seems to be precisely this observation that led to the invention of *irrational* numbers. Since we want every length to be represented by a number, we must introduce an irrational number whose square is 2.
   We are accustomed to representing numbers by decimal fractions such as $\frac{1}{8} = .125 = 1 \times 10^{-1} + 2 \times 10^{-2} + 5 \times 10^{-3}$. Even to represent as simple a fraction as $\frac{1}{3}$, we must use a nonterminating decimal, .3333 . . . . That is, no finite expression of the form $3 \times 10^{-1} + 3 \times 10^{-2} + \cdots + 3 \times 10^{-n}$ represents $\frac{1}{3}$ exactly but, by increasing n, we can make the error as small as we please. It seems natural that *every* nonterminating decimal should represent some number. As will be seen from the exercises, however, not all of these numbers are rational.
   The long division process for calculating a decimal expansion for $\frac{1}{3}$ can be thought of in this way: we try to choose each digit in the decimal expansion (call the nth digit Δ_n) as large as possible subject to the condition that each of the partial expansions $.\Delta_1$, $.\Delta_1\Delta_2$, $.\Delta_1\Delta_2\Delta_3$, . . . , should be less than $\frac{1}{3}$. For example, when we do this division, we note that $.3 < \frac{1}{3}$, but $.4 > \frac{1}{3}$, so the first digit should be $\Delta_1 = 3$. Then there is a remainder, $1.0 - 3 \otimes .3 = .1$. Since $.03 < .1/3$ and $.04 > .1/3$, we choose the next digit also to be $\Delta_2 = 3$, and so on. $\frac{1}{3}$ is at least as large as each of the numbers .3, .33, .333, ... We say then that $\frac{1}{3}$ is an *upper bound* for $\{.3, .33, .333, . . .\}$. There is no number less than $\frac{1}{3}$ that is an upper bound for $\{.3, .33, .333, . . .\}$. Thus we say that $\frac{1}{3}$ is the *least upper bound* of $\{.3, .33, .333, . . .\}$.
   Let us try the same process with $\sqrt{2}$. This will produce a practicable means for calculating a decimal expansion for $\sqrt{2}$, although *not* the most efficient one. The first digit must be 1, since $1^2 < 2$, but $2^2 > 2$. Trial-and-

Table 3

$$1^2 = 1.00000 < 2 < 4.00000 = 2^2$$
$$1.4^2 = 1.96000 < 2 < 2.25000 = 1.5^2$$
$$1.41^2 = 1.98810 < 2 < 2.01640 = 1.42^2$$
$$1.414^2 = 1.999396 < 2 < 2.002225 = 1.415^2$$

error reveals that the next digit should be 4, since $1.4^2 = 1.96 < 2 < 2.25 = 1.5^2$. Table 3 shows the results of a few steps of this process. In this way, we successively produce the digits of an **infinite decimal expansion** $1.414214\ldots$, which should represent $\sqrt{2}$. This decimal expansion is an upper bound for each of the partial expansions 1, 1.4, 1.41, 1.414, $\ldots$, and there can be no smaller upper bound for all of them. It is the *least upper bound*. Likewise, it is the *greatest lower bound* of the expansions 2, 1.5, 1.42, 1.415, 1.4143, $\ldots$, which converge to $\sqrt{2}$ from above.

3.   Consider another geometrical problem, that of determining the arc length of a circle. A procedure for performing this calculation, invented by the Greeks, is to partition the circular arc by a number of points, inscribe a polygon, and use the length of the polygon to approximate the length of the circle, as shown in Figure 2. According to the minimal length property of straight lines, "a straight line is the shortest path between two points," each of these polygons should have *smaller* length than the circle. On the other hand, it seems clear from the picture that the lengths of the inscribed polygons can be made as close as we please to the length of the circle *by making the sides of the polygons short enough*. Thus the length of the circle is an upper bound for the lengths of all possible inscribed polygons, and there ought to be no smaller upper bound. We might in fact *define* the length of the circle to be the *least upper bound* of the lengths of the inscribed polygons.

4.   In a similar way, the area of a circle may be computed by filling up the circle from the inside by rectangles (Figure 3). The least upper bound of the

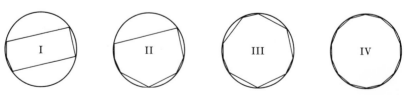

Figure 2

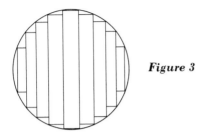

Figure 3

areas of all these inscribed rectangular polygons is the **inner Jordan content** of the circle, or more simply, its **area.**

5.   We are inclined to think that the series $1 + (x/1!) + (x^2/2!) + (x^3/3!) + \cdots$ ought to converge to some number (e^x). Generally, this number will not be rational when x is, although the partial sums of the series are rational when x is rational.

   In all these examples, we find it desirable that there be a number that represents some object arising in geometry (length, area) or analysis (decimal fractions, square roots, sums of series). I have chosen each example so that the number sought turns out to be presented naturally as a least upper bound (of lengths or areas of inscribed polygons, of decimal fractions, of partial sums of a series). The case of $\sqrt{2}$ shows that there are gaps in the system of rational numbers, so that these upper bounds will not generally be rational. It seems desirable that there ought to be enough numbers of some kind (*real numbers*) to fill the gaps.

   It is too much to hope that every set of real numbers would have a least upper bound.

6.   EXAMPLE. Consider the (empty) set of odd integers that are multiples of 2. *Any* real number is an upper bound for this set, for if x is a real number, surely no odd integer divisible by 2 is larger than x.

7.   EXAMPLE. Consider the set of all even integers. *No* real number is an upper bound for this set.

   These two examples show what more modest thing might be true—that every *nonempty* set of real numbers that *has* an upper bound should have a least upper bound. I shall try to make this assertion plausible by showing how such an upper bound could be represented by a decimal expansion. The argument, it should be noted, does *not* in general provide a practicable scheme for numerical calculation.

   Let S be a set of real numbers. Suppose that S is not empty, so that there is a number $x \in S$, and that S has an upper bound B. We shall construct a decimal expansion for the least upper bound of S by the method of trial and error used to construct a decimal expansion for $\sqrt{2}$.

First find an *integer* n that is less than x and an *integer* N that is larger than B. Then n is not an upper bound for S, but N is an upper bound for S. Among the $N - n$ integers n, $n + 1$, $n + 2$, $\ldots$, $N - 1$, there will be a largest one m, which *fails* to be an upper bound for S. $m + 1$ will be an upper bound for S. We now construct a decimal $m.\Delta_1\Delta_2\Delta_3\Delta_4 \ldots$ that is the least upper bound of S. There will be among the decimals $m.0$, $m.1$, $m.2$, $\ldots$, $m.9$ a largest one that *fails* to be an upper bound for S. Let Δ_1 be that digit among $0, 1, 2, \ldots, 9$ such that $m.\Delta_1$ is not an upper bound for S, but $m.\Delta_1 + 10^{-1}$ is an upper bound for S. Then examine the decimals $m.\Delta_1 0$, $m.\Delta_1 1$, $m.\Delta_1 2$, $\ldots$, $m.\Delta_1 9$, and pick the largest that *fails* to be an upper bound for S. That is, pick the digit Δ_2 so that $m.\Delta_1\Delta_2$ is not an upper bound for S but $m.\Delta_1\Delta_2 + 10^{-2}$ is an upper bound for S. Proceeding in this way, we get an infinite decimal expansion $m.\Delta_1\Delta_2\Delta_3\Delta_4 \ldots$.

Let us show that $m.\Delta_1\Delta_2\Delta_3\Delta_4 \ldots$ is the least upper bound of S. First of all, if $y \in S$, we know from the construction of $m.\Delta_1\Delta_2 \ldots$ that $m.\Delta_1\Delta_2 \ldots \Delta_j + 10^{-j} \geq y$ for every j, since $m.\Delta_1\Delta_2 \ldots \Delta_j + 10^{-j}$ is an upper bound for S. Taking the limit as $j \to \infty$, we find that $m.\Delta_1\Delta_2\Delta_3 \ldots \geq y$. This shows that $m.\Delta_1\Delta_2\Delta_3 \ldots$ is an upper bound for S.

It remains to be shown that if $k.\delta_1\delta_2\delta_3 \ldots$ is some real number less than $m.\Delta_1\Delta_2\Delta_3 \ldots$, then $k.\delta_1\delta_2\delta_3 \ldots$ is *not* an upper bound for S. We use the fact that $.99999 \ldots = 1$. First of all, if $k < m$, then $k.\delta_1\delta_2\delta_3 \ldots \leq k.999 \ldots \leq k + 1 \leq m$. Since m is not an upper bound for S, there is a $y \in S$ such that $y > m$. Then $y > k.\delta_1\delta_2\delta_3 \ldots$, so $k.\delta_1\delta_2\delta_3 \ldots$ is not an upper bound for S. If $k = m$, then, because $k.\delta_1\delta_2\delta_3 \ldots$ is by assumption less than $m.\Delta_1\Delta_2\Delta_3 \ldots$, there must be an i such that $k.\delta_1\delta_2 \ldots \delta_i < m.\Delta_1\Delta_2 \ldots \Delta_i$. Let j be the first such index, so that $\delta_p = \Delta_p$ if $p < j$, but $\delta_j < \Delta_j$. Then $k.\delta_1\delta_2\delta_3 \ldots \leq k.\delta_1\delta_2 \ldots \delta_j 9999 \ldots = k.\delta_1\delta_2 \ldots (\delta_j + 1) \leq m.\Delta_1\Delta_2 \ldots \Delta_j$. Since $m.\Delta_1\Delta_2 \ldots \Delta_j$ is not an upper bound for S, it follows as before that $k.\delta_1\delta_2\delta_3 \ldots$ also is not an upper bound for S. ●

It is easy, of course, to pick holes in this argument, although they could be patched up with sufficient care. I repeat, the argument is intended to make the existence of least upper bounds *plausible*, by showing that they ought to correspond to certain decimal expansions. It is not intended to be a rigorous proof.

Exercises

*1. Show that if r is rational and Hitler is irrational, then $r +$ Hitler and $r \times$ Hitler are irrational. (In the multiplicative case, assume $r \neq 0$.)

2. Show that every real number is the limit of a sequence of irrational reals. (You may assume that every real is the limit of a sequence of rationals—for example, its partial decimal expansions.)

3. Show that every repetitive decimal expansion such as .007007007007007. . . represents a rational number.

4. Show conversely that every rational number has a repetitive decimal expansion. *Hint:* Consider the long division process for finding the decimal expansion of a rational P/Q. How many distinct remainders can occur?

Schema:

$$
\begin{array}{l}
\overline{.\,*\,*\,*\,*\,*} \\
Q)\overline{P.00000.\,.\,.} \\
\underline{???} \\
\text{remainder} \\
\underline{???} \\
\text{remainder} \\
\underline{???} \\
\text{remainder} \\
\underline{???} \\
\text{etc.}
\end{array}
$$

*5. Use the result of Exercise 4 to describe a method for writing down decimal expansions for lots of irrational numbers. Show (by Cantor's Diagonal Argument) that you can actually get *uncountably many* irrationals in this way.

CHAPTER II

Axioms for the real number system and their consequences

In the introduction, I tried to make the least upper bound property plausible. Now I shall list a collection of properties that the real number system plausibly ought to have. In this book we shall give very little attention to the questions "Are there any real numbers? Are the axioms consistent? Is there anything (the real number system) that satisfies them?" These questions are worthy of investigation, but we shall take the point of view that we know what the real numbers are and that they evidently satisfy the following axioms, and concern ourselves solely with deducing the consequences of those axioms. We have to take something for granted or we could never begin.

A. The axioms

The real number system is a quadruple, $\langle \mathbf{R}, \mathbf{R}^+, +, \times \rangle$, consisting of a set, $\mathbf{R}$, a subset $\mathbf{R}^+$ of $\mathbf{R}$, and two functions $+$ and $\times$ from the set of pairs $\mathbf{R} \times \mathbf{R}$ of things in $\mathbf{R}$ into $\mathbf{R}$ with the following properties.

1. ALGEBRAIC AXIOMS. *The real numbers form a field.*

(a) *Group axioms for addition.* For every a, b, and c in $\mathbf{R}$,
   (i) *Associativity:* $(a + b) + c = a + (b + c)$;
   (ii) *Neutral element:* there is an element 0 in $\mathbf{R}$ such that $0 + a = a$;
   (iii) *Inverse:* there is an element $-a$ such that $(-a) + a = 0$;
   (iv) *Commutativity:* $a + b = b + a$.
(b) *Group axioms for multiplication.* For every a, b, and c in $\mathbf{R}$,
   (i) *Associativity:* $(a \times b) \times c = a \times (b \times c)$;
   (ii) *Neutral element:* there is an element 1 in $\mathbf{R}$ such that $1 \times a = a$;

(iii) *Inverse, 0 excepted:* if $a \neq 0$, there is an element a^{-1} in **R** such that $(a^{-1}) \times a = 1$;

(iv) *Commutativity:* $a \times b = b \times a$.

(c) *Distributive Law.* For every a, b, and c in **R**, $a \times (b + c) = (a \times b) + (a \times c)$.

(d) *Nontriviality.* $0 \neq 1$.

2. ORDER AXIOMS. (We call the elements of $\mathbf{R}^+$ *positive.*)

(a) *Closure.* If a, $b \in \mathbf{R}^+$, so are $a + b$ and $a \times b$ in $\mathbf{R}^+$.

(b) *Trichotomy.* For each a in **R**, one and *only one* of the following is true:

(i) $a \in \mathbf{R}^+$ (a is positive);

(ii) $a = 0$;

(iii) $-a \in \mathbf{R}^+$ (a is negative).

DEFINITIONS. $a < b \Leftrightarrow b > a \Leftrightarrow (b - a) \in \mathbf{R}^+$. If S is a subset of **R**, and z is a real number such that $x \leq z$ ($x < z$ or $x = z$) whenever $x \in S$, then z is an **upper bound for S** and S is **bounded above by z.** If z is an upper bound for S and $w \geq z$, then it is clear that w is also an upper bound for S. If z is less than any other upper bound, then it is a **least upper bound for S.** (Abbreviation: $z = $ **lub S** or $z = $ **sup S.** "sup" stands for **supremum.**) There are similar definitions for **lower bound** and **greatest lower bound** or **infimum** (**glb** or **inf**). If S is bounded both from above and from below, then we say simply that it is **bounded.**

3. COMPLETENESS AXIOM. Every subset of **R** that is *not empty* and *bounded above* has a least upper bound.

4. REMARKS. If z and w are least upper bounds for S, then by the definitions $z \leq w$ and $w \leq z$, so $z = w$. Thus the least upper bound is *unique.*

5. ☛ It follows immediately from 3 that *a nonempty subset S of real numbers that is bounded below has a glb.* In fact, if $x \in S$ and b is a lower bound for S, then the set T of lower bounds for S is nonempty (it contains b) and bounded above (by x). Therefore, T has a lub, L. L is a lower bound for S, for if it were not, there would be a $y \in S$ such that $y < L$. But then y would be an upper bound for T, which is less than the least one, L. By its construction, L is larger than any other lower bound for S. ●

The question whether these axioms are consistent or contain an inherent contradiction seems to be quite difficult. Whenever one deals with such a delicate logical question as the consistency of a mathematical system, the methods by which one attempts to answer it must themselves be closely scrutinized. In the case of the real number system, the problem has been approached by trying to show that there is no greater danger of inconsistency in the real number system than there is in elementary arithmetic and set theory. Of these last, set theory is most suspect. Whether there is an inconsistency in elementary arithmetic or set theory is a still more difficult question. At present, nobody knows of one, and most people doubt that there is any. Of course inconsistencies were discovered in the naive set theory invented by Cantor, as Cantor's and Russel's Paradoxes show. When I say that no one knows of an inconsistency in elementary set theory, I mean one or another of the modern axiomatic set theories such as the Zermelo-Frankel theory or the one in the back of J. L. Kelley's *General Topology* (Van Nostrand, 1955). These set theories were designed expressly to avoid the known paradoxes, and do so successfully so far as we know.

The attempt to show that the axioms for the real number system introduce no new danger of inconsistency is carried out by constructing a model for the real number system out of set-theoretical and arithmetical notions—by the use of infinite decimal expansions, for example. One then tries to prove that the axioms for the real number system can be proved as theorems *about the model*. If one could get a contradiction from the axioms, the contradiction must have been inherent in the original system in which the model was constructed.

To illustrate this procedure, let us construct within arithmetic a model for the field axioms (listed under 1 above). We start with the integers 0 and 1, but redefine addition and multiplication *modulo 2*. That is, to add or multiply in this new sense, we add or multiply as usual, divide by 2, and *keep only the remainder*. Table 4 describes the new operations $\oplus$ and $\otimes$; we could also interpret these operations by thinking of 0 as *even* and 1 as *odd*. Then the relation $1 + 1 = 0$ could be interpreted as odd + odd = even.

Because this system has only two elements, 0 and 1, the business of checking that the field axioms 1(a) to 1(d) are satisfied is reduced to a triviality. To put it differently, suppose we want a system $(\mathbf{Z}_2, \oplus, \otimes)$ that satisfies the field axioms alone, and are unconcerned with the order and completeness axioms. We could get one by putting $\mathbf{Z}_2 = \{0, 1\}$, and defining the operations $\oplus$ and $\otimes$ as above. The fact that this system does satisfy the axioms is then a theorem of (extremely) elementary arithmetic. If we could deduce a contradiction from the field axioms alone, we could deduce the same contradiction

Table 4

$\oplus$	0	1		$\otimes$	0	1
0	0	1		0	0	0
1	1	0		1	0	1

from this theorem in (extremely) elementary arithmetic, by the same chain of reasoning. Therefore, the field axioms are no more to be suspected than arithmetic, and we are quite confident of the latter.

In a similar but more complicated way, the entire axiom system for the real number system can be modeled within (less elementary) arithmetic and elementary set theory.

It is clear that the two-element field we have exhibited above is quite a different object from the real number system. Thus the field axioms alone are insufficient to characterize the real numbers. Even the addition of the order axioms does not suffice to characterize the real number system. For example, the rational number system also satisfies all the axioms 1 and 2 for an *ordered field*. However, the full axiom system for a *complete* ordered field does characterize the real number system in the following sense. If $(\mathbf{R}, \mathbf{R}^+, +, \times)$ and $\langle \mathbf{R}_1, \mathbf{R}_1^+, +_1, \times_1 \rangle$ are two complete ordered fields, then any algebraic, order, or limiting operation that can be carried out in one can be mimicked exactly by the other. That is, the two complete ordered fields cannot be distinguished by their properties.

To illustrate this point, let us show how we can distinguish the three fields $\langle \{0, 1\}, \oplus, \otimes \rangle$, the rationals (which we denote by $\mathbf{Q}$), and the reals by their properties. First of all, in any field, there is a unique multiplicative neutral element 1. The two-element field is the only one of those listed above in which $1 + 1 = 0$. It remains to distinguish the rationals from the reals, which we do by observing that the equation $x^2 = 1 + 1$ has no rational solution, but does have a real solution.

The exact sense in which any two complete ordered fields are essentially the same is that they are **isomorphic**. This means that if $\langle \mathbf{R}, \mathbf{R}^+, +, \times \rangle$ and $\langle \mathbf{R}_1, \mathbf{R}_1^+, +_1, \times_1 \rangle$ are complete ordered fields, there is a (unique) function ϕ from $\mathbf{R}$ to $\mathbf{R}_1$ with the following properties:

(a) ϕ maps $\mathbf{R}$ onto all of $\mathbf{R}_1$,
(b) ϕ is one-to-one,
(c) $\phi(a + b) = \phi(a) +_1 \phi(b)$,
(d) $\phi(a \times b) = \phi(a) \times_1 \phi(b)$,
(e) $a \in \mathbf{R}^+ \Leftrightarrow \phi(a) \in \mathbf{R}_1^+$.

(a) and (b) say that ϕ sets up a one-to-one correspondence between $\mathbf{R}$ and $\mathbf{R}_1$, (c) and (d) say that this correspondence preserves the algebraic structure, and (e) says that it preserves the order structure as well.

The construction of ϕ proceeds as follows. The *positive integers* in an ordered field are all those elements such as $1 + 1 + 1 + 1$, which can be obtained by repeated addition of the multiplicative unit element to itself. The *positive rationals* are those elements of the form pq^{-1}, where p and q are positive integers. The *rationals* are 0, the positive rationals, and the additive inverses of positive rationals. It is not hard to show that we can define a map

ϕ from the positive integers in **R** to the positive integers in $\mathbf{R}_1$ having the properties (c) and (d), by putting

$$\underbrace{\phi(1 + 1 + \cdots + 1)}_{n \text{ times}} = \underbrace{\phi(1) +_1 1 \cdots +_1 \phi(1)}_{n \text{ times}} = \underbrace{1_1 +_1 \cdots +_1 1_1}_{n \text{ times}}.$$

Then if p/q is a positive rational, we extend the definition of ϕ by putting $\phi(p/q) = p_1/q_1 = \phi(p)/\phi(q)$. If $p/q = r/s$, then $ps = qr$, so $\phi(ps) = \phi(p)\phi(s) = \phi(qr) = \phi(q)\phi(r)$, and hence $\phi(p)/\phi(q) = \phi(r)/\phi(s)$. This shows that the definition makes sense; we do not get two different values for $\phi(x)$ by representing x as a quotient of positive integers in two different ways. We then extend ϕ to all the rationals in the obvious way by putting $\phi(0) = 0_1$, $\phi(-x) = -\phi(x)$. It can be checked without great difficulty that this extension gives an isomorphism of the rationals in **R** with those in $\mathbf{R}_1$ preserving the algebraic and order structures. Finally, the way to extend ϕ to all of **R** is this: if $x \in \mathbf{R}$, let S_x be the set of all rationals that are $\leq x$. Then S_x is bounded above by some rational b and is nonempty (proof of this to follow; see Corollary C3). Therefore S_{1x}, the set of all elements of $\mathbf{R}_1$ of the form $\phi(r)$, where $r \in S_x$, is nonempty and bounded above by $\phi(b)$. We define $\phi(x)$ to be lub S_{1x}. It must then be checked (easily) that this definition makes sense, and does indeed extend ϕ to an isomorphism between $\langle \mathbf{R}, \mathbf{R}^+, +, \times \rangle$ and $\langle \mathbf{R}_1, \mathbf{R}_1^+, +_1, \times_1 \rangle$.

What will interest us in this course are the consequences of the completeness properties of the reals. However, before we take them up, let us look at the algebraic and order properties more closely. For a more detailed discussion of the topics I have treated so sketchily above, I refer you to:

(a) Birkhoff and MacLane, *A Survey of Modern Algebra* (Macmillan, 1955).
(b) van der Waerden, *Modern Algebra*, vol. 1 (Ungar, 1953).
(c) Bourbaki, *Éléments de Mathématique, Algebre, Chapitre I* (Hermann, 1958).

B. Consequences of the algebraic and order axioms

1. GROUPS. If we take just the axioms A1(a)(*i*), (*ii*), (*iii*), we have the characteristic properties of a **group**. That is, any pair $\langle G, \circ \rangle$ where G is a set and $\circ$ is a function that takes $G \times G$ into G and satisfies the axioms of associativity, (left) neutral element, and (left) inverse, is a group. If $\circ$ satisfies the commutative law as well, $\langle G, \circ \rangle$ is a **commutative** (or **Abelian**) group. Two examples follow.

(a) Let S be a set of three elements, $S = \{a, b, c\}$. Let G be the set of all permutations of S (one-to-one maps from S onto S). If f and g are permutations, their composition $f \circ g$ is defined to be the permutation that sends x into $f(g(x))$. The identity permutation leaves everything alone. The inverse of a permutation puts things back where they started. This group is *not* commutative.

(b)   The real numbers with addition $\langle \mathbf{R}, + \rangle$, the *nonzero* real numbers with multiplication $\langle \mathbf{R}^0, \times \rangle$, and the *positive* real numbers with multiplication $\langle \mathbf{R}^+, \times \rangle$ are all commutative groups.

2.   FIELDS AND ORDERED FIELDS. As we have already remarked, the entire system of axioms listed under A1 are the characteristic properties of a **field**. We have given already three examples of fields—the two-element field $\mathbf{Z}_2$, the rationals $\mathbf{Q}$, the reals $\mathbf{R}$. (Strictly speaking, a field is a *triple* $\langle F, +, \times \rangle$ consisting of a set F and two binary operations, $+$ and $\times$, on it that satisfies the field axioms. Referring to the real field by the symbol $\mathbf{R}$ rather than the symbol $\langle \mathbf{R}, +, \times \rangle$ is an abuse of language. I shall frequently be brief but abusive.)

The last two of these, the rationals and the reals, are **ordered fields** as well. That is, they also satisfy the order axioms A2. (The positive rationals $\mathbf{Q}^+$ are those rationals which, as real numbers, happen to be positive.)

A *complete* ordered field, of course, is isomorphic to the field of real numbers.

The familiar rules of arithmetic can be deduced from the axioms for a field or for an ordered field. A few examples will show how this is done. (There are a few more in the Exercises.) Since all these proofs are easy, we shall not dwell on them, but shall get on to the serious business of deducing consequences of the Completeness Axiom. Details can be found in the three texts cited above.

EXAMPLES.   (a) A *commutative group* $\langle G, \circ \rangle$ can have only one identity element. In fact, if e and e' are identity elements, then $e = e' \circ e = e \circ e'$ because e' is an identity element. But $e \circ e' = e'$ because e is an identity element. Thus $e = e'$.

Likewise an element $x \in G$ can have only one inverse, for if y and z are inverses for x and e is the identity element of G, then $y = e \circ y = (z \circ x) \circ y = z \circ (x \circ y) = z \circ (y \circ x) = z \circ e = e \circ z = z$. In particular, in a field $\langle F, +, \times \rangle$ there are only one element 0 (which is an additive identity) and only one element 1 (which is a multiplicative identity), and the additive and multiplicative inverses of an element of F are unique.

(b)   The rule "a minus times a minus is a plus," or in other words, $(-A) \times (-B) = A \times B$ is a consequence of the distributive law. First of all, for any element B of a field $\langle F, +, \times \rangle$, $0 \times B = (0 + 0) \times B = (0 \times B) + (0 \times B)$. Let A be an additive inverse for $0 \times B$. Then $0 = A + (0 \times B) = A + ((0 \times B) + (0 \times B)) = (A + (0 \times B)) + (0 \times B) = 0 + (0 \times B) = 0 \times B$, by the associative law. Now let A and B be any elements of F. Then, by the distributive law, $((-A) \times (-B)) + ((-A) \times B) = (-A) \times ((-B) + B) = (-A) \times 0 = 0 = 0 \times B = ((-A) + A) \times B = ((-A) \times B) + (A \times B)$. By adding to both sides $-((-A) \times B)$, we can cancel $(-A) \times B$ and get $(-A) \times (-B) = A \times B$.

(c)   We used above two properties of an ordered field $\langle F, F^+, +, \times \rangle$, namely, that $A \leq B, B \leq C \Rightarrow A \leq C$ and that $A \leq B, B \leq A \Rightarrow A = B$. To prove the first, note that it is trivial if $A = B$ or $B = C$. If $A < B$ and $B < C$, then $B - A$ and $C - B$ are in F^+. By Axiom A2(a), $C - A = (C - B) + (B - A)$ is also in F^+, so that $C > A$. To prove the second, note that if $A \neq B$, then according to Axiom A2(b), either $B - A \in F^+$ or $A - B \in F^+$, but *not both*. Thus either $A < B$ or $B < A$, but not both. Thus the only way both $A \leq B$ and $B \leq A$ can be true is for A to be equal to B.

Exercises

1. Check, for your own benefit, that $\langle \mathbf{Z}_2, \oplus, \otimes \rangle$ is a field, where $\mathbf{Z}_2$ is the set consisting of 0 and 1 alone, and $\oplus$ and $\otimes$ are the operations defined by Table 4.
2. Consider the system consisting of those real numbers that are of the form $p + q\sqrt{2}$, where p and q are rational. Verify that, with the usual addition and multiplication, they form a field. It is obvious (why?) that all the axioms are satisfied except possibly for the existence of multiplicative inverses. Thus, the crux of the matter is to show that $1/(p + q\sqrt{2})$ can be represented as $r + s\sqrt{2}$, where r and s are rational.
*3. Show that if S is any nonempty set, the collection $\pi(S)$ of all permutations of S (one-to-one functions from S onto S) forms a group, with the composition defined by $(f \circ g)(x) = f(g(x))$. Show that this group is commutative if, and only if, S contains at most two elements.
4. Prove the following consequences of the axioms for an ordered field, if $\langle F, F^+, +, \times \rangle$ is the ordered field. (a) For each nonzero $x \in F$, $x^2 \in F^+$. (b) Let $x/y = xy^{-1}$. Then $1/(x/y) = y/x$ and $(x/y) + (z/w) = (xw + yz)/yw$. (c) $1 + 1 \neq 0$. *Hint:* Use (a). *Conclusion:* There is no way to make the field $\langle \mathbf{Z}_2, \oplus, \otimes \rangle$ into an ordered field.

C.   Consequences of the completeness axiom

We begin with a useful characterization of least upper bounds.

1.   THEOREM.   Let S be a subset of **R**. For $B \in \mathbf{R}$ to be the least upper bound of S, it is necessary and sufficient that

(a) B be an upper bound for S, and
(b)   for every $\epsilon > 0$, there be an $x \in S$ such that $B - \epsilon < x$.

Proof. Necessity. If $B = $ lub S, then B is an upper bound for S by definition. If (b) were false, then for some $\epsilon > 0$, $B - \epsilon$ would be an upper bound for S, which is smaller than B, contradicting the fact that B is the *least* upper bound.

Sufficiency. (a) and (b) together show that S is nonempty and bounded above, so it has a least upper bound L. Certainly $L \leq B$. We have to show that $L < B$ is impossible. But if $L < B$, we could put $\epsilon = B - L > 0$. Since L is an upper bound for S, there can be no $x \in S$ such that $x > B - \epsilon = B - (B - L) = L$. This contradicts (b).   ●

As a consequence of this theorem, we get a fact that was used implicitly in the construction of the isomorphism between two complete ordered fields.

2.   THEOREM (ARCHIMEDEAN PROPERTY). Let ϵ be any positive real number, and let a and b be two real numbers such that $b - a > \epsilon$. Then there is an integer m such that $a < m\epsilon < b$.

Proof. Let S be the set of all integers n such that $n\epsilon \leq a$. Clearly S is bounded above by a/ϵ. We must first show that S is not empty. In fact, if S were empty, then a would be less than $n\epsilon$ for every integer n (by the trichotomy law). Thus a/ϵ would be a lower bound for the set of all integers. Now the set $\mathbf{Z}$ of all integers is surely nonempty, since it contains 0. Thus $\mathbf{Z}$ would have a glb M. Apply Theorem 1 (reformulated for greatest lower bounds). There must be an integer n such that $n < M + 1$. But then $n - 1$ is an integer, and $n - 1 < M$, so that M is not a lower bound for $\mathbf{Z}$, a contradiction.

Since S is thus bounded above and nonempty, it has a least upper bound B. According to Theorem 1, there is an integer n in S such that $B - 1 < n$. We already know that $n \leq a/\epsilon$. On the other hand, n must be greater than $a/\epsilon - 1$. Otherwise, if $n \leq a/\epsilon - 1$, then $n + 1$ is an integer, and $n + 1 \leq a/\epsilon$, so that $n + 1$ is in S. But $n + 1 > B$, contradicting the fact that B is an upper bound for S.

Now we know that $a/\epsilon - 1 < n \leq a/\epsilon$, or in other words that $a - \epsilon < n\epsilon \leq a$. Therefore $n + 1$ is an integer and $a < (n + 1)\epsilon \leq a + \epsilon < b$. It suffices to put $m = n + 1$.

3.   COROLLARY. If a and b are real numbers, and $b > a$, there is a rational number r such that $a < r < b$. Thus every real number is the least upper bound of a set of rational numbers.

Proof. By the preceding theorem, there is an integer $m = m \cdot 1$ between $1/(b - a)$ and $(1/(b - a)) + 2$. In particular, $m > 1/(b - a) > 0$, so that $b - a > 1/m > 0$. Now apply Theorem 2 again with $\epsilon = 1/m$. There is an integer n such that $a < n/m < b$. Finally, if b is any real number and S is the set of rationals that are $\leq b$, this result together with Theorem 1 shows that $b = \text{lub } S$. (If $\epsilon > 0$ is given, we can put $a = b - \epsilon$ to verify 1(b).)   ●

The following theorem is very useful as a tool for proving that continuous functions attain their maxima, and so on (see the Exercises).

4.   DEFINITION. If S is a set of real numbers, then x is a **cluster point** of S (or **point of accumulation** of S) if for every $\epsilon > 0$, there are *infinitely many* real numbers $y \in S$ such that $|y - x| < \epsilon$.

5.   THEOREM (BOLZANO-WEIERSTRASS PROPERTY). Every *bounded infinite* set of real numbers has a cluster point.

Proof. Let S be an infinite set of real numbers that is bounded below by b and above by B. Let C be the set of real numbers that are smaller than infinitely many of the numbers in S. Then C is bounded above by B and is nonempty, since it contains b. Thus C has a least upper bound U. I claim that U is a cluster point of S. What I have to show is that for every $\epsilon > 0$, there are infinitely many points of S between $U - \epsilon$ and $U + \epsilon$. Now $U - \epsilon$ is smaller than U, so it is not an upper bound for C. Therefore there is a $y \in C$ such that $y > U - \epsilon$. Since y lies below infinitely many of the points in C, so does $U - \epsilon$. It will be enough to show that of the infinitely many points in S that are greater than $U - \epsilon$, only finitely many are also $\geq U + \epsilon$. But if infinitely many were $\geq U + \epsilon$, $U + \epsilon$ would be in C, so U would not be an upper bound for C.   ●

One way to look at the completeness of the real number system is to observe that the Completeness Axiom shows that every sequence of real numbers that ought to have a limit actually does have one. We already have seen the importance of this fact in certain geometrical applications to length and area. Clearly it is essential in analysis as well. To be more precise, let us recall the definition of the limit of a sequence.

The **segment of the integers beginning with n** is $\{k \in \mathbf{Z} \colon k \geq n\}$. Recall that a **sequence** is simply a function whose domain is a segment of the integers. A sequence **begins with the nth term** if it is a function whose domain is the segment of the integers beginning with n. Various ways of naming such a sequence are "$k \mapsto x_k$," in which the term with which the sequence begins is not explicitly mentioned, "$\{x_k\}$," in which abusive notation it is not even explicitly shown that the terms of the sequence are ordered, "$\{x_k \colon k = n, n + 1, n + 2, \ldots\}$," and "$x_n, x_{n+1}, x_{n+2}, \ldots$."

6.   DEFINITION. The number L is the **limit** of the sequence $n \mapsto x_n$ if, and only if, for every $\epsilon > 0$, there is an integer N such that $|x_n - L| < \epsilon$ whenever $n > N$. (Abbreviation: $\lim\limits_{n \to \infty} x_n = L$.)   ●

Note that this definition is useless for recognizing limits *unless we already know what the limit is.* We must have the number L in hand before the definition can be applied. If we want to *define* L as the limit of a sequence $n \mapsto x_n$, the definition of limit alone does us no good. How do we know that there is such a limit? For example, it is customary to define e^x as the sum of the series (limit of the sequence of its partial sums) $1 + (x/1!) + (x^2/2!) + (x^3/3!) + (x^4/4!) + \cdots$. The definition would allow us to test whether or not L is the sum of this series, *if we had L in hand*, but it does not present us with any candidates for the position of limit.

What we need is some criterion that would allow us to recognize when a sequence has a limit by looking at the intrinsic properties of the sequence itself. This is afforded by the following simple observation about limits.

7.   PROPOSITION. If the sequence $n \mapsto x_n$ has a limit, then for every $\epsilon > 0$, there is an integer N such that $|x_n - x_m| < \epsilon$ whenever $n > N$ and $m > N$.

Proof. Suppose $\lim_{n \to \infty} x_n = L$. Choose N so large that $|x_n - L| < \epsilon/2$ whenever $n > N$. Then if n and m are both greater than N,

$$|x_n - x_m| = |x_n - L + L - x_m| \leq |x_n - L| + |L - x_m| < \frac{\epsilon}{2} + \frac{\epsilon}{2} = \epsilon.$$

8.   DEFINITION. The sequence $n \mapsto x_n$ is a **Cauchy sequence** or **satisfies the Cauchy criterion** if, and only if, for every $\epsilon > 0$ there is an integer N such that $|x_n - x_m| < \epsilon$ whenever $n, m > N$. ●

Proposition 7 says simply that every sequence that has a limit satisfies the Cauchy criterion. I shall now prove that the converse is true as well.

9.   THEOREM. Every Cauchy sequence of real numbers has a limit in **R**.

Proof. Let $n \mapsto x_n$ be a Cauchy sequence in **R**. By definition, we can pick an integer N such that $|x_n - x_m| < 1$ whenever $n, m > N$.

Case 1: The range of the function $n \mapsto x_n$ is an infinite set. In this case, $C = \{x_n : n > N\}$ is an infinite set of real numbers bounded below by $x_{N+1} - 1$ and above by $x_{N+1} + 1$. According to the Bolzano-Weierstrass Theorem (5), C has a point of accumulation, L. For any $\epsilon > 0$, there is an M so large that $n, m > M \Rightarrow |x_n - x_m| < \epsilon/2$, because $n \mapsto x_n$ is a Cauchy sequence. There is also an $n_0 > M$ such that $|x_{n_0} - L| < \epsilon/2$, because L is a point of accumulation of C. Then for any $m > M$, $|x_m - L| \leq |x_m - x_{n_0}| + |x_{n_0} - L| < \epsilon/2 + \epsilon/2 = \epsilon$. This shows that $\lim_{m \to \infty} x_m = L$.

Case 2: The range of the function $n \mapsto x_n$ is a finite set. Let R be the range of the function $n \mapsto x_n$. Then there must be at least one number $L \in R$ such that $x_n = L$ for infinitely many values of n. Just as in Case 1, it follows that $\lim_{m \to \infty} x_m = L$.

Exercises

*1. Let $a > 0$. Prove by induction that if n is a positive integer, $(1 + a)^n \geq 1 + na + (n(n - 1)/2)a^2$, so that $(1 + a)^n \geq na$ and $(1 + a)^n \geq (n(n - 1)/2)a^2$.

*2. Using the Archimedean Property, show that if a is any real number, $\lim_{n \to \infty} (a/n) = 0$.

*3. Show that if b is any real number and $r > 1$, there are integers n and m such that $r^n > b$ and $r^m/m > b$. Use this to show that if $0 < s < 1$, $\lim_{n \to \infty} s^n = \lim_{m \to \infty} m s^m = 0$. *Hint:* Put $r = 1 + a = 1/s$, and use the result of Problem 1.

*4. Let S be the set of all real numbers of the form $n + m\sqrt{2}$, where n and m are
   integers. Show that if a and b are any real numbers such that $a < b$, there is an
   $s \in S$ such that $a < s < b$. *Hint:* Note first that if s and t are in S, so are $s + t$,
   $s - t$, and ns for every integer n. According to Theorem 2, it is enough to show
   that for every $\epsilon > 0$, there is an $s \in S$ such that $0 < s < \epsilon$. (Why?) Prove this,
   using the Bolzano-Weierstrass Property or the **Pigeonhole Principle** (if there
   are $n + 1$ pigeons in n pigeonholes, at least one hole must contain two or more
   pigeons), and the observation that there are infinitely many points in S between
   0 and 1. To prove this last assertion, pick an integer m and try to find an integer
   n such that $0 \le n + m\sqrt{2} < 1$.

*5. For each of the following sets of real numbers, state whether it is bounded above
   or below. If it is bounded above, find its lub; if it is bounded below, find its glb.
   (a) The integers,
   (b) the positive integers,
   (c) the reciprocals of positive integers: $\frac{1}{1}, \frac{1}{2}, \frac{1}{3}, \frac{1}{4}, \ldots$,
   (d) the real numbers greater than all of the decimals $.3, .33, .333, .3333, \ldots$,
   (e) the rational numbers whose square is less than b (where $b > 0$),
   (f) the real numbers whose square is less than b ($b > 0$),
   (g) the positive real numbers whose square is less than b ($b > 0$),
   (h) the real numbers less than π.

*6. Prove that every sequence of real numbers that has a limit is bounded (more
   precisely, if $\lim\limits_{n \to \infty} x_n = L$, the *range* of the function $n \mapsto x_n$ is bounded).

7. Show that the sequence whose nth term is $1 + \dfrac{1}{2} + \dfrac{1}{3} + \cdots + \dfrac{1}{n}$ is *unbounded*.
   Hint: $1 \ge \frac{1}{2}, \frac{1}{2} \ge \frac{1}{2}, \frac{1}{3} + \frac{1}{4} \ge \frac{1}{2}, \frac{1}{5} + \frac{1}{6} + \frac{1}{7} + \frac{1}{8} \ge \frac{1}{2}$.

8. Many of the facts about continuous functions that are used in elementary calculus
   without proof can be proved using the least upper bound property (lub) or the
   Bolzano-Weierstrass Property (B-W). Here are some. Prove them.
   (a) If f is a continuous function on the interval $[a, b]$ where $a \le x \le b$, then f is
       bounded. *Hint:* If for every n there were an $x_n \in [a, b]$ such that $f(x_n) > n$,
       we could apply B-W to the set $\{x_1, x_2, x_3, \ldots\}$.
   (b) A continuous function on the interval $[a, b]$ attains its maximum on this
       interval. *Hint:* Use (a) and B-W. Choose $x_n \in [a, b]$ such that $f(x_n) > $ lub
       $\{f(x): a \le x \le b\} - 1/n$.
   (c) If f is continuous on $[a, b]$ and $f(a) < L < f(b)$, then there is a number c such
       that $a < c < b$ and $f(c) = L$. *Hint:* Apply lub to $\{y \in [a, b]: f(y) < L\}$.
   (d) Let $\{s_n\}$ be a *bounded* sequence of real numbers such that $s_1 \le s_2 \le s_3 \le$
       $s_4 \le \cdots$. Show that $\{s_n\}$ converges. *Hint:* Use B-W or lub. ●

There is more than one way to express the completeness of an ordered field.

*9. Let $\langle F, F^+, +, \times \rangle$ be an ordered field in which the Bolzano-Weierstrass Property
   is true. Prove that F has the Archimedean Property and that every Cauchy sequence in F converges to a limit in F.

*10. Let $\langle F, F^+, +, \times \rangle$ be an ordered field that has the Archimedean Property and
   in which every Cauchy sequence converges. Prove that every nonempty subset
   of F that is bounded above has a least upper bound. *Hint:* Let S be a nonempty
   subset of F that has an upper bound. Find a $b \in F$ that *is not* an upper bound for

S and a $B \in F$ which *is* an upper bound for S. Then use the **Method of Bisection** to find two sequences $\{l_n\}$ and $\{u_n\}$ in F such that for every integer $n \geq 0$,

(a) u_n is an upper bound for S, but l_n is *not*,

(b) $u_n - l_n = (B - b) \times 2^{-n}$, where $2 = 1 + 1 \in F$,

(c) $l_n \leq l_{n+1} < u_{n+1} \leq u_n$.

Do this by putting $l_0 = b$, $u_0 = B$, and choosing $\frac{1}{2}(l_n + u_n)$ to be l_{n+1} or u_{n+1} according to whether or not $\frac{1}{2}(l_n + u_n)$ *fails* to be an upper bound for S. Use induction to show that $2^n \geq n$ for every n, and the Archimedean Property to show that $\lim_{n \to \infty} 2^{-n} = \lim_{n \to \infty} (1/n) = 0$. Show then that $n \mapsto u_n$ is a Cauchy sequence and that $\lim_{n \to \infty} u_n = \text{lub } S$.

CHAPTER III

Vector spaces, inner products, and linear maps

We already have noted that the axioms for a *field* are satisfied by quite a few interesting mathematical systems, such as the rational numbers and the reals. Thus any fact we prove to be true of all fields, using the axioms for a field, is true in each of these cases. One advantage of formulating the notion of field is that it effects an economy of proof—we can prove the properties of a field once and for all, thereby avoiding repetition of the same argument for the real numbers, the rationals, and so forth. A second advantage is psychological. Suppose a certain fact is true as a consequence of the field axioms alone. If we tried to prove it in the case of the real numbers, we might well be distracted by additional properties of the reals that are not needed for the case in hand, such as that the reals are ordered and complete. These red herrings might so distract us that we would actually be unable to find a proof, or they might lead us to make an unnecessarily complicated one. Axiomatizing a notion such as that of field thus serves the psychological function of removing irrelevant and confusing distractions.

For example, suppose we want to prove that the only real numbers x such that $x^2 = x$ are 0 and 1. Because the reals are an *ordered* field, we might be distracted by the order properties so as to produce a complicated and inelegant proof by cases, as follows.

Bad proof that $x^2 = x \Leftrightarrow (x = 0 \ or \ x = 1)$. First, $0^2 = 0$ and $1^2 = 1$. What we must do is rule out all other possibilities. If x is a real number that is neither 0 nor 1, then either (a) $x < 0$, or (b) $0 < x < 1$, or (c) $x > 1$. It will be shown that in none of these cases can x satisfy the equation $x^2 = x$.

Case (a): Since every square of a nonzero number is positive, in the case that $x < 0$ we have $x < 0 < x^2$.

Case (b): If $0 < x < 1$, then $x \cdot x < x \cdot 1 = x$, since x is positive and $x < 1$.
Case (c): If $x > 1$, then as in case (b), $x \cdot x > x \cdot 1 = x$. ●

If instead we were not distracted by the ordering of the reals, but had attempted to make a proof from the field axioms alone, we might have produced the following better proof, which incidentally shows the assertion to be true *in any field*.

Good proof that $x^2 = x \Leftrightarrow (x = 0 \text{ or } x = 1)$. First, $0^2 = 0$ and $1^2 = 1$. Now $x^2 = x \Rightarrow x^2 - x = 0 \Rightarrow x(x - 1) = 0 \Rightarrow x = 0$ or $x - 1 = 0$, since the produce of two elements of a field can be zero only if at least one of them is 0. In fact, if $a \times b = 0$, but $a \neq 0$, then $b = 1 \times b = (a^{-1}) \times a \times b = a^{-1} \times 0 = 0$, since a, being different from 0, has a multiplicative inverse. ●

For the sake of efficiency, and for psychological benefits, I wish to set forth axioms for certain other algebraic systems of which we shall encounter many examples in this book, and to derive some of the consequences of the axioms. These consequences will be true, of course, in *any* system that satisfies the axioms. My program is to state the axioms first, and afterwards give examples.

A. Vector spaces

1.   DEFINITION. A **vector space over a field** F is a mathematical system $\langle V, +, \cdot \rangle$ consisting of a set V, a binary operation $+$ on V with respect to which $\langle V, + \rangle$ is an Abelian group, and a function $\cdot$, which assigns to each pair $\langle s, v \rangle \in F \times V$ a product $s \cdot v$ subject to the following axioms:

(a)   for each s and t in F, and each $v \in V$, $(st) \cdot v = s \cdot (t \cdot v)$;
(b)   for each s and t in F, and each $v \in V$, $(s + t) \cdot v = (s \cdot v) + (t \cdot v)$;
(c)   for each s in F and each v and w in V, $s \cdot (v + w) = (s \cdot v) + (s \cdot w)$;
(d)   for each $v \in V$, $1 \cdot v = v$. ●

Remark. It is common to omit the dot in the expression "$r \cdot v$."

2.   EXAMPLES. (a) The points in the plane can be made into a vector space over the field of real numbers as follows. Choose a fixed point **0**, the origin. Any point P in the plane can then be represented by an arrow from **0** to P (Figure 4, A). To add two plane vectors, use the parallelogram law as indi-

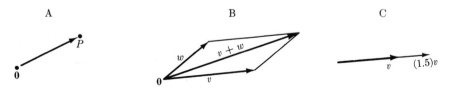

A   B   C

Figure 4

cated by Figure 4, B. To multiply a plane vector v by a *positive* real number r, produce an arrow pointing in the same direction as v whose length is r times the length of v, as in Figure 4, C. To multiply by a *negative* real number r, first multiply by $|r|$, then reverse the direction. To multiply by 0, simply produce a vector of length 0.

(b) Let S be any nonempty set, and F any field. Then the collection of all functions from S into F can be made into a vector space over F as follows. If f and g are two such functions, their sum $f + g$ is defined by $(f + g)(s) = f(s) + g(s)$; and if f is such a function and $r \in F$, then rf is the function defined by the formula $(rf)(x) = r(f(x))$.

(c)   In particular, the **n-tuples** of elements of F can be thought of as functions from the first n positive integers into F. Thus for example, a sextuple $(f_1, f_2, f_3, f_4, f_5, f_6)$ can be thought of as the function $1 \mapsto f_1$, $2 \mapsto f_2$, $3 \mapsto f_3$, $4 \mapsto f_4$, $5 \mapsto f_5$, $6 \mapsto f_6$. We add n-tuples and multiply them by elements of F according to the rule for functions:

$$(x_1, \ldots, x_n) + (y_1, \ldots, y_n) = (x_1 + y_1, \ldots, x_n + y_n),$$

$$r(x_1, \ldots, x_n) = (rx_1, \ldots, rx_n).$$

(d) Consider the collection of twice-differentiable real-valued functions f: $\mathbf{R} \to \mathbf{R}$, which satisfy the *differential equation* $f''(x) + f(x) = 0$ for all real values of f. It is an easy matter to check that the sum of two solutions to this differential equation is again a solution, as is the product of any solution by a real number. Thus the solutions to the differential equation form a vector space. In fact, I claim that every solution has the form $f(x) = a \sin x + b \cos x$, where a and b are real numbers. To see this, note that every such function is in fact a solution. Then observe that if f is any solution, $(d/dx)[(f(x))^2 + (f'(x))^2] = 2f(x)f'(x) + 2f'(x)f''(x) = 2f(x)f'(x) + 2f'(x)(-f(x)) = 0$. In particular, if $f(0) = f'(0) = 0$, then $(f(x))^2 + (f'(x))^2$ is constantly equal to zero. Since every square is nonnegative, and the square of a nonzero real number is positive, this sum of squares can be zero only if $f(x) \equiv 0$. Now if g is any solution of the differential equation, put $f(x) = g(x) - g(0) \cos x - g'(0) \sin x$. Then f is a solution, and $f(0) = f'(0) = 0$. It follows that f is constantly equal to zero, so that $g(x) \equiv g(0) \cos x + g'(0) \sin x$.

3.   DEFINITION. If V is a vector space over F, then a subset W of V is a (vector) **subspace** of V if

(a) $v + w \in W$ whenever v and w are in W, and

(b) $s \cdot V \in W$ whenever $v \in W$ and $s \in F$.

    In this case, W is a vector space in its own right, with the operations of vector addition and multiplication by scalars that it inherits from V.

    EXAMPLES. V is always a subspace of itself. The subset of V consisting of the additive neutral element **0** alone is also a subspace of V.

☛ We denote the vector space of n-tuples of elements of F by the symbol "F^n." In F^3, for example, the subset consisting of triples $(a, b, 0)$ with a zero in the last place is a subspace.

4.   DEFINITIONS. If $v_1, \ldots, v_n \in V$ and $s_1, \ldots, s_n \in F$, then the vector $(s_1 \cdot v_1) + \cdots + (s_n \cdot v_n)$ is a **linear combination** of $v_1, \ldots, v_n$. Any vector that is a linear combination of some of the vectors in a subset S of V is **linearly dependent** on S.

5.   PROPOSITION. If S is a subset of V, the collection of all vectors in V that are linearly dependent on S is a subspace W of V. (It is called the subspace **spanned by S.** We say that S **spans** W.)

Proof. If x and y are dependent on S, then there are vectors $v_1, \ldots, v_n$ and $w_1, \ldots, w_m$ in S and scalars $r_1, \ldots, r_n$ and $s_1, \ldots, s_m$ in F such that $x = r_1 \cdot v_1 + \cdots + r_n \cdot v_n$ and $y = s_1 \cdot w_1 + \cdots + s_m \cdot w_m$ (by definition). Then $x + y$ is a linear combination of vectors in S, namely $r_1 \cdot v_1 + \cdots + r_n \cdot v_n + s_1 \cdot w_1 + \cdots + s_m \cdot w_m$, and so is $t \cdot x$ for each $t \in F$ ($t \cdot x = (tr_1) \cdot v_1 + \cdots + (tr_n) \cdot v_n$).

EXAMPLES. $(1, 2) = (1, 0) + 2 \cdot (0, 1)$ is a linear combination of $(1, 0)$ and $(0, 1)$ in $\mathbf{R}^2$, but $(1, 2)$ is *not* a linear combination of the vectors $(1, 0)$ and $(3, 0)$, since any linear combination of $(1, 0)$ and $(3, 0)$ must clearly have a zero as its second component. $(1, 2, 0)$ is linearly dependent on $(1, 0, 0)$ and $(0, 1, 0)$ in $\mathbf{R}^3$, but $(1, 2, 3)$ is *not*, since it has a 3 in its last place.

6.   PROPOSITION (TRANSITIVITY OF LINEAR DEPENDENCE). Let S and T be subsets of the vector space V over the field F. Suppose every vector in T is linearly dependent on S. Then if x is linearly dependent on T, x is also linearly dependent on S.

Proof. By hypothesis there are vectors $y_1, \ldots, y_n$ in T and scalars $s_1, \ldots, s_n$ in F such that $x = s_1 y_1 + \cdots + s_n y_n$. For each $i = 1, \ldots, n$, there are vectors $z_{i1}, \ldots, z_{im(i)}$ in S and scalars $r_{i1}, \ldots, r_{im(i)}$ in F such that $y_i = r_{i1} z_{i1} + \cdots + r_{im(i)} z_{im(i)}$. We can then express x as $(s_1 r_{11}) z_{11} + \cdots + (s_1 r_{1m(1)}) z_{1m(1)} + \cdots + (s_n r_{n1}) z_{n1} + \cdots + (s_n r_{nm(n)}) z_{nm(n)}$, that is, as a linear combination of vectors in S.   ●

7.   DEFINITION. If $v_1, \ldots, v_n \in V$ and there exist $r_1, \ldots, r_n \in F$ *not all of which are zero* such that $r_1 v_1 + \cdots + r_n v_n = 0$, then $v_1, \ldots, v_n$ are **linearly dependent**. Otherwise, $v_1, \ldots, v_n$ are **linearly independent.**

EXAMPLES. $(1, 0)$ and $(0, 1)$ are independent in $\mathbf{Q}^2$, since if $r_1(1, 0) + r_2(0, 1) = (r_1, r_2) = \mathbf{0} = (0, 0)$, then $r_1 = 0$ and $r_2 = 0$. On the other hand, $(1, 0)$, $(0, 1)$, and $(1, 2)$ are linearly dependent, since $1 \cdot (1, 0) + 2 \cdot (0, 1) + (-1) \cdot (1, 2) = (0, 0)$.

8.   DEFINITION. A **basis** for a subspace W of a vector space V over F is a linearly independent set $\{v_1, \ldots, v_n\}$ of vectors that span W.

EXAMPLE. 2(d) shows that the solutions to the differential equation $f'' + f = 0$ form a subspace of the vector space of all twice-differentiable real-valued functions on **R**, and that this subspace is actually spanned by the two solutions sin x and cos x. These two solutions actually form a basis for the subspace, since they are linearly independent. In fact, if r sin $x + s$ cos $x \equiv \mathbf{0}$, then in particular, when we substitute $x = 0$ we get $0 = r \cdot \sin 0 + s \cdot \cos 0 = r \cdot 0 + s \cdot 1 = s$. It follows then that r sin x is the zero function, so $r = 0$ as well. It is not hard to see that the functions sin x + cos x and $\frac{1}{2}$(sin x − cos x) also form a basis for this subspace.

The following theorems show how these various notions of span, dependence, independence, basis, and so on are related.

9.   THEOREM. If $\{v_1, \ldots, v_n\}$ is an independent set of vectors but $\{v_1, \ldots, v_n, v_{n+1}\}$ is a dependent set, then v_{n+1} is linearly dependent on $v_1, \ldots, v_n$.

Proof. By hypothesis, there are scalars $r_1, \ldots, r_{n+1}$, not all of which are 0, such that $\mathbf{0} = r_1 v_1 + \cdots + r_n v_n + r_{n+1} v_{n+1}$. But r_{n+1} cannot be zero, or else $\mathbf{0}$ would equal $r_1 v_1 + \cdots + r_n v_n$, not all of $r_1, \ldots, r_n$ being 0, and this would contradict the presumed independence of $v_1, \ldots, v_n$. Therefore, we can divide by r_{n+1} to get $v_{n+1} = (-r_1/r_{n+1})v_1 + \cdots + (-r_n/r_{n+1})v_n$, which shows the dependence of v_{n+1} on $v_1, \ldots, v_n$.

10.   COROLLARY. $v_1, \ldots, v_n$ are linearly dependent if, and only if, some one of them is dependent on the others.

Proof. Say $v_1, \ldots, v_n$ are dependent. If $\mathbf{0}$ is among them, there is nothing to prove, since $\mathbf{0}$ is dependent on any vector ($\mathbf{0} = 0 \cdot v$). Otherwise, there is a first k such that $v_1, \ldots, v_{k+1}$ are dependent, and k must be ≥ 1. Then $v_1, \ldots, v_k$ are independent, so v_{k+1} is dependent on $v_1, \ldots, v_k$ by 9.

Conversely, suppose one of $v_1, \ldots, v_n$ is dependent on the others—say $v_n = r_1 v_1 + \cdots + r_{n-1} v_{n-1}$. Then $\mathbf{0} = r_1 v_1 + \cdots + r_{n-1} v_{n-1} + (-1)v_n$ is a dependence relation among $v_1, \ldots, v_n$. (Not all of $r_1, \ldots, r_{n-1}$, and -1 are 0. In particular, $-1 \neq 0$.)

11.   THEOREM. For $v_1, \cdots, v_n$ to be a basis for V, it is necessary and sufficient that each vector $w \in V$ can be expressed *uniquely* as $r_1 v_1 + \cdots + r_n v_n$, the coefficients $r_1, \ldots, r_n$ depending on w.

Proof. If $v_1, \ldots, v_n$ are a basis, they span V. If $w \in V$, then w must be a linear combination of them, $w = r_1 v_1 + \cdots + r_n v_n$. If w is a linear combination in two ways, $w = r_1 v_1 + \cdots + r_n v_n = s_1 v_1 + \cdots + s_n v_n$, then $\mathbf{0} =$

$(r_1 - s_1)v_1 + \cdots + (r_n - s_n)v_n$. Since the vectors in a basis are independent, this relation is impossible unless $r_1 - s_1 = r_2 - s_2 = \cdots = r_n - s_n = 0$.

Conversely, if each $w \in V$ can be expressed uniquely as a linear combination of $v_1, \ldots, v_n$, then clearly $v_1, \ldots, v_n$ span V. To show that $v_1, \ldots, v_n$ are a basis, we must show that they are independent. Suppose $0 = r_1 v_1 + \cdots + r_n v_n$. We know that $0 = 0 \cdot v_1 + \cdots + 0 \cdot v_n$. Because 0 can be expressed as a linear combination of $v_1, \ldots, v_n$ in only one way, $r_1 = 0, \ldots, r_n = 0$. Therefore, there can be no nontrivial dependence relation among $v_1, \ldots, v_n$. ●

The next theorem will allow us to deduce a number of important facts about bases.

12. EXCHANGE THEOREM. If $v_1, \ldots, v_n$ are independent vectors in V and $w_1, \ldots, w_m$ span V, then

(a) $m \geq n$, and
(b) n of the vectors $w_1, \ldots, w_m$ can be replaced by the vectors $v_1, \ldots, v_n$ in such a way that the new collection of m vectors still spans V.

Proof. We keep m fixed, and make the proof by induction on n. When $n = 0$, there are no v's at all, so the theorem is trivial in that case. We must now show that if the statement of the theorem is true about n, then it also is a true statement about $n + 1$.

Suppose we have $n + 1$ independent v's, $v_1, \ldots, v_n, v_{n+1}$. By our induction hypothesis, $m \geq n$, and we can replace n of the w's by $v_1, \ldots, v_n$ and still have a set of vectors that spans V. To avoid fancy business with the subscripts, let us call the remaining w's "$z_{n+1}, \ldots, z_m$." Then the result of exchanging n v's for w's is to get a new collection $v_1, \ldots, v_n, z_{n+1}, \ldots, z_m$ of vectors that spans V. Now v_{n+1} is dependent on $v_1, \ldots, z_m$ (since these vectors span V), but not on the vectors $v_1, \ldots, v_n$. Therefore, there must be at least one z, so $m > n$. In other words $m \geq n + 1$. This establishes part (a) with n replaced by $n + 1$.

Now the collection of vectors $v_1, \ldots, v_n, v_{n+1}, z_{n+1}, \ldots, z_m$ is dependent, since v_{n+1} is dependent on the remaining vectors. Hence, counting from the left, there must be a first vector such that it and its predecessors form a dependent set. This vector cannot be a v, since the v's are independent. Therefore, there must be a z, say z_k, such that $v_1, \ldots, v_{n+1}, z_{n+1}, \ldots, z_{k-1}$ are independent, but $v_1, \ldots, v_{n+1}, z_{n+1}, \ldots, z_k$ are not. This is precisely the situation of Theorem 9. z_k must be dependent on $v_1, \ldots, v_{n+1}, z_{n+1}, \ldots, z_{k-1}$. That is, we have $z_k = r_1 v_1 + \cdots + r_{k-1} z_{k-1}$. We can then exchange z_k for v_{n+1}.

Let $T = \{v_1, \ldots, v_{n+1}, z_{n+1}, \ldots, z_m\}$ and $S = T - \{z_k\}$. According to Proposition 6, S spans V, since every vector in T is linearly dependent on S and every vector in V is linearly dependent on T.

13. VERY IMPORTANT COROLLARY ON THE INVARIANCE OF DIMENSION.
If $v_1, \ldots, v_n$ and $w_1, \ldots, w_m$ are two bases for V, then $n = m$. In other words, any two bases for V contain the same number of elements.

Proof. If $\{v_1, \ldots, v_n\}$ and $\{w_1, \ldots, w_m\}$ are bases for V, then the v's are independent and the w's span V. According to the Exchange Theorem, $n \leq m$. By reversing this argument, we conclude also that $m \leq n$.

14. DEFINITION. If V has a finite basis, it is called **finite-dimensional,** and the number of vectors in any of its bases is its **dimension.** The dimension of a vector space consisting of $\mathbf{0}$ alone is 0. The dimension of V is denoted by "dim V." ●

Not every vector space has finite dimension. See Exercise 2 for an example.

15. THEOREM. Let V be a vector space of dimension n over F. If W is a subspace of V,

(a) W is finite-dimensional;

(b) dim $W \leq n$;

(c) if dim $W = n$, then $W = V$;

(d) if $w_1, \ldots, w_m$ is any basis for W, vectors $z_{m+1}, \ldots, z_n$ can be found such that $w_1, \ldots, w_m, z_{m+1}, \ldots, z_n$ is a basis for V.

Proof. (a) Suppose W were not finite-dimensional. Let us show by induction that for every nonnegative integer k, there must be k linearly independent vectors in W. This is clear when $k = 0$. Suppose we already have found a set $\{w_1, \ldots, w_k\}$ of k independent vectors in W. Since W is not of finite dimension, $w_1, \ldots, w_k$ do not span W. Therefore, there is a w_{k+1} in W that is independent of them. Then, according to Theorem 9, $w_1, \ldots, w_{k+1}$ is a set of independent vectors in W. But for $k = n$, this is impossible, for if $v_1, \ldots, v_n$ is a basis for V, the Exchange Theorem tells us that there cannot be more than n independent vectors in V.

(b) Since by (a), W has finite dimension, it has a basis $w_1, \ldots, w_m$. According to the Exchange Theorem, $m \leq n$.

(c) If $m = n$, then the Exchange Theorem shows that we can replace *all* of $v_1, \ldots, v_n$ by $w_1, \ldots, w_n$, thereby obtaining one independent set of vectors that spans V. Since V is thus spanned by vectors in W, W must be all of V.

(d) In any case, according to the Exchange Theorem, we can replace *some* of the vectors $v_1, \ldots, v_n$ by the vectors $w_1, \ldots, w_m$ so as to obtain a set of vectors $w_1, \ldots, w_m, z_{m+1}, \ldots, z_n$ that spans V. None of these vectors can be dependent on the others, for if one were, V could be spanned by fewer than n vectors. But the Exchange Theorem shows that if m vectors span V, then $m \geq n$.

16. COROLLARY. If dim $V = n$, then any set of n independent vectors in V spans V, and any set of n vectors in V that spans V is independent.

Proof. If $w_1, \ldots, w_n$ are independent, they span an n-dimensional subspace of V, which by 15(c) is all of V. If $w_1, \ldots, w_n$ span V, and if one were dependent on the others, then V would be spanned by fewer than n vectors. As indicated in the proof of 15(d), this is impossible.

EXAMPLE. If dim $V = 3$, every subspace of V has dimension 0, 1, 2, or 3, and the only subspace of dimension 3 is V itself.

Exercises

*1. Show that the dimension of F^n is n.
2. Let P be the vector space (over the field Q of rational numbers) of polynomials with rational coefficients. Add polynomials and multiply them by rationals in the usual way. Show that P is *infinite* dimensional. (*Hint:* If P were of finite dimension, there would be a polynomial of largest degree.)
*3. (1, 2, 3), (4, 5, 6), (1, 1, 1), and (1, 0, 2) are four vectors in $\mathbf{R}^3$. Hence they must be dependent. Express one of them as a linear combination of the other three.
*4. Show that the vectors (1, 2, 3), (1, 3, 3), and (1, 2, 4) are a basis for Q^3.

B.   Application to linear equations

The work we have just done with vector spaces has important interpretations in the setting of systems of linear equations. Consider a system of m linear equations in the n unknowns $x_1, \ldots, x_n$ in a field F, displayed in Table 5(a). If we introduce the vectors $A_1, \ldots, A_n, B$ in F^m displayed in Table 5(b),

---

Table 5

(a) $x_1 a_{11} + \cdots + x_n a_{n1} = b_1$
$\quad\;\; x_1 a_{12} + \cdots + x_n a_{n2} = b_2$

$\qquad\qquad\qquad\;\; \vdots$

$\quad\;\; x_1 a_{1m} + \cdots + x_n a_{nm} = b_m$

(b) $A_1 = (a_{11}, a_{12}, \ldots, a_{1m})$
$\quad\;\; A_2 = (a_{21}, a_{22}, \ldots, a_{2m}) \qquad\qquad B = (b_1, \ldots, b_m)$

$\qquad\qquad\qquad\;\; \vdots$

$\quad\;\; A_m = (a_{n1}, a_{n2}, \ldots, a_{nm})$

(c) $x_1 A_1 + \cdots + x_n A_n = B$

then we can rewrite the system of equations (a) as a single vector equation (Table 5(c)).

It is apparent that this equation has a solution if, and only if, B is in the subspace of F^m spanned by the vectors $A_1, \ldots, A_n$. If $A_1, \ldots, A_n$ do not span all of F^m, then the system (a) (or the single equation (c)) will have a solution for some choices of B (those in the span of $A_1, \ldots, A_n$) and not for others. Suppose that B is in the subspace of F^m spanned by $A_1, \ldots, A_n$. Call this subspace V. Then according to Theorem A 11, there is only one solution if, and only if, the vectors $A_1, \ldots, A_n$ are linearly independent.

Some important remarks are in order here about the relative sizes of n and m.

1. If $n < m$, so that there are fewer unknowns than equations, the vectors $A_1, \ldots, A_n$ can span a subspace of F^m of dimension *at most* n. In particular, $A_1, \ldots, A_n$ cannot span all of F^m, so there will certainly be choices of B for which the equation (c) has no solution.

2. If $n > m$, the vectors $A_1, \ldots, A_n$ must be dependent, since there are more of them than the dimension of F^m. In this case, (c) will always have more than one solution if it has any solution at all (that is, if B is dependent on $A_1, \ldots, A_n$). The equation will fail to have a solution for some choices of B unless some m of the n vectors $A_1, \ldots, A_n$ do form an independent set. This is the case of more unknowns than equations.

3. In case there are just as many equations as unknowns, $m = n$. If the vectors $A_1, \ldots, A_n$ are independent, they will be a basis for F^m. In this case, (c) will have one, and only one solution for each choice of B. Otherwise, if $A_1, \ldots, A_n$ are dependent, they will not span all of F^m. In this case, there will be some choices of B for which there is no solution at all, and if any solution exists, there will be more than one.

Next, I shall indicate a practical method for solving linear equations. In view of the preceding remarks, the same procedure can be used to determine whether one vector in F^m is dependent on some others, whether a given set of vectors in F^m is dependent or independent, and so on. Some readers may be familiar with *Cramer's Rule* for solving systems of n equations in n unknowns using determinants. ☛ Cramer's Rule should NOT be used in practice. It requires roughly n times as much labor as the method of **Gaussian Elimination** I am about to describe. Moreover, it is useless when $n \neq m$, although Gaussian Elimination can be used.

The basis of Gaussian Elimination is the simple remark that a system of equations in **special triangular form** as shown in Table 6(a) can be solved in a trivial way.

Table 6

(a) $\quad x_1 + x_2 a_{21} + x_3 a_{31} + \cdots + x_n a_{n1} = b_1$

$\qquad\quad x_2 \quad + x_3 a_{32} + \cdots + x_n a_{n2} = b_2$

$\qquad\qquad\quad x_3 \quad + \cdots + x_n a_{n3} = b_3$

$\qquad\qquad\qquad\qquad\qquad\quad \cdot$

$\qquad\qquad\qquad\qquad\qquad\quad \cdot$ Special

$\qquad\qquad\qquad\qquad\qquad\quad \cdot$ triangular

$\qquad\qquad\qquad\qquad\quad x_n \quad = b_n$ form

$\qquad\qquad\qquad\qquad\quad 0 \quad = b_{n+1}$

$\qquad\qquad\qquad\qquad\qquad\quad \cdot$

$\qquad\qquad\qquad\qquad\qquad\quad \cdot$

$\qquad\qquad\qquad\qquad\qquad\quad \cdot$

$\qquad\qquad\qquad\qquad\quad 0 \quad = b_m$

(b) $x_1 a_{11} + x_2 a_{21} + x_3 a_{31} + \cdots + x_n a_{n1} = b_1$

$\qquad\quad x_2 a_{22} + x_3 a_{32} + \cdots + x_n a_{n2} = b_2$

$\qquad\qquad\quad x_3 a_{33} + \cdots + x_n a_{n3} = b_3$

$\qquad\qquad\qquad\qquad\quad \cdot$

$\qquad\qquad\qquad\qquad\quad \cdot$

$\qquad\qquad\qquad\qquad\quad \cdot$ Triangular

$\qquad\qquad\qquad\quad x_n a_{nm} = b_n$ form

$\qquad\qquad\qquad\quad 0 \quad = b_{n+1}$

$\qquad\qquad\qquad\qquad\quad \cdot$

$\qquad\qquad\qquad\qquad\quad \cdot$

$\qquad\qquad\qquad\qquad\quad \cdot$

$\qquad\qquad\qquad\quad 0 \quad = b_m$

(c) $\quad x_1 + x_2 a_{21} + \cdots + \cdots + x_n a_{n1} = b_1$

$\qquad\qquad x_2 \quad + \cdots + \cdots + x_n a_{n2} = b_2$

$\qquad\qquad\qquad\qquad\quad \cdot$

$\qquad\qquad\qquad\qquad\quad \cdot$

$\qquad\qquad\qquad\qquad\quad \cdot$

$\qquad\qquad x_k \quad + \cdots + x_n a_{nk} = b_k$

$\qquad\qquad\qquad 0 x_{k+1} = b_{k+1}$ Bad

$\qquad\qquad\qquad\qquad \cdot$ triangular

$\qquad\qquad\qquad\qquad \cdot$ form

$\qquad\qquad\qquad\qquad \cdot$

$\qquad\qquad\quad 0 x_n \quad = b_n$

$\qquad\qquad\quad 0 \quad = b_{n+1}$

$\qquad\qquad\qquad\qquad \cdot$

$\qquad\qquad\qquad\qquad \cdot$

$\qquad\qquad\qquad\qquad \cdot$

$\qquad\qquad\quad 0 \quad = b_m$

Table 7

(a) $2x + 4y + 6z = -4$
(b) $3x + 3y + 12z = 0$
(c) $-2x + 8y - 11z = -6$

$$\begin{pmatrix} 2 & 4 & 6 & -4 \\ 3 & 3 & 12 & 0 \\ -2 & 8 & -11 & -6 \end{pmatrix}$$

This is the system of equations, with its coefficients arranged as a matrix

(d) $x + 2y + 3z = -2$
(b) $3x + 3y + 12z = 0$
(c) $-2x + 8y - 11z = -6$

$$\begin{pmatrix} 1 & 2 & 3 & -2 \\ 3 & 3 & 12 & 0 \\ -2 & 8 & -11 & -6 \end{pmatrix}$$

$\text{(a)} \to \tfrac{1}{2}\text{(a)} = \text{(d)}$

(d) $x + 2y + 3z = -2$
(e) $- 3y + 3z = 6$
(f) $12y - 5z = -10$

$$\begin{pmatrix} 1 & 2 & 3 & -2 \\ 0 & -3 & 3 & 6 \\ 0 & 12 & -5 & -10 \end{pmatrix}$$

$\text{(b)} \to \text{(b)} - 3\text{(d)} = \text{(e)}$
$\text{(c)} \to \text{(c)} + 2\text{(a)} = \text{(f)}$

(d) $x + 2y + 3z = -2$
(g) $y - z = -2$
(f) $12y - 5z = -10$

$$\begin{pmatrix} 1 & 2 & 3 & -2 \\ 0 & 1 & -1 & -2 \\ 0 & 12 & -5 & -10 \end{pmatrix}$$

$\text{(e)} \to -\tfrac{1}{3}\text{(e)} = \text{(g)}$

(d) $x + 2y + 3z = -2$
(g) $y - z = -2$
(h) $7z = 14$

$$\begin{pmatrix} 1 & 2 & 3 & -2 \\ 0 & 1 & -1 & -2 \\ 0 & 0 & 7 & 14 \end{pmatrix}$$

$\text{(f)} \to \text{(f)} - 12\text{(g)} = \text{(h)}$

(d) $x + 2y + 3z = -2$
(g) $y - z = -2$
(i) $z = 2$

$$\begin{pmatrix} 1 & 2 & 3 & -2 \\ 0 & 1 & -1 & -2 \\ 0 & 0 & 1 & 2 \end{pmatrix}$$

$\text{(h)} \to \tfrac{1}{7}\text{(h)} = \text{(i)}$
Triangular form

$z = 2$
$y = 0$
$x = -8$

Solution by back substitution

Case 1. $m > n$ and one of $b_{n+1}, \ldots, b_m$ is not 0. In this case the system of equations is clearly contradictory, and has no solution.

Case 2. $m > n$ and $b_{n+1} = \cdots = b_m = 0$. This case is the same as Case 3 below, with the addition of $m - n$ trivial equations of the form $0 = 0$.

Case 3. $m = n$. The solution can be found by **back substitution.** Start with the last equation, $x_n = b_n$. Substitute b_n for x_n in the second to last equation so as to get $x_{n-1} + b_n a_{n-1} = b_{n-1}$, or $x_{n-1} = b_{n-1} - b_n a_{n-1}$, then substitute for x_n and x_{n-1} in the third to last equation, and so on.

Case 4. $m < n$. In this case, the last equation has the form $x_m + x_{m+1} a_{m+1\, m} + \cdots + x_n a_{nm} = b_m$. The variables $x_{m+1}, \ldots, x_n$ can be chosen at will, and then x_m can be determined. The back substitution process then proceeds as in Case 3. The fact that the last $n - m$ variables can be chosen at will accounts for the nonuniqueness of the solution in case there are more variables than unknowns.

A system of linear equations can be put in triangular form by performing **row operations,** familiar from high school algebra. These operations consist of

(a)   rearranging the equations,
(b)   multiplying an equation by a nonzero constant,
(c)   substracting a multiple of one equation from another.

In the most pleasant situation, the resulting equations in triangular form (Table 6(b)) have nonzero **pivot elements** $a_{11}, \ldots, a_{nn}$. By dividing by these pivot elements (a row operation of type (b)), the equations can be put in special triangular form, and the solutions, if any, can be read off. In a slightly less pleasant situation, the equations can be put in triangular form with nonzero pivot elements by performing the **column operations** of permuting the unknowns $x_1, \ldots, x_n$. Finally, it may happen that the best that can be accomplished by row and column operations is to put the equations in **Bad Triangular Form** as shown in Table 6(c).

Case 5. Equations in bad triangular form, with at least one of $b_{k+1}, \ldots, b_m$ different from 0. Same as Case 1.

Case 6. Equations in bad triangular form, with $b_{k+1} = \cdots = b_m = 0$. This is just like Case 4, with $x_{k+1}, \ldots, x_n$ determined at will.

All six of these cases can occur. Case 3 occurs when there are equally many equations and unknowns, and $A_1, \ldots, A_n$ span F^n. If $n = m$ but $A_1, \ldots, A_n$ do not span F^n, then Case 5 or Case 6 will occur according to whether or not B is independent of $A_1, \ldots, A_n$. Generally, if B is dependent on $A_1, \ldots, A_n$, Case 2 occurs when $m > n$ and $A_1, \ldots, A_n$ are independent, Case 3 occurs if $m = n$ and $A_1, \ldots, A_n$ are independent, and either Case 4 or Case 6 occurs when $A_1, \ldots, A_n$ are dependent. If B is not dependent on $A_1, \ldots, A_n$, Case 1 or Case 5 occurs.

Table 7 shows an illustrative example with $n = m = 3$. The first column shows the equations, the third column gives a commentary, and the second column shows the same work in **matrix form** without the unnecessary symbols, not actually needed for the computation, which appear in the first column.

If we had wanted also to solve the equations

$$2x + 4y + 6z = 1$$
$$3x + 3y + 12z = 0$$
$$-2x + 8y - 11z = 0$$

we could have done so at the same time by just adding the column $\begin{pmatrix} 1 \\ 0 \\ 0 \end{pmatrix}$ to our initial matrix. The work is shown in Table 8. Back substitution would give the same solution to the first set of equations, $(-8, 0, 2)$, and for the second set, $\left(\dfrac{43}{14}, \dfrac{-3}{14}, \dfrac{-5}{7} \right)$.

If there are more unknowns than equations, we get a result like that shown in Table 9; then, by back substitution, $y = z - 2$, $x = -5z + 2$, which are the equations of the line through $(2, -2, 0)$ in the direction of $(-5, 1, 1)$.

Finally, here are two examples of using row operations to determine whether a set of vectors is independent.

First Case: The vectors $A_1 = (1, 2, 3)$, $A_2 = (4, 5, 6)$, $A_3 = (6, 9, 12)$. Arrange them as a matrix and perform row operations as shown in Table 10(a). These operations gave the sequence of matrices exhibited in Table

---

Table 8

$$\begin{pmatrix} 2 & 4 & 6 & -4 & 1 \\ 3 & 3 & 12 & 0 & 0 \\ -2 & 8 & -11 & -6 & 0 \end{pmatrix} \rightarrow \begin{pmatrix} 1 & 2 & 3 & -2 & \frac{1}{2} \\ 3 & 3 & 12 & 0 & 0 \\ -2 & 8 & -11 & -6 & 0 \end{pmatrix} \rightarrow \begin{pmatrix} 1 & 2 & 3 & -2 & \frac{1}{2} \\ 0 & -3 & 3 & 6 & -\frac{3}{2} \\ 0 & 12 & -5 & -10 & 1 \end{pmatrix} \rightarrow$$

$$\begin{pmatrix} 1 & 2 & 3 & -2 & \frac{1}{2} \\ 0 & 1 & -1 & -2 & \frac{1}{2} \\ 0 & 12 & -5 & -10 & 1 \end{pmatrix} \rightarrow \begin{pmatrix} 1 & 2 & 3 & -2 & \frac{1}{2} \\ 0 & 1 & -1 & -2 & \frac{1}{2} \\ 0 & 0 & 7 & 14 & -5 \end{pmatrix} \rightarrow \begin{pmatrix} 1 & 2 & 3 & -2 & \frac{1}{2} \\ 0 & 1 & -1 & -2 & \frac{1}{2} \\ 0 & 0 & 1 & 2 & -\frac{5}{7} \end{pmatrix}$$

Table 9

$$\begin{matrix} 2x - 4y + 6z = -4 \\ 3x + 3y + 12z = 0 \end{matrix} \rightarrow \begin{pmatrix} 2 & 4 & 6 & -4 \\ 3 & 3 & 12 & 0 \end{pmatrix} \rightarrow \begin{pmatrix} 1 & 2 & 3 & -2 \\ 3 & 3 & 12 & 0 \end{pmatrix} \rightarrow$$

$$\begin{pmatrix} 1 & 2 & 3 & -2 \\ 0 & -3 & 3 & 6 \end{pmatrix} \rightarrow \begin{pmatrix} 1 & 2 & 3 & -2 \\ 0 & 1 & -1 & -2 \end{pmatrix} \rightarrow \begin{matrix} x + 2y + 3z = -2 \\ y - z = -2 \end{matrix}$$

> ### Table 10
>
> (a) $\begin{pmatrix} 1 & 2 & 3 \\ 4 & 5 & 6 \\ 6 & 9 & 12 \end{pmatrix} \rightarrow \begin{pmatrix} 1 & 2 & 3 \\ 0 & -3 & -6 \\ 0 & -3 & -6 \end{pmatrix} \rightarrow \begin{pmatrix} 1 & 2 & 3 \\ 0 & -3 & -6 \\ 0 & 0 & 0 \end{pmatrix}$
>
> (b) $\begin{pmatrix} A_1 \\ A_2 \\ A_3 \end{pmatrix} \rightarrow \begin{pmatrix} A_1 \\ A_2 - 4A_1 \\ A_3 - 6A_1 \end{pmatrix} \rightarrow \begin{pmatrix} A_1 \\ A_2 - 4A_1 \\ A_3 - 6A_1 - (A_2 - 4A_1) = 0 \end{pmatrix}$

> ### Table 11
>
> $\begin{pmatrix} 1 & 2 & 3 & 4 \\ 2 & 3 & 4 & 5 \\ 3 & 4 & 6 & 6 \end{pmatrix} \rightarrow \begin{pmatrix} 1 & 2 & 3 & 4 \\ 0 & -1 & -2 & -3 \\ 0 & -2 & -3 & -6 \end{pmatrix} \rightarrow \begin{pmatrix} 1 & 2 & 3 & 4 \\ 0 & -1 & -2 & -3 \\ 0 & 0 & 1 & 0 \end{pmatrix} \rightarrow \begin{pmatrix} 1 & 2 & 3 & 4 \\ 0 & 1 & 2 & 3 \\ 0 & 0 & 1 & 0 \end{pmatrix}$

10(b), showing quickly enough a dependence relation $-2A_1 - A_2 + A_3 = A_3 - 6A_1 - (A_2 - 4A_1) = 0$.

Second Case: The vectors $B_1 = (1, 2, 3, 4)$, $B_2 = (2, 3, 4, 5)$, $B_3 = (3, 4, 6, 6)$ as shown in Table 11. Now it is clear that the vectors $C_1 = (1, 2, 3, 4)$, $C_2 = (0, 1, 2, 3)$, and $C_3 = (0, 0, 1, 0)$ are independent, since $xC_1 + yC_2 + zC_3 = \mathbf{0} \Leftrightarrow (x, 2x + y, 3x + 2y + z, 4x + 3y) = \mathbf{0}$, and then $x = 0$, $2x + y = 0$ so that $y = 0$, and $3x + 2y + z = 0$ so that $z = 0$. The vectors C_1, C_2, and C_3 are in the subspace of $\mathbf{R}^4$ spanned by B_1, B_2, and B_3—namely, $C_1 = B_1$, $C_2 = -B_2 + 2B_1$, $C_3 = (B_3 - 3B_1) - 2(B_2 - 2B_1)$. Since that subspace contains three independent vectors, it is three-dimensional. Since B_1, B_2, and B_3 span a three-dimensional space and there are only three of them, they must be independent. Note that what makes it easy to tell that C_1, C_2, and C_3 are independent is the special triangular form of the matrix $\begin{pmatrix} C_1 \\ C_2 \\ C_3 \end{pmatrix}$.

C. Inner products, the inequalities of Cauchy-Schwarz and Minkowski

The notion of vector space axiomatizes, among other things, some of the algebraic features of Euclidean geometry. The notion of *inner product* axiomatizes certain features of the Euclidean notion of distance. Once again, I shall proceed by giving first the axioms, and afterwards their motivation in Euclidean geometry.

1.   DEFINITION. Let V be a vector space over **R**. An **inner product** on V is a function that assigns a real number to each pair of vectors in V and satisfies the following axioms. (Denote the inner product of v and w by "$[v, w]$.")

(a)   For each $v \in V$, $[v, v] \geq 0$.
(b) $[v, v] = 0$ if, and only if, $v = \mathbf{0}$.
(c) $[v, w] = [w, v]$.
(d)   For each $v, w \in V$ and $t \in \mathbf{R}$, $[tv, w] = t[v, w]$.
(e)   For each $u, v, w \in V$, $[u + v, w] = [u, w] + [v, w]$.

2.   EXAMPLES. (a) Consider the vector space $\mathcal{C}([a, b])$ of continuous real-valued functions on the interval $[a, b]$. We define the inner product of two functions to be $[f, g] = \int_a^b f(x)g(x) \, dx$. The only property of inner products that is not trivial to check is that $[f, f] = 0 \Rightarrow f \equiv 0$. But $[f, f] = 0 \Rightarrow \int_a^b (f(x))^2 \, dx = 0$. $(f(x))^2$ is nonnegative and continuous. If it ever assumed a positive value, then there would be some interval I of length $\delta > 0$ on which $(f(x))^2 \geq \epsilon > 0$ for some positive ϵ (because f^2 is continuous). But then $\int_a^b (f(x))^2 \, dx \geq \int_I f^2 \geq \int_I \epsilon = \delta\epsilon > 0$.

(b) Let us choose rectangular coordinates for the Euclidean three dimensional space $\mathbf{R}^3$. The inner product of the vectors (x, y, z) and (a, b, c) is $ax + by + cz$. The expression $[(x, y, z), (x, y, z)] = x^2 + y^2 + z^2$ has a natural interpretation as the *square of the length* of the vector (x, y, z). With this in mind, we can interpret the inner product of the vectors A and B as follows: The square of the length of $A + B = [A + B, A + B] = [A, A] + 2[A, B] + [B, B]$. Also, by the law of cosines (see Figure 5), $[A + B, A + B] = [A, A] + 2[A, A]^{1/2}[B, B]^{1/2} \cos \phi + [B, B]$. By equating both expressions for $[A + B, A + B]$, we find that

3.   ☞ $[A, B] = [A, A]^{1/2}[B, B]^{1/2} \cos \phi$.

That is, the inner product of two vectors is the product of their lengths and the cosine of the angle between them.

The second example suggests the following definition.

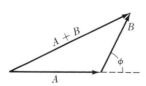

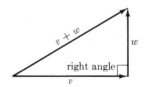

Figure 5   **Figure 6**

4. DEFINITION. If [,] is an inner product on V, the **norm** on V corresponding to it is $||v|| = [v, v]^{1/2}$.

5. THEOREM (CAUCHY-SCHWARZ INEQUALITY.) $|[v, w]| \leq ||v|| \, ||w||$.

Proof. The inequality is surely true if $v = 0$ or $w = 0$. Otherwise, we know that for every real t,
$$0 \leq ||v + tw||^2 = [v + tw, v + tw] = ||v||^2 + 2t[v, w] + t^2||w||^2.$$
In particular, if we put $t = -[v, w]/||w||^2$, we get
$$0 \leq ||v||^2 - 2[v, w]^2/||w||^2 + ||w||^2[v, w]^2/||w||^4$$
$$= ||v||^2 - [v, w]^2/||w||^2 \Rightarrow 0 \leq ||v||^2||w||^2 - [v, w]^2.$$
It follows that the absolute value of $[v, w]$ is $\leq ||v|| \, ||w||$.

6. THEOREM (MINKOWSKI OR TRIANGLE INEQUALITY.) $||v + w|| \leq ||v|| + ||w||$.

Proof. $||v + w||^2 = ||v||^2 + 2[v, w] + ||w||^2 \leq ||v||^2 + 2||v|| \, ||w|| + ||w||^2 = (||v|| + ||w||)^2$. Now take square roots.

7. THEOREM (PARALLELOGRAM LAW.) $||v + w||^2 + ||v - w||^2 = 2[||v||^2 + ||w||^2]$.

Proof. Simply calculate $||v + w||^2 + ||v - w||^2$ in terms of inner products.

EXAMPLES. There is a natural inner product in $\mathbf{R}^n$, namely
$$[(x_1, \ldots, x_n), (y_1, \ldots, y_n)] = x_1y_1 + \cdots + x_ny_n.$$
In terms of this inner product, the Cauchy and Minkowski inequalities say that
$$|x_1y_1 + \cdots + x_ny_n| \leq (x_1^2 + \cdots + x_n^2)^{1/2}(x_1^2 + \cdots + y_n^2)^{1/2}$$
and
$$((x_1 + y_1)^2 + \cdots + (x_n + y_n)^2)^{1/2} \leq (x_1^2 + \cdots + x_n^2)^{1/2} + (y_1^2 + \cdots + y_n^2)^{1/2}.$$

8. REMARK. When does $|[x, y]| = ||x|| \, ||y||$? Examination of the proof of the Schwarz inequality shows that this happens if, and only if, $x + ty = 0$ for $t = -[x, y]/||y||^2$, that is, if and only if the vectors x and y are *linearly dependent*.

9. SUMMARY OF PROPERTIES OF THE NORM:

(a) $||x|| \geq 0$,
(b) $||x|| = 0 \Leftrightarrow x = 0$,
(c) $||tx|| = |t| \, ||x||$,
(d) $||x + y|| \leq ||x|| + ||y||$,

(e) $||x||\ ||y|| \geq |[x, y]|$,

(f) $||x||^2 = [x, x]$.

Proof. Those properties listed above that are not already proved are obvious.

10. INTERPRETATION. In $\mathbf{R}^3$, the Schwarz inequality is obvious. It simply corresponds to the fact that $|\cos \phi| \leq 1$. In general, we can use the inner product to *define* the cosine of the angle between two nonzero vectors as $\cos \angle(v, w) = [v, w]/||v||\ ||w||$. It may not be clear just how to interpret the angle between two functions in $\mathcal{C}([a, b])$. The important thing is that we can reason in this case by analogy to the situation in Euclidean geometry. In fact, any two-dimensional subspace of $\mathcal{C}([a, b])$ is *isomorphic* to the Euclidean plane, in the sense that all the notions of Euclidean plane geometry—including notions of points (functions), lines, angles, lengths, circles, and so on—can be mimicked exactly in a two-dimensional subspace of $\mathcal{C}([a, b])$, using the norm for length and the inner product to define the measure of an angle.

Note in particular that two vectors in $\mathbf{R}^3$ are perpendicular when the cosine of the angle between them is 0. This suggests a definition.

11. DEFINITION. $v \perp w$ means that $[v, w] = 0$.

12. THEOREM (LAW OF PYTHAGORAS.) $v \perp w \Leftrightarrow ||v + w||^2 = ||v||^2 + ||w||^2$.

Proof. $v \perp w \Leftrightarrow 0 = 2[v, w] = ||v + w||^2 - ||v||^2 - ||w||^2$ (Figure 6).

INTERPRETATION IN TERMS OF LINEAR EQUATIONS. Consider an equation $a_1 x_1 + \cdots + a_n x_n = b$. Let P be any vector such that $[(a_1, \ldots, a_n), P] = b$. Let $A = (a_1, \ldots, a_n)$, $X = (x_1, \ldots, x_n)$. Then we can rewrite the linear equation as $[X - P, A] = 0$, or $X - P \perp A$. This says that X lies on the plane through P perpendicular to A (Figure 7).

Exercises

*1. Let g be a continuous function on $[a, b]$ with values in $\mathbf{R}$. For what function f is $\int_a^b f(x)g(x)\, dx$ a maximum subject to the condition that $\int_a^b (f(x))^2\, dx \leq 1$?

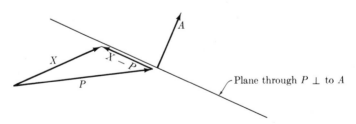

Figure 7

*2. Let $v_1, \ldots, v_n$ be a set of linearly independent vectors in the vector space V over **R**, on which there is an inner product $[\ ,\ ]$. A set of vectors $w_1, \ldots, w_n$ is **orthogonal** if $w_j \perp w_k$ whenever $j \neq k$, and **normal** if $[w_j, w_j] = 1$ for each j. It is **orthonormal** if it is both orthogonal and normal. Show that $\{v_1, \ldots, v_n\}$ can be **orthonormalized** in the following sense: an orthonormal set of vectors $w_1, \ldots, w_n$ can be found such that for each $k = 1, \ldots, n$, $\{v_1, \ldots, v_k\}$ and $\{w_1, \ldots, w_k\}$ span the same subspace of V.

Hint: Start with $w_1 = (1/\|v_1\|)v_1$. Having found $\{w_1, \ldots, w_k\}$, an orthonormal set of vectors spanning the same subspace of V as $\{v_1, \ldots, v_k\}$, express w_{k+1} as a linear combination of v_{k+1} and $w_1, \ldots, w_k$. How should the coefficients be chosen so that w_{k+1} will be orthogonal to $w_1, \ldots, w_k$? Once you have found a vector orthogonal to $w_1, \ldots, w_k$, it is easy to normalize it. Don't forget to show that $\{w_1, \ldots, w_{k+1}\}$ spans the same subspace of V as $\{v_1, \ldots, v_{k+1}\}$.

*3. Orthonormalize the functions $1, x, x^2$ with respect to the inner product $[f, g] = \int_0^1 f(x)g(x)\, dx$ on $\mathcal{C}([0, 1])$.

D.   Linear maps

Suppose we have two vector spaces V and W over the same field F. What would it mean for V and W to be *isomorphic?* An isomorphism between V and W should be a function $L: V \to W$, which is one-to-one, onto, and preserves the operations of vector addition and multiplication by scalars. That is,

(a) $L(v_1) = L(v_2) \Rightarrow v_1 = v_2$,

(b)   for every $w \in W$, there is a $v \in V$ such that $L(v) = w$,

(c) $L(v_1 + v_2) = L(v_1) + L(v_2)$,

(d) $L(sv) = sL(v)$ for each $v \in V$ and $s \in F$.

EXAMPLE. The vector space $\mathbf{R}^2$ is isomorphic to the space S of functions that satisfy the differential equation $f'' + f = 0$. One isomorphism (there are infinitely many) is the map $L: \mathbf{R}^2 \to S$ defined by $L(a, b) = a \cos x + b \sin x$.   ●

It is interesting to look at functions, which we may call *homomorphisms* ("*similar* form," as opposed to "isomorphism," meaning "*same* form") or *linear maps*, or *linear transformations*, which preserve the algebraic structure without necessarily being one-to-one or onto (that is, without necessarily preserving the set-theoretic structure).

1.   DEFINITION. A **linear map (homomorphism, linear transformation)** from a vector space V over F to a vector space W over F is a function $L: V \to W$ such that

(a) $L(v_1 + v_2) = L(v_1) + L(v_2)$ for each $v_1, v_2 \in V$,

(b) $L(rv) = rL(v)$ for each $v \in V$ and $r \in F$.

2.   EXAMPLES. (a) Let S be the vector space of functions that satisfy the differential equation $f'' + f = \mathbf{0}$. If f is a solution, so is its derivative, Df, since $(Df)'' + Df = f''' + f' = (d/dx)(f'' + f) = (d/dx)\mathbf{0} = \mathbf{0}$. Then D is a linear map from S to S.

(b)   Let $v_1, \ldots, v_n$ be a basis for V and $w_1, \ldots, w_m$ a basis for W. Each $m \times n$ matrix

$$M = \begin{pmatrix} a_{11} & \cdots & a_{1n} \\ a_{21} & \cdots & a_{2n} \\ & \vdots & \\ a_{m1} & \cdots & a_{mn} \end{pmatrix}$$

determines a linear map L_M in terms of these bases, as follows. If $v \in V$, v can be expressed uniquely as $x_1 v_1 + \cdots + x_n v_n$. We then put

$$L_M(v) = x_1(a_{11}w_1 + \cdots + a_{m1}w_m) + \cdots + x_n(a_{1n}w_1 + \cdots + a_{mn}w_m)$$
$$= (a_{11}x_1 + \cdots + a_{1n}x_n)w_1 + \cdots + (a_{m1}x_1 + \cdots + a_{mn}x_n)w_m.$$

It is easy to verify that this map is linear. For example, if $v' = y_1 v_1 + \cdots + y_n v_n$, then

$$v + v' = (x_1 + y_1)v_1 + \cdots + (x_n + y_n)v_n,$$

and

$$L_M(v + v') = [a_{11}(x_1 + y_1) + \cdots + a_m(x_n + y_n)]w_1 + \cdots$$
$$= [(a_{11}x_1 + \cdots + a_{1n}x_n)w_1 + \cdots] + [(a_{11}y_1 + \cdots + a_{1n}y_n)w_1 + \cdots]$$
$$= L_M(v) + L_M(v').$$

☛ *Note that the definition of L_M depends not only on the matrix M, but also on the choice of bases for V and W.*

(c)   In case L is a linear map from a vector space V over F into F (which is a one-dimensional vector space over itself) we call L a **linear functional.** For instance, if V is the space $\mathcal{C}([a, b])$ of continuous functions from $[a, b]$ to **R** and $a \le c \le b$, then the **evaluation at c,** $e_c(f) = f(c)$, and the **integral,** $\int f = \int_a^b f(x)\, dx$, are linear functionals on V.

3.   REMARK. If V and W are *finite-dimensional* vector spaces over F, every linear map from V into W can be expressed as in 2(b) by a matrix. Let us choose a basis $v_1, \ldots, v_n$ for V and a basis $w_1, \ldots, w_m$ for W. Let $L: V \to W$ be a linear map. For each j, $L(v_j)$ can be expressed as a linear combination of $w_1, \ldots, w_m$, say $L(v_j) = a_{1j}w_1 + \cdots + a_{mj}w_m$. Then, if $v \in V$, we can represent v uniquely as $x_1 v_1 + \cdots + x_n v_n$. Because L is a linear map,

$$L(x_1 v_1 + \cdots + x_n v_n) = x_1 L(v_1) + \cdots + x_n L(v_n)$$
$$= x_1(a_{11}w_1 + \cdots + a_{m1}w_m) + \cdots + x_n(a_{1n}w_1 + \cdots + a_{mn}w_m),$$

which is exactly the formula for the linear map determined in terms of the given bases for V and W by the matrix

$$M = \begin{pmatrix} a_{11} & \cdots & a_{1n} \\ a_{21} & \cdots & a_{2n} \\ & \vdots & \\ a_{m1} & \cdots & a_{mn} \end{pmatrix} \text{ whose } j\text{th column } \begin{pmatrix} a_{1j} \\ a_{2j} \\ \vdots \\ a_{mj} \end{pmatrix}$$

lists the coefficients for $L(v_j)$ in terms of the basis $w_1, \ldots, w_m$ for W.

☛ It is very important to note that *the particular matrix that represents L depends on the choice of bases in V and W.*

In fact, for fixed bases for V and W, the correspondence we have introduced between linear maps $L: V \to W$ and matrices is one-to-one and onto. If $v_1, \ldots, v_n$ is a basis for V and $w_1, \ldots, w_m$ is a basis for W, then we have shown that every m (rows) $\times$ n (columns) matrix represents a unique linear map with respect to these bases, and that every linear map has such a representation. Thus the function that assigns to each linear map $L: V \to W$ its matrix representation in terms of the given bases for V and W is one-to-one and onto.

If a linear map is not one-to-one, it will send a nonzero vector into $\mathbf{0}$. The amount by which a linear map fails to be one-to-one can be measured by observing the size of its *null-space*, the set of vectors mapped into $\mathbf{0}$.

4. DEFINITION. The **kernel** or **null-space** of a linear map $L: V \to W$ is $\ker L = \{v \in V : L(v) = 0\} = L^{-1}\{0\}$. The **range** of L is

$$\{w \in W : \text{there is a } v \in V \text{ such that } L(v) = w\} = \text{range } L = L(V).$$

5. THEOREM. Let L be a linear map from the vector space V over F to the vector space W over F.

(a) $\ker L$ is a linear subspace of V and range L is a linear subspace of W.
(b) L is one-to-one if, and only if, $\dim \ker L = 0$.
   Suppose in addition that V is finite-dimensional. Then
(c) $\dim V = \dim \ker L + \dim \text{range } L$,
(d) L is an isomorphism of V onto W if, and only if,
   (i) $\dim \ker L = 0$ and
   (ii) $\dim \text{range } L = \dim W$,
      or equivalently, if and only if,
   (iii) $\dim \text{range } L = \dim W = \dim V$.

In particular, there is an isomorphism between V and W if, and only if, they have the same dimension.

Proof. The verification of (a) is trivial.

(b) If $\dim \ker L \neq 0$, there is a nonzero vector $v \in \ker L$. Then $L(v) = L(0) =$

0, so L is not one-to-one. If dim ker $L = 0$, then ker L consists of **0** alone. Then $L(v_1) = L(v_2) \Rightarrow L(v_1) - L(v_2) = L(v_1 - v_2) = \mathbf{0}$, so $v_1 - v_2 \in$ ker L, and therefore $v_1 - v_2 = \mathbf{0}$.

(c)   Choose a basis $v_1, \ldots, v_d$ for ker L, and extend it to a basis $v_1, \ldots, v_d$, $v_{d+1}, \ldots, v_n$ for all of V. To prove (c), it will be enough to show that dim range $L = n - d$, and in order to show this, it will be enough to prove that $L(v_{d+1}), \ldots, L(v_n)$ form a basis for range L. First, these vectors span range L, since if $w \in$ range L, there is a vector $x_1 v_1 + \cdots + x_n v_n \in V$ such that

$$w = L(x_1 v_1 + \cdots + x_n v_n) = x_1 L(v_1) + \cdots + x_d L(v_d)$$
$$+ x_{d+1} L(v_{d+1}) + \cdots + x_n L(v_n) = x_{d+1} L(v_{d+1}) + \cdots + x_n L(v_n),$$

since $L(v_1) = \cdots = L(v_d) = \mathbf{0}$. Next, $L(v_{d+1}), \ldots, L(v_n)$ are independent, since

$$x_{d+1} L(v_{d+1}) + \cdots + x_n L(v_n) = \mathbf{0} \Rightarrow L(x_{d+1} v_{d+1} + \cdots + x_n v_n) = \mathbf{0}$$
$$\Rightarrow x_{d+1} v_{d+1} + \cdots + x_n v_n \in \text{ker } L \Rightarrow x_{d+1} v_{d+1} + \cdots + x_n v_n$$
$$= -x_1 v_1 - \cdots - x_d v_d$$

for some $x_1, \ldots, x_d \in F$, since $v_1, \ldots, v_d$ span ker L. Then

$$x_1 v_1 + \cdots + x_d v_d + x_{d+1} v_{d+1} + \cdots + x_n v_n = \mathbf{0}.$$

Since $v_1, \ldots, v_n$ are a basis for V, $x_1 = x_2 = \cdots = x_d = x_{d+1} = \cdots = x_n = 0$.

(d)   The fact that (i) and (ii) together are equivalent to (iii) follows from (c). Now (i) implies that L is one-to-one, and (ii) that L maps V onto all of W. Since L is then linear, one-to-one, and onto, it is by definition an isomorphism. Conversely, if L is an isomorphism, it is one-to-one, so dim ker $L = 0$, and onto, so that $W = $ range L, or dim $W = $ dim range L.

If there is an isomorphism between V and W, this shows that they have the same dimension. Conversely, if dim $V = $ dim W, we can choose bases $v_1, \ldots, v_n$ for V and $w_1, \ldots, w_n$ for W. Then $L(x_1 v_1 + \cdots + x_n v_n) = x_1 w_1 + \cdots + x_n w_n$ defines a linear map between V and W, which is one-to-one and onto because $\{v_1, \ldots, v_n\}$ and $\{w_1, \ldots, w_n\}$ are bases.

6.   APPLICATION TO LINEAR EQUATIONS. A system of linear equations such as (a) or (c) of Table 5, namely, $x_1 A_1 + \cdots + x_n A_n = B$, can be reformulated in terms of linear maps. We can use the matrix form of the equation, for example, to define the linear map $L((x_1, \ldots, x_n)) = x_1 A_1 + \cdots + x_n A_n$ from F^n to F^m. The linear equation then can be written as $L(x) = B$. The condition for this equation to have a solution is that $B \in$ range L. For the solution, when it exists, to be unique, it is necessary and sufficient that dim ker $L = 0$. If $n < m$, dim range $(L) = n - $ dim ker $L < m$, so range L is not all of F^m. If $m > n$, dim ker $L = n - $ dim range $L \geq n - m > 0$, so ker L is not trivial. If ker L is not trivial, the space of solutions to the equation $L(x) = \mathbf{0}$ has dimension $d = $ dim ker L. The solutions to the equation $L(x) = B$, supposing one, say y, to exist, is obtained by translating this d-dimensional subspace of F^n by the vector y (Figure 8).

Figure 8

In particular, if the equation $L(x) = B$ has more than one solution for a particular choice of B, it has more than one solution for every $B \in$ range L.

E. Operations with linear maps

Linear maps from V to W, being functions, can be added and multiplied by scalars in the usual way. That is, for each $v \in V$, $r \in F$, and each two linear maps L and L', $(L + L')(v) = L(v) + L'(v)$ and $(rL)(v) = r(L(v))$. One has to check that the sum of linear maps is linear, but this is trivial to do. Indeed, it is easy to check the following points.

1. *The linear maps from V to W form a vector space over F.* If V and W are finite-dimensional, so is the space $\mathcal{L}(V, W)$ of linear maps between V and W. This can be seen from the correspondence between $\mathcal{L}(V, W)$ and the $m \times n$ matrices over F ($m = \dim W$, $n = \dim V$). In fact, to add L to L', one has only to add the corresponding entries in their matrix representations in terms of any fixed bases, and likewise to multiply L by a scalar. For example, if we choose the basis $(1, 0)$ and $(0, 1)$ for $\mathbf{R}^2$, and the basis $\cos x$, $\sin x$ for the space of functions that satisfy the differential equation $f'' + f = \mathbf{0}$, then the linear map $L((a, b)) = a \cos x + b \sin x$ has the matrix representation $\begin{pmatrix} 1 & 0 \\ 0 & 1 \end{pmatrix}$. If we follow L by differentiation, we get the linear map DL defined by

$$DL((a, b)) = (d/dx)(a \cos x + b \sin x)$$
$$= -a \sin x + b \cos x = b \cos x - a \sin x.$$

It has the matrix representation $\begin{pmatrix} 0 & 1 \\ -1 & 0 \end{pmatrix}$. Then $L + DL$ is given by the matrix

$$\begin{pmatrix} 1 & 0 \\ 0 & 1 \end{pmatrix} + \begin{pmatrix} 0 & 1 \\ -1 & 0 \end{pmatrix} = \begin{pmatrix} 1+0 & 0+1 \\ 0-1 & 1+0 \end{pmatrix} = \begin{pmatrix} 1 & 1 \\ -1 & 1 \end{pmatrix}.$$

Likewise $3L$ has the matrix representation

$$3\begin{pmatrix} 1 & 0 \\ 0 & 1 \end{pmatrix} = \begin{pmatrix} 3 & 0 \\ 0 & 3 \end{pmatrix}.$$

Now it is easy to see that the $m \times n$ matrices form a vector space (with these operations) that is isomorphic to F^{mn}.

2.   *If* dim $V = n$ *and* dim $W = m$, dim $\mathcal{L}(V, W) = mn$. In particular, a basis for $\mathcal{L}(V, W)$ consists of the linear maps E_{ij} defined by $E_{ij}(v_j) = w_i$, $E_{ij}(v_k) = \mathbf{0}$ if $k \neq j$. Note that, unsurprisingly, this choice of a basis for $\mathcal{L}(V, W)$ depends on the bases chosen for V and W.

I have already illustrated the notion of the *product* of two linear maps with the maps D and L above.

3.   DEFINITION. If U, V, and W are vector spaces, and $L: U \to V$, $M: V \to W$ are linear maps, their **product** ML is defined by $(ML)(v) = M(L(v)) = (M \circ L)(v)$.   ●

It is again trivial to check that the product of linear maps is linear. We also can note the following algebraic properties.

4.   PROPOSITION. If L and L' are linear maps from U to V, M and M' are linear maps from V to W, and N is a linear map from W to Y, then

(a)   *(Associativity)* $N(ML) = (NM)L$,
(b)   *(Right distributive law)* $(M + M')L = (ML) + (M'L)$,
(c)   *(Left distributive law)* $M(L + L') = (ML) + (ML')$.

Proof. The proofs are routine. Take (c) as an example. For each $u \in U$, $M(L + L')(u) = M(L(u) + L'(u)) = M(L(u)) + M(L'(u))$ (because M is linear) $= (ML)(u) + (ML')(u) = [(ML) + (ML')](u)$.

5.   REMARK. The product is *not* generally commutative. For one thing, if we have maps $U \xrightarrow{L} V \xrightarrow{M} W$, then the product ML from U to W is defined, but the product LM makes no sense, since M maps V into W, but L is not even defined on W. Now even when L and M map V into itself, so that both LM and ML are defined, LM may not equal ML. For example, let V be the vector space of real-valued functions on $\mathbf{R}$ that have derivatives of all orders. For $f \in V$, let $(Df)(x) = f'(x)$, and $\left(\int^x f\right)(x) = \int_0^x f(t)\, dt$. Then $\left(D \int^x f\right)(x) = (d/dx) \int_0^x f(t)\, dt = f(x)$, but $\left(\int^x Df\right)(x) = \int_0^x Df(t)\, dt = \int_0^x f'(t)\, dt = f(x) - f(0)$. Thus $\left(D \int^x - \int^x D\right)f$ is not $\mathbf{0}$ unless $f(0) = 0$. Likewise, if we consider the linear maps from $\mathbf{R}^2$ to $\mathbf{R}^2$ defined in terms of the basis $\{(0, 1), (1, 0)\}$ by the matrices

$$L \leftrightarrow \begin{pmatrix} 0 & 1 \\ 1 & 0 \end{pmatrix}, \qquad M \leftrightarrow \begin{pmatrix} 0 & -1 \\ 1 & 0 \end{pmatrix},$$

then LM has the matrix representation $\begin{pmatrix} 1 & 0 \\ 0 & -1 \end{pmatrix}$, but ML has the matrix representation $\begin{pmatrix} -1 & 0 \\ 0 & 1 \end{pmatrix}$.   ●

This suggests a general question: given two linear maps L and M represented by matrices

$$\begin{pmatrix} a_{11} & \cdots & a_{1n} \\ \cdots\cdots\cdots\cdots \\ a_{m1} & \cdots & a_{mn} \end{pmatrix}, \quad \begin{pmatrix} b_{11} & \cdots & b_{1m} \\ \cdots\cdots\cdots\cdots \\ b_{k1} & \cdots & b_{km} \end{pmatrix},$$

what is the matrix representation of ML? More precisely, suppose L maps U into V, and M maps V into W. Suppose we have bases for these spaces $\{u_1, \ldots, u_n\}$, $\{v_1, \ldots, v_m\}$, $\{w_1, \ldots, w_k\}$, and that L and M are represented in terms of these bases by matrices as above. What is the representation of ML in terms of the bases for U and W? Let us compute what ML does to the basis vector u_j:

$$ML(u_j) = M(a_{1j}v_1 + \cdots + a_{mj}v_m) = (b_{11}a_{1j} + \cdots + b_{1m}a_{mj})w_1 + $$
$$\cdots + (b_{k1}a_{1j} + \cdots + b_{km}a_{mj})w_k.$$

That is, if

$$\begin{pmatrix} c_{11} & \cdots & c_{1n} \\ \cdots\cdots\cdots\cdots \\ c_{k1} & \cdots & c_{kn} \end{pmatrix}$$

is the matrix representation for ML, then

☛ $$\boxed{c_{ij} = b_{i1}a_{1j} + \cdots + b_{im}a_{mj}.}$$

This can be expressed as follows. Call the matrix

$$\begin{pmatrix} c_{11} & \cdots & c_{1n} \\ \cdots\cdots\cdots\cdots \\ c_{k1} & \cdots & c_{kn} \end{pmatrix}$$

for ML the matrix product

$$\begin{pmatrix} b_{11} & \cdots & b_{1m} \\ \cdots\cdots\cdots\cdots \\ b_{k1} & \cdots & b_{km} \end{pmatrix} \times \begin{pmatrix} a_{11} & \cdots & a_{1n} \\ \cdots\cdots\cdots\cdots \\ a_{m1} & \cdots & a_{mn} \end{pmatrix}$$

of the matrices for M and for L (in terms of appropriate bases). Then we have the following

6. ☛ To obtain the entry in the ith row and jth column of the product BA of the matrix B by the matrix A, form the **inner product**

$$\boxed{b_{i1}a_{1j} + \cdots + b_{im}a_{mj}}$$

of the ith row of B by the jth column of A (Figure 9, A).

In particular, let M be an $m \times n$ matrix with entry a_{ij} in the ith row and jth column. Then M determines a linear map $L_M \colon F^n \to F^m$ by the usual formula $L_M((x_1, \ldots, x_m)) = (a_{11}x_1 + \cdots + a_{1n}x_n, \ldots, a_{m1}x_1 + \cdots + a_{mn}x_n)$.

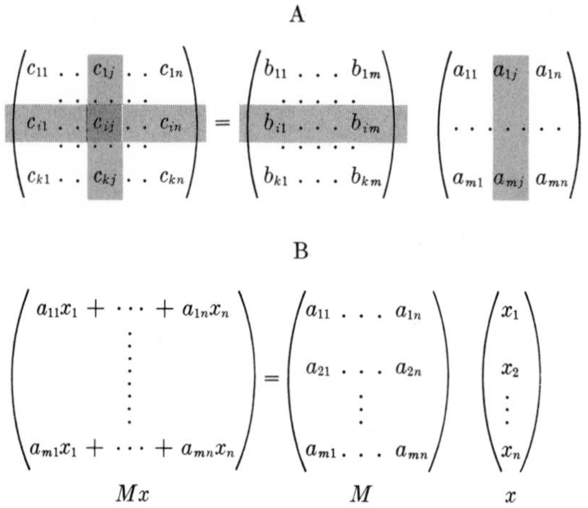

A

$$\begin{pmatrix} c_{11} & .. & c_{1j} & .. & c_{1n} \\ & . & . & . & \\ c_{i1} & .. & c_{ij} & .. & c_{in} \\ & . & . & . & \\ c_{k1} & .. & c_{kj} & .. & c_{kn} \end{pmatrix} = \begin{pmatrix} b_{11} & . & . & . & b_{1m} \\ & . & . & . & \\ b_{i1} & . & . & . & b_{im} \\ & . & . & . & \\ b_{k1} & . & . & . & b_{km} \end{pmatrix} \begin{pmatrix} a_{11} & a_{1j} & a_{1n} \\ & . & . & . & . \\ a_{m1} & a_{mj} & a_{mn} \end{pmatrix}$$

B

$$\begin{pmatrix} a_{11}x_1 + \cdots + a_{1n}x_n \\ \vdots \\ \vdots \\ a_{m1}x_1 + \cdots + a_{mn}x_n \end{pmatrix} = \begin{pmatrix} a_{11} & . & . & . & a_{1n} \\ a_{21} & . & . & . & a_{2n} \\ & & \vdots & & \\ a_{m1} & . & . & . & a_{mn} \end{pmatrix} \begin{pmatrix} x_1 \\ x_2 \\ \vdots \\ x_n \end{pmatrix}$$

$$\qquad Mx \qquad\qquad\qquad M \qquad\qquad x$$

Figure 9. Rule for forming matrix products.

If we think of the vector $(x_1, \ldots, x_n)$ as a **column vector** x, then $L_M((x_1, \ldots, x_n))$ is just the matrix product Mx (Figure 9, B). For this reason it is common to omit parentheses in the expression "$L(x)$" when L is a linear map and x is a vector, and write "Lx" as if $L(x)$ were the product of the vector x by the linear map L.

There is a special linear map of importance, the **identity map,** defined by the formula $Iv = v$. We can also define the inverse of a linear transformation.

7.   DEFINITION. Let T and S be linear transformations from V to V such that $ST = TS = I$. Then S is the **inverse** of T and we write $S = T^{-1}$.

8.   THEOREM. (a) A linear map has at most one inverse.

(b) The following conditions are equivalent if T is a linear map from the *finite-dimensional* space V into V:

   (i) $S = T^{-1}$,
   (ii) $ST = I$,
   (iii) $TS = I$.

(c) If V is *finite-dimensional* and T is a linear map from V into V, then the following conditions are equivalent:

   (i) T has an inverse,
   (ii) T maps V onto V,
   (iii) T is one-to-one.

Proof. (a) If $ST = I = TS'$, then $S = SI = STS' = IS' = S'$ (compare Example II B2(a)).

(b) (i) $\Rightarrow$ (ii). If $S = T^{-1}$, then by definition $ST = I$. (ii) $\Rightarrow$ (iii). If $ST = I$, then S maps V onto V. In fact, if $v \in V$, then $v = S(w)$, where $w = T(v)$. Now by Theorem D5 dim ker $S = $ dim (V) − dim range $(S) = 0$. Therefore, S is also one-to-one. Now $STS = IS = S = SI$, so $S(TS − I) = \mathbf{0}$. If $v \in V$, then $S(TS − I)(v) = \mathbf{0}$. Since S is one-to-one, $(TS − I)(v) = \mathbf{0}$. That is, $(TS − I)(v) = \mathbf{0}$ for every $v \in V$, so $TS − I$ is the zero map, and $TS = I$. (iii) $\Rightarrow$ (i). If $TS = I$, we will show that $S = T^{-1}$ if we show that $ST = I$ as well. But $STS = SI = IS$, so $(ST − I)S = \mathbf{0}$. Now S is one-to-one, since $S(v) = \mathbf{0} \Rightarrow \mathbf{0} = T\mathbf{0} = TS(v) = Iv = v$. Therefore, dim range $S = $ dim V − dim ker $(S) = $ dim V, so S is onto. Then, if v is any vector in V, there is a $w \in V$ such that $v = S(w)$. Then $(ST − I)(v) = (ST − I)S(w) = \mathbf{0}$. That is, $ST − I$ is the zero map, so $ST = I$.

(c)   We have shown in the course of proving (b) that if T has an inverse, it is both one-to-one and onto, and that if T is either one-to-one or onto, then it is both one-to-one and onto. Thus all that remains to be shown is that if T is both one-to-one and onto, it has a linear inverse. *This is true even without the hypothesis that V be finite-dimensional.* In fact, if T is one-to-one and onto, there is an inverse map $T^{-1}: V \to V$, and it has only to be shown that T^{-1} is a linear map. But if $v, w \in V$, $T(T^{-1}(v) + T^{-1}(w)) = TT^{-1}(v) + TT^{-1}(w) = v + w = TT^{-1}(v + w)$. Since T is one-to-one, it follows that $T^{-1}(v) + T^{-1}(w) = T^{-1}(v + w)$. Likewise, if $r \in F, v \in V$, $T(T^{-1}(rv)) = TT^{-1}(rv) = rv = rTT^{-1}(v) = T(rT^{-1}(v))$, and since T is one-to-one, $T^{-1}(rv) = rT^{-1}(v)$.   ●

Exercises

1. Give an example of a linear map from $\mathbf{R}^2$ into $\mathbf{R}^2$ that has no inverse, but is *not* identically $\mathbf{0}$.

2. Let P be the vector space of polynomials with real coefficients, and let D be the differentiation operator, which is a linear map from P into $P: Df = f'$ for each polynomial $f \in P$. Show that D maps P onto P, but that D is not one-to-one. In fact, dim ker $D = 1$. Show that there is an operator J from P into P that is one-to-one but not onto, such that $DJ = I$. *Hint:* If $f(x) \in P$, let $Jf(x)$ be the polynomial defined by $Jf(x) = \int_0^x f(t)\, dt$. Show that this is indeed a polynomial. Why doesn't Theorem E8(c) apply?

3. If dim $V = n$, and L is a linear functional on V which is not the identically zero functional, what is dim ker L?

*4. Show that if L is a linear functional on F^n, there is a vector $v = (x_1, \ldots, x_n) \in F^n$ such that $L(y_1, \ldots, y_n) = (y_1, \ldots, y_n) \cdot v = y_1 x_1 + \cdots + y_n x_n$, and conversely.

*5. Consider the linear map D of Example D2(a), which differentiates functions satisfying the differential equation $f'' + f = 0$.

   (a)   Show that with respect to the basis $\cos x$, $\sin x$ for the space S of solutions to this differential equation, the matrix of D is $\begin{pmatrix} 0 & 1 \\ -1 & 0 \end{pmatrix}$.

   (b)   The functions $\cos x + \sin x$ and $\sin x$ also form a basis for S. What is the matrix for D with respect to this new basis?

*6. Let L be the linear map whose matrix is $\begin{pmatrix} 1 & 2 \\ 3 & 4 \end{pmatrix}$ *with respect to the basis* $(1, 0)$, $(0, 1)$ for $\mathbf{R}^2$. What is the matrix for L *with respect to the basis* $(1, 1)$, $(1, -1)$?

F. Determinants

The following material on determinants and the succeeding section on eigenvalues and eigenvectors will not be used elsewhere in this book, but it is presented because it fits in naturally with the preceding discussion of linear maps, and because it may be useful to students of applied mathematics. The most useful thing this section could accomplish is to persuade the reader to *avoid using determinants whenever possible.* Almost all computations that can be made using determinants can better be made without them. This is true in particular of Cramer's Rule for solving linear equations, of the use of the Wronskian determinant in the elementary theory of linear differential equations, and of the use of determinants for the calculation of eigenvalues of matrices of large dimension. Determinants are useful primarily as a theoretical device, in algebra (the theory of integral dependence), in the theory of analytic functions of several complex variables, in algebraic geometry, in making changes of variable in a multiple integral, in differential geometry.

Throughout this section, V will be an *n-dimensional* vector space over the field F.

1. DEFINITION. An **alternating multilinear form** on V is a function ϕ that maps the set V^n of n-tuples of elements in V into F, and has the following properties.

(a) *Alternating property.* If $v_1, \ldots, v_n \in V$ and $v_i = v_j$ for some *distinct* pair of indices i, j, then $\phi(v_1, \ldots, v_n) = 0$.

(b) *Multilinear property.* ϕ is a linear function of each argument separately. That is, for each $i = 1, \ldots, n$,

$$\phi(v_1, \ldots, sv_i + tw_i, v_{i+1}, \ldots, v_n) = s\phi(v_1, \ldots, v_i, v_{i+1}, \ldots, v_n)$$
$$+ t\phi(v_1, \ldots, w_i, v_{i+1}, \ldots, v_n). \quad \bullet$$

2. REMARK. It is a consequence of the alternating property that

$$\phi(v_1, \ldots, \overset{i\text{th place}}{v_i}, \ldots, \overset{j\text{th place}}{v_j}, \ldots, v_n) = -\phi(v_1, \ldots, \overset{i\text{th place}}{v_j}, \ldots, \overset{j\text{th place}}{v_i}, \ldots, v_n).$$

That is, interchanging two of the arguments of ϕ changes the sign of ϕ. In fact,

$$\phi(v_1, \ldots, \overset{i\text{th place}}{v_i + v_j}, \ldots, \overset{j\text{th place}}{v_i + v_j}, \ldots, v_n)$$
$$= \phi(v_1, \ldots, v_i, \ldots, v_i, \ldots, v_n) + \phi(v_1, \ldots, v_i, \ldots, v_j, \ldots, v_n)$$
$$+ \phi(v_1, \ldots, v_j, \ldots, v_i, \ldots, v_n) + \phi(v_1, \ldots, v_j, \ldots, v_j, \ldots, v_n)$$

by the multilinearity of ϕ. But the expression on the left side of this equation, as well as the first and last terms on the right, are all 0 because ϕ is alternating. Thus

$$0 = \phi(v_1, \ldots, v_i, \ldots, v_j, \ldots, v_n) + \phi(v_1, \ldots, v_j, \ldots, v_i, \ldots, v_n).$$

If F is a field in which $1 + 1 \neq 0$, then a multilinear form that has this property is alternating. In fact if $v_i = v_j$, then

$$\phi(v_1, \ldots, v_i, \ldots, v_j, \ldots, v_n) = \phi(v_1, \ldots, v_j, \ldots, v_i, \ldots, v_n).$$

But because interchanging v_i and v_j changes the sign of ϕ,

$$\phi(v_1, \ldots, v_i, \ldots, v_j, \ldots, v_n) = -\phi(v_1, \ldots, v_j, \ldots, v_i, \ldots, v_n).$$

Adding these two equations, we find that

$$(1 + 1)\phi(v_1, \ldots, v_i, \ldots, v_j, \ldots, v_n) = 0.$$

If $1 + 1 \neq 0$, this implies that

$$\phi(v_1, \ldots, v_i, \ldots, v_j, \ldots, v_n) = 0,$$

so that ϕ is alternating. $\bullet$

3. EXAMPLE. Alternating multilinear forms arise naturally in geometry as **oriented lengths, areas, and volumes.** There is a familiar interpretation of $\int_a^b f(x)\,dx$ as the area of the figure bounded by the curve $y = f(x)$, the x-axis, and the lines $x = a$ and $x = b$ (Figure 10, A). To use this interpretation when $f(x) < 0$ for $x \in [a, b]$, we must introduce the notion of *oriented area*. Areas of figures lying above the x-axis are counted as positive; areas of figures lying below the x-axis are counted as negative (Figure 10, B).

The notion of oriented area does not only appear as a device for interpreting definite integrals. Here is a physical interpretation (Figure 11). The figure shows a fluid flowing with velocity C across a rectangular surface perpendicular to the fluid flow. Two sides of the rectangle are spanned by the vectors A and B. During t units of time, a parallelopiped of fluid with edges A, B, and tC will flow across the rectangle. The volume of fluid that crosses the rectangle is the product of the length $\|tC\|$ by the *area* of the rectangle spanned by A and B. If we want to measure the amount of fluid flowing *into* Box 2 *out of* Box 1, we should consider this quantity of fluid to be positive. Alternatively, we could consider it to be a negative quantity of fluid flowing *into* Box 1 *out of* Box 2. (Perhaps this is easier to see if you think of the fluid as positive electrical charge. A positive charge flowing *out of* Box 1 has the same effect on the total charge in Box 1 as a negative charge flowing *into* Box 1.) We can consider the quantity of fluid flowing *out of* Box 1 across the rectangle to be the product of the length $\|tC\|$ by the *oriented area* of the rectangle. This rectangle on the surface of Box 1 can be considered to be *positively oriented* with respect to the fluid flow when the fluid flows *out of* Box 1. The

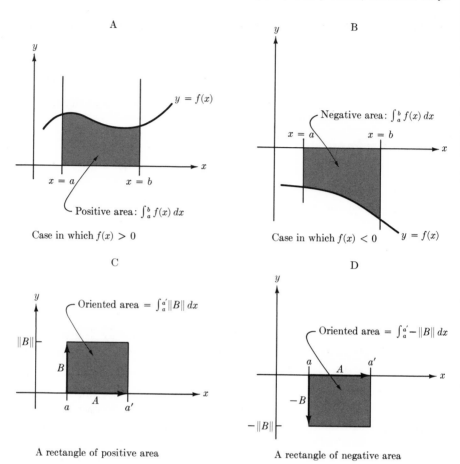

A

$y = f(x)$

$x = a$ $x = b$

Positive area: $\int_a^b f(x)\,dx$

Case in which $f(x) > 0$

B

Negative area: $\int_a^b f(x)\,dx$

$x = a$ $x = b$

Case in which $f(x) < 0$ $y = f(x)$

C

Oriented area $= \int_a^{a'} \|B\|\,dx$

$\|B\|$

B

a A a'

A rectangle of positive area

D

Oriented area $= \int_a^{a'} -\|B\|\,dx$

a A a'

$-B$

$-\|B\|$

A rectangle of negative area

Figure 10

negative orientation of the rectangle with respect to the flow corresponds to flow *into* Box 1.

If we turn the boxes upside down, by a 180° rotation about the axis through A, without disturbing the fluid flow, the orientation of the rectangle

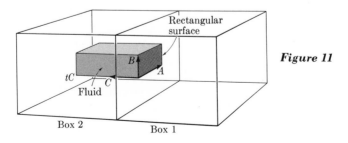

Rectangular surface

B

tC

C

A

Fluid

Box 2

Box 1

Figure 11

changes. The positively oriented rectangle is shown in Figure 10, C; the negatively oriented rectangle is shown in Figure 10, D. Note the correspondence between the orientations of the rectangles and the signs of the corresponding definite integrals.

Any two vectors A and B in the plane span a parallelogram, as shown in Figure 12, A. We define an alternating multilinear form $\phi(A, B)$ to be the oriented area of the parallelogram spanned by A and B. To compute this area, rotate the coordinate axes so that the positive x-axis lies along the vector A (Figure 12, B). Then compute the area of the parallelogram as usual, with a plus sign if B lies above the x-axis, and a minus sign if B lies below the

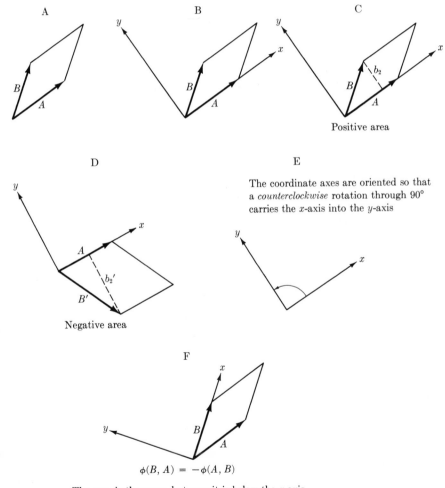

A B C

Positive area

D   E

Negative area

The coordinate axes are oriented so that a *counterclockwise* rotation through 90° carries the x-axis into the y-axis

F

$\phi(B, A) = -\phi(A, B)$

The area is the same, but now it is below the x-axis

Figure 12

x-axis. Suppose that when the coordinates have been rotated so that the positive x-axis lies along A, the components of the vector B are (b_1, b_2). Then $\phi(A, B) = ||A|| b_2$ (Figure 12, C). This makes it clear that ϕ is a *linear* function of B. Clearly the rectangle spanned by A and A has width 0, so $\phi(A, A) = 0$. That is, ϕ is *alternating*. As shown in Figure 12, F, $\phi(B, A) = -\phi(A, B)$. Thus $\phi(A + A', B) = -\phi(B, A + A') = -\phi(B, A) - \phi(B, A') = \phi(A, B) + \phi(A', B)$, and similarly $\phi(tA, B) = t\phi(A, B)$. That is, $\phi(A, B)$ is a *linear* function of A as well. In sum, ϕ is *multilinear*.   ●

The notion of oriented lengths, areas, and volumes is essential to a precise statement of such theorems in vector analysis as Stokes' Theorem and the Divergence Theorem. A general treatment of this subject starts with the use of alternating multilinear forms to *define* oriented areas, volumes, and so on, rather than using the intuitive geometric approach sketched above.   ●

It is easy to see that the sum of two alternating multilinear forms is again an alternating multilinear form, as is the product of an alternating multilinear form by a scalar. That is, *the alternating multilinear forms on V form a vector space $AM(V)$ over F.*

4.   THEOREM. If $n > 0$, the dimension of $AM(V)$ is 1.

Proof. There are two major parts to the proof. First it will be shown that there is a nonzero element ϕ of $AM(V)$. Then it will be shown that every element of $AM(V)$ is a multiple of ϕ. These two facts together prove that ϕ is a basis for $AM(V)$. Since $AM(V)$ has a basis consisting of one element, its dimension is 1. The proof is complicated, so it will be broken up into several lemmas. (A lemma is a *branch* of a proof, that is, a little theorem proved on the way to establishing a big one.)

Let $b_1, \ldots, b_n$ be a basis for V. It is easy to construct a multilinear form μ on V. For each $i = 1, \ldots, n$ let L_i be the linear functional $L_i: x_1 b_1 + \cdots + x_n b_n \mapsto x_i$. L_i has the $1 \times n$ matrix representation $(0, \ldots, 1, \ldots, 0)$ with 1 in the ith place. Then put $\mu(v_1, \ldots, v_n) = L_1(v_1) L_2(v_2) \ldots L_n(v_n)$. The trouble is that μ is not alternating (unless $n = 1$, in which case the alternating property is trivial). An alternating form has the property that a permutation of its arguments multiplies its value by ± 1. Since μ does not have this property, it must be modified so as to give each permutation of its arguments equal influence on the final result.

5.   DEFINITION. A **permutation** of the integers $1, \ldots, n$ is a function mapping the set $\{1, \ldots, n\}$ one-to-one onto itself. π_n is the set of all permutations of $\{1, \ldots, n\}$.

6.   LEMMA. There is a function sgn from π_n into F with the following properties.

(a) sgn $(\sigma) = +1$ or sgn $(\sigma) = -1$ for each $\sigma \in \pi_n$.

(b) $(\text{sgn }(\sigma))(\text{sgn }(\tau)) = \text{sgn }(\sigma \circ \tau)$. That is, sgn is a *homomorphism* from the group π_n into the multiplicative group $\{1, -1\} \subseteq F$. (Compare Section IIB.)

(c) If σ interchanges just two integers and leaves the others fixed, sgn $(\sigma) = -1$. (Such a permutation is called a **transposition**.)

Proof. Consider the polynomial $p(X_1, \ldots, X_n)$ in n indeterminates with coefficients in F defined by

$$p(X_1, \ldots, X_n) = (X_n - X_{n-1}) \ldots (X_n - X_1)(X_{n-1} - X_{n-2}) \ldots$$
$$(X_{n-2} - X_1) \ldots (X_2 - X_1) = \underset{i<j}{\text{Product }} (X_j - X_i).$$

If $\sigma \in \pi_n$, then $p(X_{\sigma(1)}, \ldots, X_{\sigma(n)})$ is the product of the same factors $(X_j - X_i)$ with $i < j$, except that the order of the factors may be interchanged, and the signs of some factors may be changed. For example, if σ transposes 1 and 2, then the factor $(X_{\sigma(2)} - X_{\sigma(1)})$ is just $(X_1 - X_2) = -(X_2 - X_1)$. Thus $p(X_{\sigma(1)}, \ldots, X_{\sigma(n)}) = \pm p(X_1, \ldots, X_n)$. Put

$$\text{sgn }(\sigma) = \frac{p(X_{\sigma(1)}, \ldots, X_{\sigma(n)})}{p(X_1, \ldots, X_n)}.$$

This proves (a). Note that

$$p(X_{\sigma \circ \tau(1)}, \ldots, X_{\sigma \circ \tau(n)}) = \text{sgn }(\sigma)p(X_{\tau(1)}, \ldots, X_{\tau(n)})$$
$$= \text{sgn }(\sigma) \text{ sgn }(\tau)p(X_1, \ldots, X_n).$$

This proves (b).

Proof of (c). Suppose σ interchanges the integers i and j $(i < j)$ and leaves the other integers from 1 to n fixed. Classify the factors $X_r - X_s$ of p $(s < r)$ into these four classes.

(i) r and s are both different from i and from j.

(ii) One of r and s is i or is j, the other is less than i or greater than j.

(iii) One of r and s is i or is j, the other lies between i and j.

(iv) $r = j$, $s = i$.

In the case of a factor $X_r - X_s$ of type (i), $X_{\sigma(r)} - X_{\sigma(s)} = X_r - X_s$. Factors of type (ii) occur in pairs: if $r > j$, the pair is $(X_r - X_j)(X_r - X_i)$. Then

$$(X_{\sigma(r)} - X_{\sigma(j)})(X_{\sigma(r)} - X_{\sigma(i)}) = (X_r - X_i)(X_r - X_j) = (X_r - X_j)(X_r - X_i).$$

The same is true if $s < i$. Thus the effect of σ on such a pair is just to interchange the pair of factors with no change in sign. Factors of type (iii) also occur in pairs: $(X_j - X_r)(X_r - X_i)$. Then

$$(X_{\sigma(j)} - X_{\sigma(r)})(X_{\sigma(r)} - X_{\sigma(i)}) = (X_i - X_r)(X_r - X_j) = (X_j - X_r)(X_r - X_i).$$

Thus the effect of σ on such a pair is to interchange the pair with two canceling changes of sign. Finally, the single pair $(X_j - X_i)$ has its sign changed by σ:

$$(X_{\sigma(j)} - X_{\sigma(i)}) = (X_i - X_j) = -(X_j - X_i). \quad \bullet$$

7.   LEMMA. The form

$$\phi: (v_1, \ldots, v_n) \rightarrow \sum_{\sigma \in \pi_n} \text{sgn } (\sigma)\mu(v_{\sigma(1)}, \ldots, v_{\sigma(n)})$$

is alternating and multilinear and $\phi(b_1, \ldots, b_n) = 1$.

Proof. For each permutation σ, the form

$$(v_1, \ldots, v_n) \mapsto \mu(v_{\sigma(1)}, \ldots, v_{\sigma(n)})$$

is multilinear. Thus ϕ, the sum of multilinear forms, is multilinear. Since $L_i(b_j) = L_i(0b_1 + \cdots + 1b_j + \cdots + 0b_n)$, $L_i(b_j) = 0$ if $i \neq j$, and $L_i(b_i) = 1$. It follows that $\mu(b_1, \ldots, b_n) = 1$ but that $\mu(b_{\sigma(1)}, \ldots, b_{\sigma(n)}) = 0$ unless σ is the *identity* permutation I, which leaves each of the integers $1, \ldots, n$ fixed. Since sgn $I = 1$, obviously, it follows that $\phi(b_1, \ldots, b_n) = 1$. In particular, ϕ is not the zero form.

It remains to be shown that ϕ is alternating. Suppose then that $v_i = v_j$ for some pair of indices $i \neq j$. Then in particular n must be at least 2 (ϕ is trivially alternating if $n = 1$). Let τ be the transposition that interchanges i and j and leaves every other integer from 1 to n fixed. $\tau \circ \tau$ is the identity permutation, If $\sigma \in \pi_n$, there is only one permutation ω different from σ such that $\omega(k) = \sigma(k)$ for each $k \in \{1, \ldots, n\} - \{i, j\}$. In fact, if ω is such a permutation, $\sigma^{-1} \circ \omega$ is not the identity permutation, but it leaves fixed every integer from 1 to n except i and j. Thus $\sigma^{-1} \circ \omega = \tau$, $\omega = \sigma \circ \tau$. If $\{\omega, \omega \circ \tau\}$ contains an element of $\{\sigma, \sigma \circ \tau\}$, then either $\omega = \sigma$ or $\omega = \sigma \circ \tau$. ($\omega \circ \tau = \sigma \Rightarrow \omega = \sigma \circ \tau^{-1} = \sigma \circ \tau$, and $\omega \circ \tau = \sigma \circ \tau \Rightarrow \omega = \sigma$.) If $\omega = \sigma$, then $\omega \circ \tau = \sigma \circ \tau$. If $\omega = \sigma \circ \tau$, then $\omega \circ \tau = \sigma \circ (\tau \circ \tau) = \sigma$. Thus $\{\omega, \omega \circ \tau\} = \{\sigma, \sigma \circ \tau\}$ or else $\{\omega, \omega \circ \tau\}$ and $\{\sigma, \sigma \circ \tau\}$ have no element in common. We can express π_n as the union of such pairs of permutations. Thus to show that

$$\sum_{\sigma \in \pi_n} \text{sgn } (\sigma)\mu(v_{\sigma(1)}, \ldots, v_{\sigma(n)}) = 0,$$

it is enough to prove that each pair of terms

$$\text{sgn } (\sigma)\mu(v_{\sigma(1)}, \ldots, v_{\sigma(n)}) + \text{sgn } (\sigma \circ \tau)\mu(v_{\sigma\circ\tau(1)}, \ldots, v_{\sigma\circ\tau(n)})$$

adds up to 0. But this is true because $v_i = v_j \Rightarrow v_{\sigma(k)} = v_{\sigma\circ\tau(k)}$ for each $k = 1, \ldots, n \Rightarrow \mu(v_{\sigma(1)}, \ldots, v_{\sigma(n)}) = \mu(v_{\sigma\circ\tau(1)}, \ldots, v_{\sigma\circ\tau(n)})$, and sgn $(\sigma \circ \tau) = $ sgn (σ) sgn $(\tau) = -$sgn σ. ●

8.   LEMMA. If $\sigma \in \pi_n$, then σ is the product of at most $n - 1$ transpositions.

Proof, by induction on n. The Lemma is trivial if $n = 1$, since π_1 contains only the identity permutation. Suppose every element of π_{n-1} is the product of at most $n - 2$ transpositions. If $\sigma \in \pi_n$ and $\sigma(n) \neq n$, let τ be the transposition that interchanges n and $\sigma(n)$. Then $\tau \circ \sigma$ leaves n fixed. We can thus define a permutation ω of $\{1, \ldots, n - 1\}$ by the formula $\omega(k) = \tau \circ \sigma(k)$ for $k = 1, \ldots, n - 1$. By the induction hypothesis, there are m transpositions $\tau_1, \ldots, \tau_m \in \pi_{n-1}$ ($m \leq n - 2$) such that $\omega = \tau_1 \circ \tau_2 \circ \cdots \circ \tau_m$. Let τ_j' be the permutation of $\{1, \ldots, n\}$ defined by $\tau_j'(k) = \tau_j(k)$ if $k < n$, $\tau_j'(n) = n$. Then $\tau \circ \sigma = \tau_1' \circ \tau_2' \circ \cdots \circ \tau_m'$, and $\sigma = \tau \circ \tau_1' \circ \tau_2' \circ \cdots \circ \tau_m'$. If $\sigma(n) = n$, the same argument works without the intervention of τ: $\sigma = \tau_1' \circ \tau_2' \circ \cdots \circ \tau_m'$. ●

9.   LEMMA. If $\psi \in AM(V)$ and $\sigma \in \pi_n$, then $\psi(v_{\sigma(1)}, \ldots, v_{\sigma(n)}) = \text{sgn } (\sigma)\psi(v_1, \ldots, v_n)$.

Proof. Express σ as a product of m transpositions $\tau_1 \circ \cdots \circ \tau_m$. According to Remark 2, and Lemma 6,

$$\psi(v_{\sigma(1)}, \ldots, v_{\sigma(n)}) = (-1)\psi(v_{\tau_1 \circ \cdots \circ \tau_{m-2}(1)}, \ldots, v_{\tau_1 \circ \cdots \circ \tau_{m-2}(n)})$$
$$= \cdots = (-1)^m\psi(v_1, \ldots, v_n) = \text{sgn } (\tau_1) \ldots \text{sgn } (\tau_m)\psi(v_1, \ldots, v_n)$$
$$= \text{sgn } (\tau_1 \circ \cdots \circ \tau_m)\psi(v_1, \ldots, v_n) = \text{sgn } (\sigma)\psi(v_1, \ldots, v_n).$$

10.   LEMMA. $\psi \in AM(V) \Rightarrow \psi = \psi(b_1, \ldots, b_n)\phi$.

Proof. Let $\chi = \psi - \psi(b_1, \ldots, b_n)\phi$. Then $\chi \in AM(V)$, and since $\phi(b_1, \ldots, b_n) = 1$, $\chi(b_1, \ldots, b_n) = 0$. If $v_1, \ldots, v_n \in V$, we can express each v_j as $a_{j1}b_1 + \cdots + a_{jn}b_n$, where $a_{ji} \in F$. Then

$$\chi(v_1, \ldots, v_n) = \sum_{1 \leq i_1, \ldots, i_n \leq n} a_{1i_1}a_{2i_2} \ldots a_{ni_n}\chi(b_{i_1}, \ldots, b_{i_n})$$

by the multilinearity of χ. Note that the indices $i_1, \ldots, i_n$ run from 1 to n independently of one another. In case $i_k = i_j$ for some $j \neq k$, $\chi(b_{i_1}, \ldots, b_{i_n}) = 0$ because χ is multilinear. Otherwise, the map $\sigma: k \mapsto i_k$ is a permutation. Then $\chi(b_{i_1}, \ldots, b_{i_n}) = \chi(b_{\sigma(1)}, \ldots, b_{\sigma(n)}) = \text{sgn } (\sigma)\chi(b_1, \ldots, b_n) = 0$. Thus every term of the sum above is 0. This shows that χ is the 0 element of $AM(V)$, so that $\psi = \psi(b_1, \ldots, b_n)\phi$. ●

Lemma 10 completes the proof of Theorem 4. ●

If ϕ is now any nonzero alternating multilinear form on V, and L is a linear map from V into V, we can define a new alternating multilinear form $L\phi$ by the formula $L\phi(v_1, \ldots, v_n) = \phi(L(v_1), \ldots, L(v_n))$. Because $AM(V)$ is one-dimensional, there is a unique scalar $s \in F$ such that $L\phi = s\phi$. This scalar depends only on L, not on the choice of ϕ, since if ψ is another form in $AM(V)$, then there is a scalar t such that $\psi = t\phi$. Then

$$L\psi(v_1, \ldots, v_n) = \psi(L(v_1), \ldots, L(v_n)) = t\phi(L(v_1), \ldots, L(v_n))$$
$$= ts\phi(v_1, \ldots, v_n) = st\phi(v_1, \ldots, v_n) = s\psi(v_1, \ldots, v_n),$$

so the same scalar that works for ϕ works for ψ as well.

11.   DEFINITION. If L is a linear map from V to V, its **determinant** det (L) is the unique scalar s such that $L\phi = s\phi$ for each $\phi \in AM(V)$.

12. REMARK. This is a somewhat unconventional definition of the determinant function. Its advantage is that it leads very quickly to proofs of the fundamental properties of determinants, as well as providing a natural interpretation of the determinant (see below) and the interesting information contained in Theorem 4 about alternating multilinear forms. In particular, it is clear at once from this definition that det (L) depends only on the linear map L, and not on any choice of basis for V.

13. INTERPRETATION. If we interpret an alternating multilinear form as giving an oriented n-dimensional volume as in 3, then the determinant has a natural interpretation as follows. Take the two-dimensional case as an example. If we give two vectors v and w determining a parallelogram P (Figure 13, A) and then operate on them by a linear map L, we get two new vectors $L(v)$ and $L(w)$ determining a new parallelogram $L(P)$ (Figure 13, B). The determinant of L is the ratio of the oriented areas Area $(L(P))/$Area (P). Theorem 4 shows that this ratio depends only on L, not on P. Thus det (L) tells us how much L expands oriented areas. In n-dimensions, det (L) is the expansion factor for oriented n-dimensional volumes. $|$det $(L)|$ is then the expansion factor for ordinary (nonoriented) volumes.

14. METHOD FOR COMPUTATION. Our definition of det (L) immediately affords a method for computing it. Take any nonzero alternating multilinear form such as the one constructed in the proof of Theorem 4. Take any basis $b_1, \ldots, b_n$ for V. Then det $(L) = \phi(L(b_1), \ldots, L(b_n))/\phi(b_1, \ldots, b_n)$. If we write out L in terms of a matrix using this basis, L corresponds to the matrix

$$\begin{pmatrix} a_{11} & \cdots & a_{1n} \\ \cdots\cdots\cdots\cdots \\ \cdots\cdots\cdots\cdots \\ a_{n1} & \cdots & a_{nn} \end{pmatrix},$$

where $L(b_j) = a_{ij}b_1 + \cdots + a_{nj}b_n$. Suppose we adjust ϕ so that $\phi(b_1, \ldots, b_n) = 1$. Then

$$\text{det } (L) = \phi(a_{11}b_1 + \cdots + a_{n1}b_n, \ldots, a_{1n}b_1 + \cdots + a_{nn}b_n).$$

To evaluate det (L), then, one needs only to know its matrix representation in terms of any basis for V. It is common, indeed, to speak of the determi-

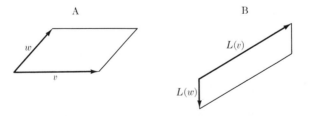

Figure 13

nant of a matrix. If M is a matrix, $\{b_1, \ldots, b_n\}$ is a basis for V, and L_M is the linear map corresponding to M with respect to this basis, then by definition det $(M) = $ det (L_M). The preceding remarks show that det (M) depends only on M, not on the choice of $b_1, \ldots, b_n$.

15. EXAMPLE. Let us calculate the determinant of the linear map D that differentiates a function satisfying the differential equation $f'' + f = 0$. Pick a basis cos x, sin x for the space of solutions. One alternating multilinear form on this space is defined by

$$\phi(a \cos x + b \sin x, c \cos x + d \sin x) = ad - bc.$$

Since $D \cos x = 0 \cos x - 1 \sin x$, $D \sin x = 1 \cos x + 0 \sin x$, det $(D) = 0 \cdot 0 - (-1)(1) = 1$.

16. FURTHER REMARKS ON COMPUTATION. In practice, the most efficient way to calculate a determinant is to use the following three

☞ *rules for column operations on a multilinear form.*

17. THEOREM. Let ϕ be any alternating multilinear form on V.

(a) If two of the arguments in the expression $\phi(v_1, \ldots, v_n)$ are interchanged, ϕ is multiplied by -1. That is,

$$\phi(v_1, \ldots, v_i, \ldots, v_j, \ldots, v_n) = -\phi(v_1, \ldots, v_j, \ldots, v_i, \ldots, v_n).$$

(b) If one argument in the expression $\phi(v_1, \ldots, v_n)$ is multiplied by t, ϕ is multiplied by t. That is,

$$\phi(v_1, \ldots, tv_i, \ldots, v_n) = t\phi(v_1, \ldots, v_i, \ldots, v_n).$$

(c) If a multiple of one argument is added to another, ϕ is unchanged. That is,

$$\phi(v_1, \ldots, v_i, \ldots, v_j + tv_i, \ldots, v_n) = \phi(v_1, \ldots, v_i, \ldots, v_j, \ldots, v_n).$$

Proof. (a) and (b) are simply part of the definition of multilinearity (see Remark 2). As to (c), we have

$$\phi(v_1, \ldots, v_i, \ldots, v_i, \ldots, v_n) = 0$$

because the two arguments are the same. Then

$$\phi(v_1, \ldots, v_i, \ldots, tv_i, \ldots, v_n) = t\phi(v_1, \ldots, v_i, \ldots, v_i, \ldots, v_n) = 0,$$

and

$$\phi(v_1, \ldots, v_i, \ldots, v_j + tv_i, \ldots, v_n)$$
$$= \phi(v_1, \ldots, v_i, \ldots, v_j, \ldots, v_n) + \phi(v_1, \ldots, v_i, \ldots, tv_i, \ldots, v_n)$$
$$= \phi(v_1, \ldots, v_i, \ldots, v_j, \ldots, v_n),$$

by linearity in the jth argument. ●

When we represent L in terms of a matrix, the coefficients of the arguments in the expression

$$\det (L) = \frac{\phi(L(b_1), \ldots, L(b_n))}{\phi(b_1, \ldots, b_n)}$$

$$= \frac{\phi(a_{11}b_1 + \cdots + a_{n1}b_n, \ldots, a_{1n}b_1 + \cdots + a_{nn}b_n)}{\phi(b_1, \ldots, b_n)}$$

appear as the columns of the matrix representing L. Theorem 11 tells us how to simplify the calculation of $\phi(L(b_1), \ldots, L(b_n))$ by *using column operations*.
☛ In fact, *to evaluate* $\det (L)$, *we never have to know that ϕ is!*

EXAMPLE. Let us pick the basis $(1, 0, 0)$, $(0, 1, 0)$, $(0, 0, 1)$ for $\mathbf{R}^3$. Let L be the linear map whose matrix is

$$\begin{pmatrix} 1 & 2 & 3 \\ 4 & 8 & 6 \\ 7 & 7 & 9 \end{pmatrix}$$

Let us find $\det (L)$, after first performing column operations on the matrix as indicated below.

(i) $\begin{pmatrix} 1 & 2 & 3 \\ 4 & 8 & 6 \\ 7 & 7 & 9 \end{pmatrix}$ Original matrix.

(ii) $\begin{pmatrix} 1 & 0 & 0 \\ 4 & 0 & -6 \\ 7 & -7 & -12 \end{pmatrix}$ Subtract $2 \times$ first column from second column, and subtract $3 \times$ first column from third column to get two zeros in first row.

(iii) $\begin{pmatrix} 1 & 0 & 0 \\ 4 & 0 & 1 \\ 7 & 1 & 2 \end{pmatrix}$ Multiply second column by $-\frac{1}{7}$ and third by $-\frac{1}{6}$.

(iv) $\begin{pmatrix} 1 & 0 & 0 \\ 4 & 0 & 1 \\ 0 & 1 & 0 \end{pmatrix}$ Subtract $7 \times$ second column from first and $2 \times$ second column from third to get two zeros in third row.

(v) $\begin{pmatrix} 1 & 0 & 0 \\ 0 & 0 & 1 \\ 0 & 1 & 0 \end{pmatrix}$ Subtract $4 \times$ third column from first to get two zeros in second row.

(vi) $\begin{pmatrix} 1 & 0 & 0 \\ 0 & 1 & 0 \\ 0 & 0 & 1 \end{pmatrix}$ Interchange second and third columns to get identity matrix.

(v alternate)

$\begin{pmatrix} 1 & 0 & 0 \\ 4 & 1 & 0 \\ 0 & 0 & 1 \end{pmatrix}$ Interchange second and third columns to get triangular matrix.

At the step (vi) we have the matrix whose columns give the coefficients for the vectors $b_1 = 1 \times b_1 + 0 \times b_2 + 0 \times b_3$, $b_2 = 0 \times b_1 + 1 \times b_2 + 0 \times b_3$, and $b_3 = 0 \times b_1 + 0 \times b_2 + 1 \times b_3$. If we apply ϕ to these three vectors, we get $\phi(b_1, b_2, b_3)$. How did we reach this stage from the original three vectors $v_1 = (1, 4, 7)$, $v_2 = (2, 8, 7)$, $v_3 = (3, 6, 9)$? Steps (i) $\rightarrow$ (ii) and (iii) $\rightarrow$ (iv) $\rightarrow$ (v) are of type (c). They do not change the value of ϕ. Step (ii) $\rightarrow$ (iii) is of type (b). It multiplies the value of ϕ by $(-\frac{1}{7}) \times (-\frac{1}{6})$. Step (v) $\rightarrow$ (vi) is of type (a). It multiplies the value of ϕ by -1. Thus $\phi(b_1, b_2, b_3) = (-1)(-\frac{1}{7}) \times (-\frac{1}{6})\phi(V_1, V_2, V_3)$, or $\phi(v_1, v_2, v_3) = -42\phi(b_1, b_2, b_3)$. Thus det $(L) = \phi(v_1, v_2, v_3)/\phi(b_1, b_2, b_3) = -42\phi(b_1, b_2, b_3)/\phi(b_1, b_2, b_3) = -42$. Notice that we didn't have to make any particular choice of ϕ to get this result. Notice also that any triangular matrix with 1's on its diagonal has the determinant 1 (as in the matrix (v alternate)), because it can be made into an

identity matrix $\begin{pmatrix} 1 & 0 & 0 \\ 0 & 1 & 0 \\ 0 & 0 & 1 \end{pmatrix}$ by operations of type (c), which don't change the

value of the determinant. Therefore, we could have saved labor by stopping at (v alternate) when we had achieved triangular form.

If at some point we persuaded one of the columns to be zero, then the original determinant would have to have been zero, because the operations performed multiply ϕ only by nonzero factors. If they produce a zero result, ϕ must have been zero at the beginning.

Finally, let us compare the method used above with a calculation based directly on a formula for a multilinear form. Here is one alternating trilinear form:

$$\phi\begin{pmatrix} a & d & g \\ b & e & h \\ c & f & i \end{pmatrix} = aei - ahf + dhc - dbi + gbf - gec.$$

To use this formula directly would have required 12 multiplications. The method of row operations required only 9. In general, in n dimensions, the direct use of the formula given by Lemma 7 for an alternating multilinear form requires $(n - 1)n!$ multiplications, as opposed to $((n - 1)/3)(n^2 + n + 3)$ for the method of column operations. The direct method involves fewer operations only when $n = 2$. As n increases, the advantage of the method of column operations increases rapidly. For example, for $n = 5$ the direct method requires 480 multiplications as against 44 for the method of column operations. The familiar technique of expansion by minors is essentially equivalent to direct evaluation of ϕ, and should be avoided. It is useful only in conjunction with the use of column operations. (Only multiplications have been counted. The number of additions is comparable.)

Finally, let us prove some of the basic properties of determinants.

18. ☛ IMPORTANT THEOREM ON DETERMINANTS.

(a) det $(I) = 1$, where I is the identity map.

(b) If L and M are both linear maps from V into V, det $(LM) = $ det (L) det (M) (*homomorphism property*).

(c) det $(L) = 0$ if, and only if, L is *not* an isomorphism. That is, if det $(L) = 0$, L is neither one-to-one nor onto, and conversely.

Proof. (a) follows at once from the definition, since if $\phi \in AM(V)$, $\phi(Ib_1, \ldots, Ib_n) = \phi(b_1, \ldots, b_n)$.

(b) Let $\phi \in AM(V)$, and let $v_1, \ldots, v_n \in V$. Then, by the definition of det, det $(LM)\phi(v_1, \ldots, v_n) = \phi(LMv_1, \ldots, LMv_n) = $ det $(L)\phi(Mv_1, \ldots, Mv_n) = $ det (L) det $(M)\phi(v_1, \ldots, v_n)$. To prove (b), it is enough to choose ϕ and $v_1, \ldots, v_n$ so that $\phi(v_1, \ldots, v_n) \neq 0$.

(c) If L is an isomorphism, it has an inverse, L^{-1}. Then $1 = $ det $(I) = $ det $(LL^{-1}) = $ det (L) det (L^{-1}). This equation would be impossible if det $(L) = 0$, since $0 \times$ det $(L^{-1}) = 0$. Conversely, suppose L is not an isomorphism. Then there is a nonzero vector v_1 such that $L(v_1) = 0$. We can choose vectors $v_2, \ldots, v_n$ so that $v_1, v_2, \ldots, v_n$ is a basis for V. Then

$$\text{det } (L) = \frac{\phi(Lv_1, \ldots, Lv_n)}{\phi(v_1, \ldots, v_n)} = \frac{\phi(0, Lv_2, \ldots, Lv_n)}{\phi(v_1, \ldots, v_n)} = 0,$$

for any nonzero $\phi \in AM(V)$.

(The relevant theorem on isomorphisms is E8(c).) ●

The definition I used for determinants was chosen to facilitate the proof of Theorem 18.

G. Eigenvectors and eigenvalues

A particularly simple formula for a linear map is

$$L(x_1b_1 + \cdots + x_nb_n) = \lambda_1x_1b_1 + \cdots + \lambda_nx_nb_n.$$

The corresponding matrix is

$$\begin{pmatrix} \lambda_1 & 0 & 0 & \cdots & 0 \\ 0 & \lambda_2 & 0 & \cdots & 0 \\ 0 & 0 & \lambda_3 & \cdots & 0 \\ & & \cdots & & \\ 0 & 0 & 0 & \cdots & \lambda_n \end{pmatrix}.$$

Such a linear map simply stretches the jth basis vector by the factor λ_j. We may ask under what circumstances can a basis for V be chosen so that L has this convenient form.

1. DEFINITION. A vector v is an **eigenvector (proper vector)** for the linear map L if

(a) $v \neq \mathbf{0}$, and

(b) there is a scalar, λ, the **eigenvalue (proper value, characteristic value)** corresponding to v such that $Lv = \lambda v$. ●

Obviously, the condition that it be possible to express L in **diagonal form** as above is that the eigenvectors of L span V.

2.   THEOREM. The following conditions are equivalent.

(a) λ is an eigenvalue of L.
(b) $L - \lambda I$ is *not* an isomorphism.
(c) $\det (L - \lambda I) = 0$.

Proof. λ is an eigenvalue of $L \Leftrightarrow$ there is a nonzero vector $v \in V$ such that $Lv = \lambda v \Leftrightarrow$ there is a nonzero vector $v \in V$ such that $(L - \lambda I)v = 0 \Leftrightarrow L - \lambda I$ is not one-to-one $\Leftrightarrow L - \lambda I$ is not an isomorphism $\Leftrightarrow \det (L - \lambda I) = 0$ (compare Theorem F18).

3.   THEOREM. Let V be a vector space of dimension n over F, and let L be a linear map from V into V. Then $\det (L - \lambda I) = P_L(\lambda)$ is a polynomial in λ of degree n whose leading coefficient is $(-1)^n$. L has an eigenvalue in F if, and only if, its **characteristic polynomial** P_L has a root in F. In fact, the eigenvalues of L are exactly the roots of the characteristic polynomial.

Proof. Let $\phi \in AM(V)$, and let $b_1, \ldots, b_n$ be a basis for V such that $\phi(b_1, \ldots, b_n) = 1$. Then

$$\det (L - \lambda I) = \phi((L - \lambda I)b_1, \ldots, (L - \lambda I)b_n)$$
$$= \phi(Lb_1 - \lambda b_1, \ldots, Lb_n - \lambda b_n).$$

We now prove by induction on k that if $A_n, A_{n-1}, \ldots, A_{k+1}$ are any vectors in V,

$$\phi(Lb_1 - \lambda b_1, \ldots, Lb_k - \lambda b_k, A_{k+1}, \ldots, A_n)$$

is a polynomial in λ of degree k with leading coefficient $\phi(-b_1, \ldots, -b_k, A_{k+1}, \ldots, A_k)$. When $k = 1$, our polynomial is just

$$\phi(Lb_1 - \lambda b_1, A_2, \ldots, A_n) = \phi(Lb_1, A_2, \ldots, A_n) + \lambda\phi(-b_1, A_2, \ldots, A_n).$$

Suppose the assertion is true about k. Then

$$\phi(Lb_1 - \lambda b_1, \ldots, Lb_{k+1} - \lambda b_{k+1}, A_{k+2}, \ldots, A_n)$$
$$= \phi(Lb_1 - \lambda b_1, \ldots, Lb_k - \lambda b_k, Lb_{k+1}, A_{k+2}, \ldots, A_n)$$
$$+ \lambda\phi(Lb_1 - \lambda b_1, \ldots, Lb_k - \lambda b_k, -b_{k+1}, A_{k+2}, \ldots, A_n).$$

By the induction hypothesis, the first term above is a polynomial in λ of degree k, and the second is λ times a polynomial of degree k whose leading coefficient is $\phi(-b_1, \ldots, -b_k, -b_{k+1}, A_{k+2}, \ldots, A_n)$. Thus the whole is a polynomial of degree $k + 1$ with that leading coefficient. This completes the induction step. $P_L(\lambda)$ is the polynomial that results in case $k = n$. Its leading coefficient is thus $\phi(-b_1, \ldots, -b_n) = (-1)^n\phi(b_1, \ldots, b_n) = (-1)^n$.

The rest of the theorem is simply the observation that λ is an eigenvalue of $L \Leftrightarrow \det (L - \lambda I) = P_L(\lambda) = 0$.   ●

4. COMPUTATIONAL EXAMPLES. (a) Take the basis $(1, 0)$, $(0, 1)$ for $\mathbf{R}^2$, and consider the linear map L whose matrix is $\begin{pmatrix} 1 & 2 \\ 5 & 4 \end{pmatrix}$. In this case the calculation of $\det (L - \lambda I) = \det \begin{pmatrix} 1 - \lambda & 2 \\ 5 & 4 - \lambda \end{pmatrix}$ is trivial. The result is

$$P_L(\lambda) = (1 - \lambda)(4 - \lambda) - 2 \times 5 = 4 - 5\lambda + \lambda^2 - 10$$
$$= \lambda^2 - 5\lambda - 6 = (\lambda + 1)(\lambda - 6).$$

The eigenvalues are -1 and 6. To find the corresponding eigenvectors, we must find the nontrivial solutions to the equations $(L + I)(x, y) = \mathbf{0}$ and $(L - 6I)(x, y) = \mathbf{0}$. Here is how we do so, by row operations. The equation $(L + I)(x, y) = \mathbf{0}$ has the matrix $\begin{pmatrix} 2 & 2 \\ 5 & 5 \end{pmatrix}$. By row operations, we get the matrices $\begin{pmatrix} 1 & 1 \\ 5 & 5 \end{pmatrix}$, $\begin{pmatrix} 1 & 1 \\ 0 & 0 \end{pmatrix}$. The last corresponds to the equations $\begin{matrix} x + y = 0 \\ 0 + 0 = 0 \end{matrix}$. A solution is $(1, -1)$. In fact, $L(1, -1) = (1 - 2, 5 - 4) = (-1, 1) = (-1)(1, -1)$. $(L - 6I)(x, y) = \mathbf{0}$ has the matrix $\begin{pmatrix} -5 & 2 \\ 5 & -2 \end{pmatrix}$. Row operations yield at once $\begin{pmatrix} -5 & 2 \\ 0 & 0 \end{pmatrix}$, or $-5x + 2y = 0$. A solution is $(2, 5)$. In fact, $L(2, 5) = (2 + 10, 10 + 20) = (12, 30) = 6(2, 5)$. With respect to the new basis, $(1, -1)$, $(2, 5)$, L has the matrix $\begin{pmatrix} -1 & 0 \\ 0 & 6 \end{pmatrix}$

(b) Consider the operator D on the space of functions f such that $f'' + f = \mathbf{0}$. With respect to the basis $\cos x$, $\sin x$, it has the matrix $\begin{pmatrix} 0 & 1 \\ -1 & 0 \end{pmatrix}$, so the characteristic polynomial is $\lambda^2 + 1 = 0$. There are no *real* solutions, so there are no real eigenvalues. However, if we permit complex-valued solutions to the differential equation and admit complex scalars, we can find the complex eigenvalues $\pm i$. The corresponding complex eigenvectors are the complex-valued solutions $\cos x + i \sin x$, $\cos x - i \sin x$.

5. REMARK. We shall prove later that every nonconstant polynomial with complex coefficients has a complex root. It follows that every linear map on a finite-dimensional vector space over the complex number field has an eigenvalue. It need not have more than one, and indeed the eigenvectors need not span $\mathbf{C}^n$. ($\mathbf{C}$ is the field of complex numbers.) For example, the characteristic polynomial of the matrix $L \leftrightarrow \begin{pmatrix} 2 & 1 \\ 0 & 2 \end{pmatrix}$ is $(\lambda - 2)^2 = 0$. The only eigenvalue is $\lambda = 2$. If the eigenvectors spanned $\mathbf{C}^2$, the kernel of $L - 2I \leftrightarrow \begin{pmatrix} 0 & 1 \\ 0 & 0 \end{pmatrix}$ would have to be all of $\mathbf{C}^2$. In fact, this kernel is one-dimensional.

Exercises on determinants and eigenvectors

1. Show that if ϕ is an alternating bilinear form on $\mathbf{R}^2$, and $\phi((1, 0), (0, 1)) = 1$, then $\phi \begin{pmatrix} a & c \\ b & d \end{pmatrix} = ad - bc$.

*2. Find all the eigenvalues and eigenvectors of the linear map whose matrix in terms of the basis $(1, 0, 0)$, $(0, 1, 0)$, $(0, 0, 1)$ for $\mathbf{R}^3$ is

$$\begin{pmatrix} 1 & 1 & -1 \\ -2 & 0 & 2 \\ -2 & 0 & 2 \end{pmatrix}.$$

3. Prove **Cramer's Rule** for the solution of linear equations.
   Let $L: V \to V$ be an isomorphism, let $b_1, \ldots, b_n$ be a basis for V, and $c \in V$. For each $j = 1, \ldots, n$, let L_j be the linear map defined by $L_j(b_i) = L(b_i)$ if $i \neq j$, and $L_j(b_j) = c$. Then if $x_j = \det (L_j)/\det (L)$, the solution to the linear equation $L(X) = c$ is $X = x_1 b_1 + \cdots + x_n b_n$.

   Hint: Let ϕ be the alternating multilinear form such that $\phi(b_1, \ldots, b_n) = 1$. Choose $x_1, \ldots, x_n$ so that $c = L(x_1 b_1 + \cdots + x_n b_n) = x_1 L(b_1) + \cdots + x_n L(b_n)$, and evaluate $\det (L_j) = \phi(L_j(b_1), \ldots, L_j(b_n))$.
   Do *not* use Cramer's Rule for the practical solution of linear equations. See paragraph B3 for remarks on this question.

CHAPTER IV

Metric and normed spaces

We now begin a general study of limits and continuity. Our program is once more to axiomatize certain properties of a familiar system, in this case the real number system, which can be recognized in other settings. In this case, the properties in question are those involving convergence. The fundamental operation of convergence is taking the limit. We can say that $\lim_{n \to \infty} x_n = L$ if the *distance* from x_n to L can be made as small as we like by making n large enough. The distance is measured by means of the *absolute value*. That is, the distance from the real number a to the real number b is the length $|b - a|$ of the line segment from a to b.

Exactly the same thing can be done in a vector space having an inner product. From the inner product, we get a length, or *norm*, and the distance from a vector v to a vector w is the length of the vector $v - w$ from w to v, namely $||v - w||$ (Figure 14). There are other ways of getting a norm on a vector space besides using an inner product. In general, norms or lengths will have only some of the properties of the norm in an inner product space, but not all of them. As usual, I shall list those properties of the norm in an inner product space that interest me generally, and then give examples.

A. Normed spaces

1. DEFINITION. A **normed linear space** (or simply **normed space**) is a pair $\langle\langle V, +, \cdot \rangle, || \; || \rangle$ consisting of a vector space $\langle V, +, \cdot \rangle$ over the *real* numbers, and a function $|| \; ||$, the **norm,** from V into $\mathbf{R}$ subject to the following axioms.

(a) $||v|| \geq 0$ for each $v \in V$.
(b) $||v|| = 0 \Leftrightarrow v = \mathbf{0}$.

(c) *Positive homogeneity.* If $t \in \mathbf{R}$, $||tv|| = |t| \, ||v||$.

(d) *Triangle inequality.* For each $v, w \in V$, $||v + w|| \leq ||v|| + ||w||$.

2. REMARK. Mathematicians often are interested in *complex* normed spaces, in which V would be a vector space over the *complex* numbers, but the norm would still take *real* values. All the axioms are the same in this case. When there is a distinction to be made, we speak of *real normed spaces* and *complex normed spaces*. Conceivably, one could also study normed spaces over other fields, such as the rationals. In this book, our interest will usually be in the real case.

3. EXAMPLES. (a) The real numbers are a normed space, with the absolute value as norm.

(b) Any *inner product space*, that is, any real vector space with an inner product, is a normed space, with $||v|| = [v, w]^{1/2}$. The relevant properties of the norm are parts (a) through (d) of Theorem IIIC9.

(c) If $[a, b]$ is a finite interval in $\mathbf{R}$, the space $\mathcal{C}([a, b])$ can be normed not only by an inner product (see Example IIIC2(a)), but also by the uniform norm $|\ |_\infty$ defined as follows. Each continuous function $f: [a, b] \to \mathbf{R}$ is bounded (Exercise IIC8(a) and Theorem VIIB5). It follows that $\{|f(x)|: a \leq x \leq b\}$ has a least upper bound, which is by definition $|f|_\infty$. The axioms 1(a) to (c) are trivial to verify. As to (d), for any $x \in [a, b]$, $|f(x) + g(x)| \leq |f(x)| + |g(x)| \leq \mathrm{lub}\ \{f(y): a \leq y \leq b\} + \mathrm{lub}\ \{g(y): a \leq y \leq b\} = |f|_\infty + |g|_\infty$. This shows that $|f|_\infty + |g|_\infty$ is an upper bound for the numbers $|f(x) + g(x)|$, so it is at least as large as the lub, $|f + g|_\infty$.

(d) $\mathcal{C}([a, b])$ also can be normed by the $\mathcal{L}^1$ norm, $|f|_1 = \int_a^b |f(x)|\ dx$.

(e) There are similar norms on the space $\mathbf{R}^n$:

$$||(x_1, \ldots, x_n)||_\infty = \max_{1 \leq k \leq n} |x_k|,$$

$$||(x_1, \ldots, x_n)||_1 = |x_1| + \cdots + |x_n|.$$

(The subscripts 1 and ∞ are part of a system of subscripts for the l_p-norms. The inner product norm is one of them: $(x_1^2 + \cdots + x_n^2)^{1/2} = ||(x_1, \ldots, x_n)||_2$.) Note that none of the spaces described under (c), (d), or (e) can be an inner product space, since none of the norms satisfy the parallelogram law. For example,

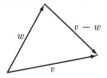

Figure 14

$2[||(1, 0)||_\infty{}^2 + ||(0, 1)||_\infty{}^2]$

$$= 2(1 + 1) = 4 \neq ||(1, 0) + (0, 1)||_\infty{}^2 + ||(1, 0) - (0, 1)||_\infty{}^2$$

$$= 1 + 1 = 2.$$

(f)   Here is an important example. Let $\langle V, ||\ ||_v \rangle$ and $\langle W, ||\ ||_w \rangle$ be normed spaces. (By abuse of notation, I have not indicated the vector operations $+$ and $\cdot$ in denoting the vector space $\langle V, +, \cdot \rangle$ by the single symbol V.) A linear map L from V to W is **bounded** if there exists a constant B such that $||L(x)||_w \leq B||x||_v$ for every $x \in V$. B is called a *bound* for L. *The smallest bound* for L is the number $|\ L\ | = $ glb $\{B\colon ||L(x)||_w \leq B||x||_v$ for every $x \in V\}$. Let us show that the set $B(V, W)$ of bounded linear maps from $\langle V, ||\ ||_v \rangle$ to $\langle W, ||\ ||_w \rangle$ is a vector space, and that $|\ |$ is a norm on this space.

   It is clear that $|\ L\ | \geq 0$ for every $L \in B(V, W)$, and that $|\ L\ | = 0 \Leftrightarrow$ $L(x) = \mathbf{0}$ for every $x \in V \Leftrightarrow L = \mathbf{0}$. If $t \in \mathbf{R}$, then for every $x \in V$, $||tL(x)||_w = |t|\ ||L(x)||_w \leq |t|\ |\ L\ |\ ||x||_v$. Thus $|t|\ |\ L\ |$ is a bound for tL. It is larger than the smallest bound for tL: $|\ tL\ | \leq |t|\ |\ L\ |$. If $t = 0$, it is clear that $|\ tL\ | = |t|\ |\ L\ |$. Otherwise, we can apply this reasoning to $1/t$ to conclude that $|\ L\ | = |\ (1/t)tL\ | \leq |1/t|\ |\ tL\ |$, so that $|t|\ |\ L\ | \leq |\ tL\ |$. This shows that $|\ tL\ | = |t|\ |\ L\ |$ even when $t \neq 0$. Finally, if $M \in B(V, W)$ as well,

$$||(L + M)(x)||_v = ||L(x) + M(x)||_v \leq ||L(x)||_v + ||M(x)||_v \leq |\ L\ |\ ||x||_w$$

$$+ |\ M\ |\ ||x||_w = (|\ L\ | + |\ M\ |)||x||_w,$$

so that $|\ L\ | + |\ M\ |$ is a bound for $L + M$. $|\ L\ | + |\ M\ |$ is at least as large as the smallest bound for $L + M$, $|\ L + M\ |$. We have shown that scalar multiples and sums of bounded operators are bounded, and have proved that $|\ |$ is a norm.

   In particular, let $||\ ||$ be any norm on $\mathbf{R}^n$ such that $||(x_1, \ldots, x_n)|| \geq |x_i|$ for $i = 1, \ldots, n$. Each of the norms $||\ ||_1, ||\ ||_2$, and $||\ ||_\infty$ has this property. If L is *any* linear map from $\mathbf{R}^n$ into a normed space $\langle V, ||\ ||_v \rangle$, then

$$||L(x_1, \ldots, x_n)||_v \leq |x_1|\ ||L(1, 0, \ldots, 0)||_v + \cdots + |x_n|\ ||L(0, \ldots, 1)||_v$$

$$\leq ||(x_1, \ldots, x_n)||(||L(1, 0, \ldots, 0)||_v + \cdots + ||L(0, 0, \ldots, 1)||_v).$$

This shows that L is bounded by $||L(1, 0, \ldots, 0)||_v + \cdots + ||L(0, 0, \ldots, 1)||_v$. Thus every linear map from $\langle \mathbf{R}^n, ||\ ||\rangle$ to $\langle V, ||\ ||_v \rangle$ is bounded. Later we shall see that this is true for *any* norm on $\mathbf{R}^n$.

   The norm on $B(V, V)$ has the following important property: $|\ LM\ | \leq$ $|\ L\ |\ |\ M\ |$.

Proof. For any $x \in V$, $||LMx||_v \leq |\ L\ |\ ||Mx||_v \leq |\ L\ |\ |\ M\ |\ ||x||_v$.

Equality need not hold. For example, let L and M be the linear maps from $\mathbf{R}^2$ to $\mathbf{R}^2$ defined by $L(a, b) = (a, 0)$, $M(a, b) = (0, b)$. Then, although neither L nor M is $\mathbf{0}$, $LM = \mathbf{0}$. Thus $|\ L\ |\ |\ M\ | > 0 = |\ LM\ |$.

EXAMPLE. Let V be the vector space of all functions on $[0, 1]$ that have continuous derivatives of all orders. V is a subspace of $\mathcal{C}([0, 1])$, so we can norm it with the uniform norm. The differentiation operator $D: f \to f'$ is a linear operator on V, which is *not* bounded. In particular, if we look at the function e^{-nx}, we find that $| e^{-nx} |_\infty = e^{-n \cdot 0} = 1$, but $| De^{-nx} |_\infty = | -ne^{-nx} |_\infty = n | e^{-nx} |_\infty = n$. Thus the ratio $| Df |_\infty / | f |_\infty$ can be made as large as we please.

B. Metric spaces

We can generalize the notion of distance still further as follows.

1. DEFINITION. A **metric space** is a pair $\langle M, d \rangle$ consisting of a set M and a function $d: M \times M \to \mathbf{R}$ satisfying the following axioms.

(a) For every $x, y \in M$, $d(x, y) \geq 0$.
(b) $d(x, y) = 0 \Leftrightarrow x = y$.
(c) $d(x, y) = d(y, x)$ for every $x, y \in M$.
(d) *Triangle inequality.* $d(x, z) \leq d(x, y) + d(y, z)$ for every $x, y, z \in M$. ●

Intuitively, (a) says the distance from x to y is at least 0, and (c) says it is the same as the distance from y to x. (b) says that if x is unequal to y, the distance is positive, and conversely. (d) says you must go at least as far to get from x to z by way of y as to go directly.

2. EXAMPLES. (a) Let $\langle V, \| \ \| \rangle$ be a normed space. Define a distance d on V by putting $d(x, y) = \|x - y\|$. Then properties (a) and (b) of the distance correspond to properties (a) and (b) of the norm. $d(y, x) = \|y - x\| = \|(-1)(x - y)\| = |-1| \|x - y\| = \|x - y\| = d(x, y)$. As to (d), $d(x, y) + d(y, z) = \|x - y\| + \|y - z\| \geq \|(x - y) + (y - z)\| = \|x - z\| = d(x, z)$.
(b) If $\langle M, d \rangle$ is a metric space, and S is any subset of M, then we get a metric d_S on S simply by restricting d to $S \times S$. We call the pair $\langle S, d_S \rangle$ a **metric subspace** of $\langle M, d \rangle$. By abuse of notation, we shall usually omit the subscript from the symbol "d_S." For example, the interval $[a, b]$ and the rational numbers are metric spaces by virtue of being metric subspaces of $\mathbf{R}$, which is a metric space because $\mathbf{R}$ is normed.
(c) Let M be any set, and define $d(x, y)$ to be 0 if $x = y$, and to be 1 otherwise. Then d is a metric, the **discrete metric,** on M.
(d) Let $\langle M, d \rangle$ be any metric space. If we put $\delta(x, y) = \min(d(x, y), 1)$, then δ is also a metric on M, *which never assumes a value larger than 1.*

3. PROPOSITION. Let $\langle M, d \rangle$ be a metric space. If $x, y, z \in M$, then $|d(x, y) - d(x, z)| \leq d(y, z)$. In particular, if $\langle V, \| \ \| \rangle$ is a normed space, $| \|y\| - \|z\| | \leq \|y - z\|$.

Proof. By the triangle inequality, $d(x, y) \leq d(y, z) + d(z, x) = d(y, z) + d(x, z)$, so $d(x, y) - d(x, z) \leq d(y, z)$. Similarly, $d(x, z) \leq d(x, y) + d(y, z)$, so $d(x, z) - d(x, y) \leq d(y, z)$. But $|d(x, y) - d(x, z)|$ equals $d(x, y) - d(x, z)$ or $-(d(x, y) - d(x, z)) = d(x, z) - d(x, y)$ depending on which is nonnegative. In particular, in the normed case, if we put $x = \mathbf{0}$, we get $\big|\, ||y|| - ||z||\,\big| \leq ||y - z||$. ●

Now let us show how to carry over various notions of limits and continuity to the setting of metric spaces.

4.   DEFINITIONS. Let $\langle M, d \rangle$ be a metric space.

(a) If $n \mapsto x_n$ is a sequence of points in M, its **limit** is the point L in M if, and only if, for every $\epsilon > 0$, there is an integer N such that $d(x_n, L) < \epsilon$ whenever $n > N$.

(b) $n \mapsto x_n$ is a **Cauchy sequence** if, for every $\epsilon > 0$, there is an integer N such that $d(x_n, x_m) < \epsilon$ whenever $n, m > N$.

(c) $\langle M, d \rangle$ is **complete** if every Cauchy sequence in M has a limit in M.

(d) If $\langle N, e \rangle$ is a second metric space and f is a function from M into N, we say that f is **continuous at the point x_0** in M if, for every $\epsilon > 0$, there is a $\delta > 0$ such that $e(f(x), f(x_0)) < \epsilon$ whenever $d(x, x_0) < \delta$. f is **continuous** if it is continuous at every point of M.

EXAMPLES. The **usual metric** for the real numbers is $d(x, y) = |x - y|$. We have shown that **R**, with the usual metric, is a complete metric space. The metric subspace of rational numbers is not. For example, if we put $x_0 = 1$, $x_{n+1} = \frac{1}{2}(x_n + (2/x_n))$, then $x \mapsto x_n$ is a Cauchy sequence of rationals, which does *not* converge to a rational number. (It converges to $\sqrt{2}$.)

5.   PROPOSITION. Let $\langle V, ||\ ||_v \rangle$ and $\langle W, ||\ ||_w \rangle$ be normed spaces, and let L be a linear map. If L is continuous at any one point in V, it is bounded. If L is bounded, then it is continuous everywhere.

Proof. Say L is continuous at x. Then there is a $\delta > 0$ such that $||y - x||_v < \delta \Rightarrow ||Ly - Lx||_w < 1$. Now, given any $z \in V$, if $z \neq \mathbf{0}$, we can put $\alpha = \delta/(2||z||_v)$. Then $||\alpha z||_v = \delta/2 < \delta \Rightarrow ||L(x + \alpha z) - L(x)||_w = \alpha ||L(z)||_w < 1 \Rightarrow$ (since L is linear) $||L(\alpha z)||_w = \alpha ||L(z)||_w < 1 \Rightarrow ||L(z)||_w < (1/\alpha) = (2/\delta)||z||_v$, so that L is bounded by $2/\delta$. If L is bounded and $y \in V$, then $||L(x) - L(y)||_w = ||L(x - y)||_w \leq |\,L\,|\ ||x - y||_v$. Given $\epsilon > 0$, put $\delta = \epsilon/|\,L\,|$. Then if $||x - y||_v < \delta$, $||L(x) - L(y)||_w < |\,L\,|\,\delta = \epsilon$. This shows that L is continuous at y. (If $|\,L\,| = 0$, it is obvious that L is continuous.)

C. Open and closed sets

Some definitions will facilitate our study of convergence.

1.   DEFINITION. A subset C of a metric space $\langle M, d \rangle$ is **closed** if, whenever $n \mapsto x_n$ is a sequence of points *in* C that has a limit *in* M, the limit is actually in C.

2.   THEOREM. (a) Each point in a metric space is closed.

(b)   The empty subset of M is closed, as is all of M.

(c)   The union of finitely many closed subsets of M is closed.

(d)   The intersection of *any number* of closed subsets of M is closed.

Proof. (a) If S consists of a single point x, every sequence in S is constantly equal to x, so its limit is the point x in S.

(b)   If S is empty, there are no sequences in S, so the condition for closure could not fail to be true. That M itself is closed is trivial.

(c)   Let $S_1, \ldots, S_p$ be closed subsets of M. If $n \mapsto x_n$ is a sequence in their union, then infinitely many of its terms must lie in some one of them, say S_1. Let us count off those terms that are in S_1; say they are $x_{n_1}, x_{n_2}, x_{n_3}, \ldots$. If $\lim\limits_{n \to \infty} x_n = L$, then surely $\lim\limits_{k \to \infty} x_{n_k} = L$. In fact, if $\epsilon > 0$, there is an N so large that $n > N \Rightarrow d(x_n, L) < \epsilon$. Just choose K so large that when $k > K$, $n_k > N$. Then $k > K \Rightarrow d(x_{n_k}, L) < \epsilon$. But then $\{x_{n_k}\}$ is a sequence in S_1 that has a limit in M, so the limit, L, is in S_1. Therefore, L is certainly in $S_1 \cup S_2 \cup \cdots \cup S_p$.

(d)   Let $\mathcal{S}$ be any collection of closed subsets of M. If $n \mapsto x_n$ is a sequence that is in every one of the sets in $\mathcal{S}$, and it has the limit L, then because each of the sets in $\mathcal{S}$ is closed, L is in every one of them. That is, the intersection of all the sets in $\mathcal{S}$ is closed.

3.   THEOREM. If $\langle M, d \rangle$ and $\langle N, e \rangle$ are metric spaces, f is a continuous function from M to N, and $\lim\limits_{n \to \infty} x_n = L$ in M, then $\lim\limits_{n \to \infty} f(x_n) = f(L)$ in N. That is, f takes convergent sequences into convergent sequences. In fact, $\lim\limits_{n \to \infty} f(x_n) = f(L)$ if f is continuous at L.

Proof. Let $\epsilon > 0$ be given. Then there is a $\delta > 0$ such that $d(x_n, L) < \delta \Rightarrow e(f(x_n), f(L)) < \epsilon$. Now choose N so large that $d(x_n, L) < \delta$ whenever $n > N$. Then $e(f(x_n), f(L)) < \epsilon$ whenever $n > N$.

4.   COROLLARY. If $f \colon M \to N$ is continuous and S is a closed subset of N, $f^{-1}(S)$ is closed in M.

Proof. Let $\{x_n\}$ be a sequence in $f^{-1}(S)$ that converges to L. Since f is continuous, $\lim\limits_{n \to \infty} f(x_n) = f(L)$, so $f(L) \in S$ because S is closed. But then $L \in f^{-1}(S)$.

5.   EXAMPLES. Proposition 3 of Section B shows that for each fixed x in M, the distance function $d(x, y)$ is a continuous function of y. Therefore, if

$\lim_{n \to \infty} y_n = L$, $\lim_{n \to \infty} d(x, y_n) = d(x, L)$. In particular, if S is the set of points y in M such that $d(x, y) \leq \epsilon$, S is *closed*, since if $n \mapsto y_n$ is a sequence in S converging to z, $d(x, z) = \lim d(x, y_n) \leq \epsilon$, so that $z \in S$. Similarly, the set of points y in M such that $d(x, y) \geq \epsilon$ is closed. The set of rationals is *not* closed in the set of reals, since the limit of a sequence of rationals need not be rational. Note, however, that the set of rationals *is* the union of an *infinite* collection of closed sets, since each individual rational is closed.

It is convenient to have names for the sets introduced in 5.

6.   DEFINITION. In a metric space $\langle M, d \rangle$, the **open ball of radius r about x** is the set $B_r(x) = \{y \in M : d(y, x) < r\}$, and the **closed ball of radius r about x** is the set $B_r^c(x) = \{y \in M : d(x, y) \leq r\}$.   ●

Let S be any subset of M. There is certainly at least one closed subset of M that contains S—namely, all of M. According to Theorem 2(d), the intersection of *all* the closed subsets of M that contain S is a closed subset S^c of M, and clearly S^c contains S. Clearly, S^c is the *smallest* closed subset of M containing S, in the sense that it is contained in any other closed subset of M that contains S.

7.   DEFINITION. The *closure* S^c of S in the metric space $\langle M, d \rangle$ is the smallest closed subset of M which contains S.

8.   THEOREM. The following three conditions are equivalent.

(a) $x \in S^c$.
(b)   There is a sequence $\{y_n\}$ of points in S that converges to x.
(c)   For every $r > 0$, the intersection of $B_r(x)$ with S is not empty.

Proof. (b) $\Rightarrow$ (a). Suppose there is a sequence y_n in S converging to x. Then y_n is also in S^c, since S^c contains S. Since S^c is closed, the limit $x = \lim_{n \to \infty} y_n$ is in S^c by the definition of *closed set*.

(a) $\Rightarrow$ (c). Suppose $x \in S^c$. If there were an $r > 0$ such that $B_r(x) \cap S$ were empty, then $M - B_r(x) = \{y \in M : d(y, x) \geq r\}$ would be a closed subset of M (see 5) that does not contain x, but does contain S. Since S^c is the smallest closed subset of M containing S, $S^c \subseteq M - B_r(x)$. But then x would not be in S^c, a contradiction.

(c) $\Rightarrow$ (b). Suppose $B_r(x) \cap S$ is not empty for any $r > 0$. Then, in particular, for each integer $n > 0$, there is a y_n in $B_{1/n} \cap S$, so that $y_n \in S$ and $d(y_n, x) < 1/n$. Then $\lim_{n \to \infty} y_n = x$.

9.   THEOREM. The closure operation has the following properties.

(a) S is closed $\Leftrightarrow S = S^c$.
(b) $(S^c)^c = S^c$.

(c) $(S \cup T)^c = S^c \cup T^c$.

(d) If $S \subseteq T$ then $S^c \subseteq T^c$.

(e) $(S \cap T)^c \subseteq S^c \cap T^c$.

Proof. (a) If S is closed, it is obviously itself the smallest closed set containing S, so $S = S^c$. Since the closure of a set is closed, if $S = S^c$, then S is closed.

(b) Since S^c is closed, it is equal to its closure $(S^c)^c$ by (a).

(c) Since by 2(c), the union of the closed sets S^c and T^c is closed, and since $S^c \cup T^c$ contains S and contains T, it therefore contains their union— $(S \cup T)^c \subseteq S^c \cup T^c$. On the other hand, $(S \cup T)^c$ is a closed set containing S and containing T, so $S^c \subseteq (S \cup T)^c$ and $T^c \subseteq (S \cup T)^c$, and therefore $S^c \cup T^c \subseteq (S \cup T)^c$. These two inclusions together imply that $S^c \cup T^c = (S \cup T)^c$.

(d) If $S \subseteq T$, then $S \subseteq T^c$, since $T \subseteq T^c$. Then $S^c \subseteq T^c$ as well, because T^c is closed.

(e) $S \cap T \subseteq S \subseteq S^c$ and $S \cap T \subseteq T \subseteq T^c$. Therefore $S \cap T \subseteq S^c \cap T^c$, so $(S \cap T)^c \subseteq S^c \cap T^c$ because the latter set is closed. (The reverse inclusion, $S^c \cap T^c \subseteq (S \cap T)^c$ is *not* always true. For an example, see the Exercises.)

10. THEOREM. A closed subset of a complete metric space is complete.

Proof. Let $\langle M, d \rangle$ be complete, and let S be a closed subset of M. If $n \mapsto x_n$ is a Cauchy sequence in S, it is also a Cauchy sequence in M, and therefore has a limit L in M. But since $\lim_{n \to \infty} x_n = L$ and S is closed, L is in S as well.

11. DEFINITION. A subset U of a metric space $\langle M, d \rangle$ is **open** if its *complement*, the set $M - U$ of points in M that are *not* in U, is closed.

12. THEOREM. In order for U to be open, it is necessary and sufficient that for each $x \in U$, there be an $r > 0$ such that $B_r(x) \subseteq U$.

Proof. Suppose U is open. If $x \in U$, then $x \notin M - U$. Since $M - U$ is closed, there is a ball $B_r(x)$ of positive radius whose intersection with $M - U$ is empty, according to 8(c). Then $B_r(x) \subseteq U$. Conversely, suppose that for every $x \in U$, there is an $r > 0$ such that $B_r(x) \subseteq U$. Then $M - U \subseteq M - B_r(x)$, and the latter set is closed and does not contain x. Thus $(M - U)^c \subseteq M - B_r(x)$, so that $x \notin (M - U)^c$. In other words, $x \in U \Rightarrow x \notin (M - U)^c$. Turned around, this says that $x \in (M - U)^c \Rightarrow x \notin U \Rightarrow x \in M - U$. That is, $(M - U)^c \subseteq M - U$. Since $M - U \subseteq (M - U)^c$, it follows that $M - U = (M - U)^c$, so that $M - U$ is closed, and U is open.

13. THEOREM. (a) The empty subset of M is open, and is all of M.

(b) The intersection of *finitely many* open sets is open.

(c) The union of *any number* of open sets is open.

Proof. (a) The complement of the empty subset of M is all of M, which is closed. The complement of M is the empty subset of M, which is closed.

(b) If $U_1, \ldots, U_n$ are open sets, then by I B 17(e), $M - (U_1 \cap \cdots \cap U_n) = (M - U_1) \cup \cdots \cup (M - U_n)$, and the latter set, being a finite union of closed sets, is closed.

(c) If $\mathfrak{U}$ is a collection of open sets, then by I B 17(d), $M - \bigcup \mathfrak{U} = \bigcap \{M - U : U \in \mathfrak{U}\}$, which is an intersection of closed sets. Thus $M - \bigcup \mathfrak{U}$ is closed. ●

If S is any subset of M, the union of all the open subsets of M contained in S is clearly open, and indeed is the largest open subset of M contained in S.

14. DEFINITION. The **interior** S^0 of S is the largest open subset of S. The **boundary** of S, ∂S, is the set $S^c - S^0$ of those points in the closure of S that are *not* in its interior.

15. THEOREM. $\partial S = S^c \cap (M - S)^c$. Therefore, ∂S is closed.

Proof. $(M - S)^c$ is the smallest closed subset of M containing $M - S$, or in other words, the smallest closed subset of M whose complement is contained in S. That is, $M - (M - S)^c = S^0$. Thus, a point is in S^c but not in $S^0 \Leftrightarrow$ it is in S^c but not in $M - (M - S)^c \Leftrightarrow$ it is in S^c and $(M - S)^c$. ●

16. EXAMPLES. (a) We know that $B_r^c(x)$ and $M - B_r(x)$ are closed (compare 5). Therefore $B_r(x)$ and $M - B_r^c(x)$ are open.

(b) Let M be $\mathbf{R}^2$ with the metric arising from the usual inner product. Then $B_1(0)$ is the disk of radius one about 0, *not* including the circle of radius 1 (Figure 15, A). $B_1^c(0)$ is $B_1(0)$ *plus* the points on the circle of radius 1 about 0 (Figure 15, B). Let S be the set consisting of $B_1(0)$ plus those points on the circle whose y coordinate is nonnegative (Figure 15, C). Then $S^c = B_1^c(0)$, $S^0 = B_1(0)$, and ∂S is the circle of radius 1 about 0. S is ☞ *neither* open nor closed.

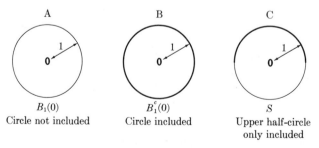

A	B	C
$B_1(0)$	$B_1^c(0)$	S
Circle not included	Circle included	Upper half-circle only included

Figure 15

17. THEOREM. Let $\langle M, d \rangle$ be a metric space, and let $S \subseteq M$, so that $\langle S, d \rangle$ is a subspace of $\langle M, d \rangle$. For a subset C of S to be closed in $\langle S, d \rangle$, it is necessary and sufficient that there be a closed subset C^* of $\langle M, d \rangle$ such that $C^* \cap S = C$. For a subset U of S to be open in $\langle S, d \rangle$, it is necessary and sufficient that there be an open subset U^* of $\langle M, d \rangle$ such that $U^* \cap S = U$.

Proof. Suppose C^* is closed in $\langle M, d \rangle$. Let $C = C^* \cap S$. If $n \mapsto x_n$ is a sequence of elements of C that has a limit L *in* S, then $L \in C^*$ because C^* is closed. Since $L \in S$ and $L \in C^*$, L is actually in $C = C^* \cap S$. This shows that C is closed in $\langle S, d \rangle$ (see Definition 1). Conversely, let C be a closed subset of $\langle S, d \rangle$. Let C^* be the closure of C *in* $\langle M, d \rangle$. Clearly $C \subseteq C^* \cap S$. Let us show that $C^* \cap S \subseteq C$. If $L \in C^* \cap S$, then $L \in S$, and according to Theorem 8 (replace S by C in the statement of that theorem) there is a sequence $n \mapsto y_n$ of elements of C such that $\lim_{n \to \infty} y_n = L$. Since C is closed in S, it follows that $L \in C$.

Now suppose U^* is an open subset of $\langle M, d \rangle$. Then $C^* = M - U^*$ is closed in $\langle M, d \rangle$. It follows that $U^* \cap S = (M - C^*) \cap S = S - C^* = S - (S \cap C^*)$. Since $S \cap C^*$ is closed in S, $U^* \cap S$ is open in S. Conversely, suppose U is an open subset of $\langle S, d \rangle$. Let $C = S - U$, and let C^* be a closed subset of M such that $C^* \cap S = C$. Put $U^* = M - C^*$. Then U^* is open in M and $U^* \cap S = (M - C^*) \cap S = S - C^* = S - (S \cap C^*) = S - C = U$. ●

Exercises

1. Let $\langle \mathbf{R}, d \rangle$ be the metric space of real numbers with the usual metric. Let $S = \{x \in \mathbf{R} : x < 0\}$ and $T = \{x \in \mathbf{R} : x > 0\}$. Show that $(S \cap T)^c \neq S^c \cap T^c$.
2. Again let $\langle \mathbf{R}, d \rangle$ be the metric space of reals. Let $\mathbf{Q}$ be as usual the subspace of rationals. Show that $\mathbf{Q}^c = \mathbf{R}$, $\mathbf{Q}^0$ is empty, and $\partial \mathbf{Q} = \mathbf{R}$.
3. Let $\langle M, d \rangle$ be any set with the *discrete metric*. Show that *every* subset of M is *both* open and closed.
4. Let $\langle \mathbf{R}, d \rangle$ be the reals with the usual metric. Show that a subset S of $\mathbf{R}$ is *both* open and closed if, and only if, S is empty or S is all of $\mathbf{R}$. *Hint:* If S is neither empty nor all of $\mathbf{R}$, then there are an $x \in S$ and a $y \notin S$. We may as well assume that $x < y$. Show that $\{z \in S : x \leq z \leq y\}$ has a lub, w, and deduce a contradiction by showing that w must be simultaneously in S because S is closed and not in S because S is open.
5. Show that there is a *countable* family $\mathcal{C}$ of open subsets of the reals such that every open subset U of the reals is the union of those open sets in $\mathcal{C}$ that are contained in U. *Hint:* Let $\mathcal{C}$ be the collection of all open intervals of the form $r < x < s$, where r and s are rational.
6. Show that if $\lim_{n \to \infty} x_n = L$, then the set consisting of L and the points x_1, x_2, x_3, x_4, ... is closed.
7. Consider the *closed interval* $[0, 1] = \{x \in \mathbf{R} : 0 \leq x \leq 1\}$ as a subspace of the metric space $\langle \mathbf{R}, \text{usual metric} \rangle$. Show that the *half-open* interval $]a, 1] = \{x \in \mathbf{R} : a < x \leq 1\}$ is an *open* subset of $\langle [0, 1], \text{usual metric} \rangle$, although it is *not* an open subset of $\langle \mathbf{R}, \text{usual metric} \rangle$.

D. Continuous functions

According to C 4, if $\langle M, d \rangle$ and $\langle N, e \rangle$ are metric spaces, $f: M \to N$ is continuous, and S is closed in N, then $f^{-1}(S)$ is closed in M. It follows that if U is open in N, $f^{-1}(U)$ is open in M. In fact, $N - U$ is closed, so $M - f^{-1}(U) = f^{-1}(N - U)$ is closed in M (Proposition I B 2). Since $M - f^{-1}(U)$ is closed, $f^{-1}(U)$ is open. The main thing to be proved in this section is that this property *characterizes* the continuous functions from M to N. First, a local form of the result.

1.   DEFINITION. U is a **neighborhood** of x in M if $x \in U^0$.

2.   THEOREM. Let f be a function from M to N. Then f is continuous at x if and only if $f^{-1}(U)$ is a neighborhood of x whenever U is a neighborhood of $f(x)$.

Proof. Suppose f is continuous at x. If U is a neighborhood of $f(x)$, Theorem C 12 shows that there is an $\epsilon > 0$ such that $B_\epsilon(f(x)) \subseteq U^0$, since U^0 is open and $f(x) \in U^0$. By the definition of continuity, there is a $\delta > 0$ such that $d(x, y) < \delta \Rightarrow e(f(x), f(y)) < \epsilon$, or in other words, $B_\delta(x) \subseteq f^{-1}(B_\epsilon(f(x)) \subseteq f^{-1}(U)$. Now $B_\delta(x)$ is an open subset of $f^{-1}(U)$ containing x, so x is in the interior of $f^{-1}(U)$. That is, $f^{-1}(U)$ is a neighborhood of x.

Conversely, suppose $f^{-1}(U)$ is a neighborhood of x whenever U is a neighborhood of $f(x)$. In particular, for each $\epsilon > 0$, $B_\epsilon(f(x))$ is a neighborhood of $f(x)$. Thus $f^{-1}(B_\epsilon(f(x)))$ is a neighborhood of x. Again, by C 12, there is a $\delta > 0$ such that $B_\delta(x) \subseteq f^{-1}(B_\epsilon(f(x)))$. But that says that if $d(x, y) < \delta$, then $e(f(x), f(y)) < \epsilon$. Thus f is continuous at x.

3.   THEOREM. In order for a function $f: M \to N$ to be continuous (on all of M) it is necessary and sufficient that $f^{-1}(U)$ be open whenever U is open, or equivalently that $f^{-1}(S)$ be closed whenever S is closed.

Proof. Suppose f is continuous and U is open. Then for each $x \in f^{-1}(U)$, U is a neighborhood of $f(x)$ because $U = U^0$. Then $f^{-1}(U)$ is a neighborhood of x. Thus every point in $f^{-1}(U)$ is in the interior of $f^{-1}(U)$, or $f^{-1}(U) = (f^{-1}(U))^0$. Since the interior of a set is open, $f^{-1}(U)$ is open. Conversely, if $f^{-1}(V)$ is open whenever V is open, then whenever $x \in M$ and U is a neighborhood of $f(x)$, $V = U^0$ is open, and $f^{-1}(U) \supseteq f^{-1}(U^0)$, so $f^{-1}(U)$ contains an open set $f^{-1}(U^0)$ containing x. Thus x is in the interior of $f^{-1}(U)$. It follows that f is continuous. We have already remarked that since $f^{-1}(N - U) = M - f^{-1}(U)$, it is the same thing to state that $f^{-1}(U)$ is open when U is open and to state that $f^{-1}(S)$ is closed when S is closed.

4.   PROPOSITION. Let $\langle M_1, d_1 \rangle$, $\langle M_2, d_2 \rangle$, and $\langle M_3, d_3 \rangle$ be metric spaces, and let $f: M_1 \to M_2$ be continuous *at* x, and $g: M_2 \to M_3$ be continuous *at* $f(x)$.

Then the function $h: M_1 \to M_3$ defined by $h(y) = g(f(y)) = (g \circ f)(y)$ is continuous *at* x. Consequently, if f and g are continuous, so is h.

Proof. Let U be a neighborhood of $h(x) = g(f(x))$. Then $g^{-1}(U)$ is a neighborhood of $f(x)$, and hence $f^{-1}(g^{-1}(U))$ is a neighborhood of x. But $f^{-1}(g^{-1}(U)) = h^{-1}(U)$, so $h^{-1}(U)$ is a neighborhood of x.

5. DEFINITION. Let $\langle M_i, d_i \rangle$ be a metric space for $i = 1, \ldots, n$. The **product space** $M = M_1 \times \cdots \times M_n$ is the set of all n-tuples $(x_1, \ldots, x_n)$ such that $x_i \in M_i$. The **product metric** $d = d_1 \times \cdots \times d_n$ on M is defined by $d(x, y) = \max_{1 \le i \le n} d_i(x_i, y_i)$. The **ith projection** is the function $\pi_i: M \to M_i$, defined by $\pi_i(x_1, \ldots, x_n) = x_i$.

REMARK. There are other ways the product metric could be defined. It would have worked just as well to put $d(x, y) = d_1(x_1, y_1) + \cdots + d_n(x_n, y_n)$. One can also make sense of the product of countably many metric spaces, $\langle M_1, d_1 \rangle, \langle M_2, d_2 \rangle, \ldots$. In this case, an appropriate product metric would be not the lub of the factor metrics, but rather something like $\sum_{n=1}^{\infty} 2^{-n} \min(d_n, 1)$.

6. THEOREM.
(a) For each $i = 1, \ldots, n$, $d_i(\pi_i(x), \pi_i(y)) \le d(x, y)$. Hence each of the projections π_i is continuous.

(b) For a function f from a metric space $\langle N, e \rangle$ into the product space $\langle M_1, d_1 \rangle \times \cdots \times \langle M_n, d_n \rangle$ to be continuous, it is necessary and sufficient that $\pi_i \circ f: N \to M_i$ be continuous for each i.

Proof. (a) If $x = (x_1, \ldots, x_n)$ and $y = (y_1, \ldots, y_n)$, then $d_i(\pi_i(x), \pi_i(y)) = d_i(x_i, y_i) \le \max_{1 \le j \le n} d_j(x_j, y_j) = d(x, y)$.

(b) If $f: N \to M$ is continuous, so is $\pi_i \circ f$ by (a) and Proposition 4. Conversely, suppose $\pi_i \circ f$ is continuous for $i = 1, \ldots, n$. We can write $f(x)$ as $(f_1(x), \ldots, f_n(x))$. Then $\pi_i \circ f = f_i$. Now for each $x \in N$ and $\epsilon > 0$, there are $\delta_1, \ldots, \delta_n > 0$ such that $e(y, x) < \delta_i \Rightarrow d_i(f_i(y), f_i(x)) < \epsilon$. Put $\delta = \min(\delta_1, \ldots, \delta_n)$. Then $\delta > 0$, and $e(y, x) < \delta \le \delta_i \Rightarrow d_i(f_i(y), f_i(x)) < \epsilon$ for every $i = 1, \ldots, n$. Therefore, $d(f(x), f(y)) = \max_{1 \le i \le n} d_i(f(y), f(x)) < \epsilon$ when $e(y, x) < \delta$. ●

We can think of $\mathbf{R}^2$ as being $\mathbf{R} \times \mathbf{R}$. The product metric $d((x, y), (a, b)) = \max(|x - a|, |y - b|)$ is exactly the metric arising from the norm $||(x, y)||_\infty = \max(|x|, |y|)$.

7. THEOREM. Let d be the product metric on $\mathbf{R}^2 = \mathbf{R} \times \mathbf{R}$. Let S, D, P, and Q be the functions from $\mathbf{R}^2$ to $\mathbf{R}$ defined by $S(x, y) = x + y$, $D(x, y) = x - y$, $P(x, y) = xy$, $Q(x, y) = x/y$. (Note that Q is actually defined only on

$\mathbf{R} \times (\mathbf{R} - \{0\})$.) Then each of these functions is continuous. (Q is continuous only on the set $\mathbf{R} \times (\mathbf{R} - \{0\})$, where the second coordinate is not 0.)

Proof. Suppose $d((a, b), (x, y)) < \epsilon/2$. Then $|S(a, b) - S(x, y)| = |(a+b) - (x+y)| \leq |a - x| + |b - y| < \epsilon$. Likewise, $|D(a, b) - D(x, y)| < \epsilon$. Let $B = 1 + |x| + |y|$, and let $\delta = \min(1, \epsilon/B)$. Then

$$d((a, b), (x, y)) < \delta \Rightarrow |a - x| < \epsilon/B, \quad |b - y| < \epsilon/B, \quad |a - x|, |b - y| < 1,$$

and finally, $|a| \leq |x| + |a - x| < |x| + 1$. Then

$$|P(a, b) - P(x, y)| = |ab - xy| = |a(b - y) + y(a - x)| \leq |a| \, |b - y|$$
$$+ |y| \, |a - x| < (1 + |x|)\epsilon/B + |y|\epsilon/B = (1 + |x| + |y|)\epsilon/B = \epsilon.$$

Finally, suppose that $y \neq 0$. Put

$$C = 2(|x| + |y| + 2)/|y|^2.$$

Let $\delta = \min(\epsilon/C, 1, |y|/2)$. Then

$$d((a, b), (x, y)) < \delta \Rightarrow |a - x|, |y - b| < \epsilon/C; \qquad |a - x|, |y - b| < 1,$$

so that $|a| < |x| + 1$, $|b| < |y| + 1$; and finally $|b - y| < |y|/2$, so that $|b| > |y|/2$. Then

$$|Q(a, b) - Q(x, y)| = \left| \frac{a}{b} - \frac{x}{y} \right| = \left| \frac{ay - bx}{by} \right| = \left| \frac{a(y - b) + b(a - x)}{by} \right|$$

$$\leq \frac{|a|}{|b| \, |y|} |y - b| + \frac{|b|}{|b| \, |y|} |a - x| < \frac{|x| + 1}{|y|^2/2} \frac{\epsilon}{C} + \frac{|y| + 1}{|y|^2/2} \frac{\epsilon}{C}$$

$$= \frac{\epsilon}{C} 2 \frac{|x| + |y| + 2}{|y|^2} = \epsilon. \quad \bullet$$

8.   COROLLARY. If f and g are real-valued functions on a metric space $\langle N, e \rangle$ that are continuous at x, then $f + g$, $f - g$, and fg are continuous at x, and f/g is continuous at x if $g(x) \neq 0$.

Proof. Let H be the function from N to $\mathbf{R} \times \mathbf{R}$ defined by $H(x) = (f(x), g(x))$. According to 6(b), H is continuous at x. But the functions $f + g$, $f - g$, fg, and f/g are respectively the composite functions $S \circ H$, $D \circ H$, $P \circ H$, and $Q \circ H$, which are continuous by Theorem 7 and the fact that the composite of two continuous functions is continuous.

Exercises

1. We know that if the function $f \colon M \to N$ is continuous, then f carries convergent sequences into convergent sequences (see C3). Show that the converse is also true: if f carries convergent sequences into convergent sequences, then it is continuous.

Hint: Let S be a closed set in N. Let $n \mapsto x_n$ be a convergent sequence in $f^{-1}(S)$. Using the facts that $n \mapsto x_n$ converges and that S is closed, show that $\lim\limits_{n \to \infty} x_n$ is in $f^{-1}(S)$. Conclude that $f^{-1}(S)$ is closed.

Remark. By a trick, we can show that if the sequence $n \mapsto f(x_n)$ has a limit whenever $\lim_{n \to \infty} x_n = L$, then $\lim_{n \to \infty} f(x_n)$ can only be $f(L)$. Thus it is unnecessary to assume that $f(x_n) \to f(L)$ whenever $x_n \to L$; we need only assume that $f(x_n)$ approaches some limit whenever $x_n \to L$. Here is the trick. If $x_n \to L$, then so does the sequence $n \to y_n$ defined by $y_{2n-1} = x_n$, $y_{2n} = L$ for $n \geq 1$. $n \mapsto y_n$ is the sequence $x_1, L, x_2, L, x_3, L, \ldots$. Therefore the sequence $n \mapsto f(y_n)$ $(f(x_1), f(L), f(x_2), f(L), f(x_3), f(L), \ldots)$ must converge to something λ. It is clear that λ can only be $f(L)$, since $\lambda = \lim_{n \to \infty} f(y_n) = \lim_{n \to \infty} f(y_{2n}) = \lim_{n \to \infty} f(L) = f(L)$. But then $\lim_{n \to \infty} f(x_n) = \lim_{n \to \infty} f(y_{2n-1}) = \lim_{n \to \infty} f(y_n) = \lambda = f(L)$ as well.

*2. Let J be the set consisting of the positive integers and one other element, ∞. Let d be the function from $J \times J$ to $\mathbf{R}$ defined by $d(n, m) = \left| \dfrac{1}{n} - \dfrac{1}{m} \right|$, $d(n, \infty) = d(\infty, n) = 1/n$, $d(\infty, \infty) = 0$.

   (a) Show that d is a metric on J.

   (b) Show that a sequence $k \mapsto n_k$ of positive integers converges to ∞ in $\langle J, d \rangle$ if, and only if for every N, there is an integer k such that $n_k > N$ whenever $k > K$.

   (c) Let $\langle M, e \rangle$ be a metric space and $n \mapsto x_n$ a sequence in M. Let L be an element of M. Define a function $f: J \to M$ by putting $f(n) = x_n$, $f(\infty) = L$. Show that f is a continuous function from $\langle J, d \rangle$ to $\langle M, e \rangle$ if and only if $\lim_{n \to \infty} x_n = L$.

   (d) Use (c) and Corollary 8 to show that if $n \mapsto x_n$ and $n \mapsto y_n$ are sequences of real numbers, $\lim_{n \to \infty} x_n = X$, and $\lim_{n \to \infty} y_n = Y$, then $\lim_{n \to \infty} (x_n + y_n) = X + Y$ and $\lim_{n \to \infty} x_n y_n = XY$.

E.   The Baire category theorem

The theorem I am about to prove is one of the most important theorems in the theory of metric spaces. Anyone seriously interested in analysis should master its use.

1.   DEFINITIONS. A subset of a metric space $\langle M, d \rangle$ is **nowhere dense** in M if the interior of its closure is empty. A subset of a metric space is of the **first category** if it is the union of countably many nowhere dense subsets. It is of the **second category** if it is *not* of the first category. It is a **residual set** if its complement is of the first category.

2.   BAIRE CATEGORY THEOREM. Every nonempty complete metric space is of the second category. That is, a nonempty complete metric space cannot be the union of countably many of its nowhere dense subsets.

Proof. Let $\langle M, d \rangle$ be a complete metric space. If M were of the first category, there would be countably many nowhere dense subsets $C_1, C_2, C_3, \ldots$ whose union is M. Then their closures $C_1^c, C_2^c, C_3^c, \ldots$ would have no interior, but the union of these closed sets C_1^c, C_2^c, ... would also be M. Let x_0 be any point in M. The ball $B_1(x_0)$ cannot be included in C_1^c, since C_1^c has no interior. Therefore $B_1(x_0) - C_1^c = B_1(x_0) \cap (M - C_1^c)$ is nonempty, and it is open since it is the intersection of two open sets. Choose a point $x_1 \in B_1(x_0) - C_1^c$. Since $B_1(x_0) - C_1^c$ is open, there is a positive number $r_1 < 2^{-1}$ such that $B_{2r_1}(x_1) \subseteq B_1(x_0) - C_1^c$. Similarly, there are a point $x_2 \in B_{r_1}(x_1) - C_2^c$, and a positive number $r_2 < 2^{-2}$ such that $B_{2r_2}(x_2) \subseteq B_{r_1}(x_1) - C_2^c$. In general, proceeding in this way, we can obtain a sequence of points $x_0, x_1, x_2, \ldots$ and a sequence of balls $B_{2r_n}(x_n)$ such that $r_n < 2^{-n}$ and $B_{2r_n}(x_n) \subseteq B_{r_{n-1}}(x_{n-1}) - C_n^c$. If $m > n$, $x_m \in B_{r_{m-1}}(x_{m-1}) \subseteq B_{2r_{m-1}}(x_{m-1}) \subseteq B_{r_{m-2}}(x_{m-2}) \subseteq \cdots \subseteq B_{r_{n+1}}(x_{n+1}) \subseteq B_{r_n}(x_n)$, so that $d(x_m, x_n) < 2^{-n}$. It follows that $n \mapsto x_n$ is a Cauchy sequence, so it has a limit L in M. We have shown that if $m > n$, $x_m \in B_{r_n}(x_n) \subseteq B_{r_n}^c(x_n)$. Since this last set is closed, $\lim_{m \to \infty} x_m = L \in B_{r_n}^c(x_n) \subseteq B_{2r_n}(x_n)$ for every n. For each n, L is in $B_{2r_n}(x_n) \subseteq B_{r_{n-1}}(x_{n-1}) - C_n^c$. Therefore, L is not in C_n^c. That is, L is not in any of the sets C_1^c, C_2^c, C_3^c, ..., contradicting the (evidently false) assumption that their union is all of M.   ●

As a rather spectacular application of the Baire Category Theorem, let us prove the existence of a continuous real-valued function on **R** that is not differentiable anywhere. Let P be the set of all real-valued functions f on **R** with the following two properties.

(a) f is periodic of period $1: f(x + 1) = f(x)$ for every $x \in$ **R**.
(b) f is *uniformly continuous*: for every $\epsilon > 0$, there is a $\delta > 0$ such that $|x - y| < \delta \Rightarrow |f(x) - f(y)| < \epsilon$. Later we shall be in a position to show that *every continuous* periodic function is uniformly continuous, but for the present we do not need this fact.

It is clear that P is nonempty, since it contains every constant function. Indeed, P is a vector space. Moreover, if $f \in P$, then f is bounded on $[0, 1]$ (see Example A3(c)), and therefore, because f is periodic, it is bounded on all of **R**. We can then norm P with the uniform norm: $|f|_\infty = \text{lub } \{|f(x)| : x \in \mathbf{R}\}$. We shall show later, and it is not hard to see (Exercise 1) that $\langle P, |\ |_\infty \rangle$ is actually *complete*.

3.   DEFINITION. A *complete* normed vector space is a **Banach space.** A *complete* inner product space is a **Hilbert space.**

4.   DEFINITION. A continuous function $g: [a, b] \to$ **R** is **piecewise linear** if there are points $a = x_0 < x_1 < x_2 < \cdots < x_n = b$ such that g is linear on each of the intervals $[x_i, x_{i+1}]$ for $i = 0, \ldots, n - 1$. A function $g: \mathbf{R} \to \mathbf{R}$ is **piecewise linear** if g is piecewise linear on every finite interval (see Figure 16).

5.   LEMMA. If $f \in P$ and $\epsilon > 0$, there is a piecewise linear function $g \in P$ such that $|f - g|_\infty \leq \epsilon$.

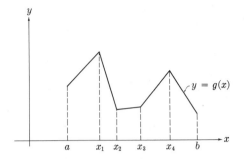

Figure 16. A piecewise linear function g.

Proof. Because f is uniformly continuous, we can choose a $\delta > 0$ such that $|x - y| < \delta \Rightarrow |f(x) - f(y)| \le \epsilon/2$. Choose an integer $n > 1/\delta$. Define g as follows: for every integer k, if $k/n < x \le (k+1)/n$, then $g(x) = (k+1-nx)f(k/n) + (nx-k)f((k+1)/n)$. It is easy to see that $g(k/n) = f(k/n)$ for every integer k, and that g is continuous and piecewise linear. A routine calculation verifies that the fact that f is periodic of period 1 implies that g is also periodic of period 1. To see that g is uniformly continuous, put $x_k = k/n$, and observe that $x_k \le x \le y \le x_{k+1} \Rightarrow |g(y) - g(x)| = |n(y-x)(f(x_{k+1}) - f(x_k))| \le n(y-x)\epsilon/2$. It follows that if $x_k \le x \le x_{k+1} \le x_j \le y \le x_{j+1}$, then

$$|g(y) - g(x)| \le n\epsilon/2[(y - x_j) + (x_j - x_{j-1}) + \cdots + (x_{k+2} - x_{k+1}) + (x_{k+1} - x)]$$
$$= (n\epsilon/2)[y - x].$$

Given any $x, y \in \mathbf{R}$, $x \le y$, either $x_k \le x \le y \le x_{k+1}$ for some k, or $x_k \le x \le x_{k+1} \le x_j \le y \le x_{j+1}$ for some k and j. Thus for any $x, y \in \mathbf{R}$, $|g(y) - g(x)| \le (n\epsilon/2)|y - x|$, which clearly shows that g is uniformly continuous.

Finally, if $x_k \le x \le x_{k+1}$, then $|f(x) - g(x)| \le |f(x) - f(x_{k+1})| + |g(x_{k+1}) - g(x)|$, since $g(x_{k+1}) = f(x_{k+1})$. But $|x - x_{k+1}| < \delta$, so that $|f(x) - f(x_{k+1})| \le \epsilon/2$ and $|g(x_{k+1}) - g(x)| \le (n\epsilon/2)|x - x_{k+1}| < (n\epsilon/2) \cdot (1/n) = \epsilon/2$. Thus $|g(x) - f(x)| < \epsilon$. ●

6.   THEOREM. The set of functions in P that are nowhere differentiable is a residual set in P. In particular, there is a nowhere differentiable function in P.

Proof. If $f \in P$ and f is differentiable at x, then

$$\lim_{h \to \infty} |(f(x+h) - f(x))/h - f'(x)| = 0.$$

In particular, there is a $\delta > 0$ such that

$$|(f(x+h) - f(x))/h - f'(x)| < 1$$

whenever $0 < h < \delta$. Then

$$0 < h < \delta \Rightarrow |(f(x+h) - f(x))/h| \le |f'(x)| + 1.$$

If $h \geq \delta$, $|(f(x + h) - f(x))/h| \leq 2 \, | \, f \, |_\infty/h \leq 2 \, | \, f \, |_\infty/\delta$. Thus if $n \geq \max \, [|f'(x)| + 1, 2 \, | \, f \, |_\infty/\delta]$, then $|f'(x + h) - f(x)|/h \leq n$ for all $h > 0$. Let

$$E_n = \{f \in P \colon \text{there is an } x \in \mathbf{R} \text{ such that } |f(x + h) - f(x)|/h \leq n \text{ for all } h > 0\}.$$

If $f \in P$ is differentiable at any $x \in \mathbf{R}$, then $f \in E_n$ for some n. If it can be shown that E_n is nowhere dense in P, it will follow that the set S of functions in P that are differentiable at at least one point of $\mathbf{R}$ is of the first category, since $S \subseteq \bigcup_{n=1}^\infty E_n$. Since P is a complete metric space, $P - S$ will have to be nonempty according to the Baire Category Theorem.

Let us show that E_n is closed. Let $k \mapsto f_k$ be a sequence of functions in E_n converging to an element f of P. Since $f_k \in E_n$, there is a $y_k \in \mathbf{R}$ such that $|f_k(y_k + h) - f_k(y_k)|/h \leq n$ for all $h > 0$. Choose $x_k \in [0, 1]$ such that $y_k - x_k$ is an integer. (If $m \leq y_k < m + 1$, $x_k = y_k - m$.) Then because f_k is periodic,

$$\frac{|f_k(x_k + h) - f_k(x_k)|}{h} = \frac{|f_k(y_k + h) - f_k(y_k)|}{h} \leq n$$

for all $h > 0$. Either $\{x_k \colon k = 1, 2, 3, \ldots\}$ is a finite set, so that there is some $x \in [0, 1]$ such that $x_k = x$ for infinitely many values of k, or else $\{x_k\}$ is an infinite set, which has an accumulation point $x \in [0, 1]$ by the Bolzano-Weierstrass Theorem II C 5. In either case, for every $\delta > 0$ and every integer K, there is an integer $k > K$ such that $|x_k - x| < \delta$. I claim that $|f(x + h) - f(x)|/h \leq n$ for all $h > 0$, so that $f \in E_n$. If this were false, we could choose an $h > 0$ such that $|f(x + h) - f(x)|/h = n + \epsilon$ for some $\epsilon > 0$. Choose K so large that $| \, f_k - f \, |_\infty < \epsilon h/4$ whenever $k > K$. Because f is uniformly continuous, we can choose $\delta > 0$ so small that $|f(y + h) - f(x + h)|$ and $|f(y) - f(x)|$ are less than $\epsilon h/4$ whenever $|y - x| < \delta$. Finally, we could choose $k > K$ such that $|x_k - x| < \delta$. Then

$$\begin{aligned}
|f(x + h) - f(x)| \\
\leq |f(x + h) - f(x_k + h)| + |f(x_k + h) - f_k(x_k + h)| \\
+ |f_k(x_k + h) - f_k(x_k)| + |f_k(x_k) - f(x_k)| + |f(x_k) - f(x)| \\
< \frac{\epsilon h}{4} + \frac{\epsilon h}{4} + nh + \frac{\epsilon h}{4} + \frac{\epsilon h}{4} = (n + \epsilon)h.
\end{aligned}$$

This shows that $|f(x + h) - f(x)|/h < n + \epsilon$, a contradiction, which completes the proof that $f \in E_n$.

Finally, let us show that E_n is nowhere dense. Since E_n is closed, this means that E_n has no interior. If E_n^0 were nonempty, it would contain a ball $B_r^c(f)$ in $\langle P, | \, | \, |_\infty \rangle$ for some $r > 0$. Lemma 5 shows that there is a piecewise linear function $g \in B_{r/2}^c(f)$. Moreover, the proof of Proposition 5 shows that there is a bound B such that $|g(y) - g(z)| \leq B|y - z|$ for all $y, z \in \mathbf{R}$. Choose an integer m so large that $mr/2 > B + n$. The function $G \colon x \mapsto \sin (4\pi mx)$ is an element of P, and $| \, G \, |_\infty = 1$. Therefore

$|f - (g + (r/2)G)|_\infty \le |f - g|_\infty + (r/2)|G|_\infty \le r/2 + r/2 = r$, so that $g + (r/2)G \in B_r^c(f)$. Take any $x \in \mathbf{R}$. $\sin 4\pi my$ varies from -1 to 1 on $[x + (1/(2m)), x + (1/m)]$. Therefore, we can certainly find an $h \in [1/(2m), 1/m]$ such that $|\sin (4\pi m(x + h)) - \sin 4\pi mx| \ge 1$. Then

$$|(g + (r/2)G)(x + h) - (g + (r/2)G)(x)|$$
$$= |g(x + h) - g(x) + r/2(G(x + h) - G(x))| \ge r/2|G(x + h) - G(x)|$$
$$- |g(x + h) - g(x)| \ge r/2 - Bh.$$

That is,

$$|(g + (r/2)G)(x + h) - (g + (r/2)G)(x)|/h \ge r/(2h) - B \ge mr/2 - B > n.$$

This shows that $g + (r/2)G \notin E_n$ and that $B_r^c(f) \not\subseteq E_n$. ●

There are shorter proofs of the existence of continuous nowhere differentiable functions (see *Functional Analysis*, by F. Riesz and B. Sz.-Nagy, Ungar, New York, 1955, pages 3–5), but this proof has the advantage of showing that *most* functions in P are nowhere differentiable, in the following sense. A set of the first category in a complete metric space $\langle M, d \rangle$ is *small*, in the sense that even countably many such sets cannot fill up the space. Indeed, the closure of a residual set must be all of M (see Exercise 5). In this sense, a residual set in M is *large*, because its complement is small.

Theorem 6 is typical of the applications of the Baire Category Theorem in that category is applied to a metric space of functions to show that there exists a function with a certain property. There are a few other applications given as Exercises. It is amusing to keep the Baire Category Theorem in mind while reading older texts on analysis. Many of the proofs in such texts merely establish special cases of the Baire Category Theorem, using more complicated versions of the general argument (compare the remarks at the beginning of Chapter III).

Exercises

1. Prove that $\langle P, |\,|_\infty \rangle$ is complete. *Hint:* If $n \mapsto f_n$ is a Cauchy sequence in P, then for each $x \in \mathbf{R}$, $n \mapsto f_n(x)$ is a Cauchy sequence of real numbers. Let $f(x) = \lim_{n \to \infty} f_n(x)$. Show that $f(x + 1) = f(x)$ for all $x \in \mathbf{R}$, that f is uniformly continuous, and that $\lim_{n \to \infty} |f - f_n|_\infty = 0$.

2. Let $\langle M, d \rangle$ be a complete metric space in which no point is open. Show that M is uncountable, using Baire's Theorem.

3. Use 2 to show that $\mathbf{R}$ is uncountable.

4. Show that the Euclidean plane $\mathbf{R}^2$ is not the union of countably many lines or circles.

5. Show that the proof of the Baire Category Theorem implies the following apparently stronger assertion: if $\langle M, d \rangle$ is a complete metric space and $\mathfrak{F}$ is a countable family of open subsets of M such that $U^c = M$ for each $U \in \mathfrak{F}$, then $(\cap \mathfrak{F})^c = M$.

6. Let C be a countable set and $\langle B, \| \ \| \rangle$ a Banach space of real-valued functions on C with the following two properties.

(a)   If $f \in B$, $\{f_n\} \subseteq B$, and $\lim\limits_{n \to \infty} \|f - f_n\| = 0$, then $\lim\limits_{n \to \infty} f_n(x) = f(x)$ for each $x \in C$.

(b)   For each $x, y \in C$ such that $x \neq y$, there is an $f \in B$ such that $f(x) \neq f(y)$. Prove that B contains a one-to-one function from C to $\mathbf{R}$. *Hint:* For each pair $\{x, y\}$ of distinct points of C, let $S_{xy} = \{f \in B : f(x) = f(y)\}$.

CHAPTER V

Introduction to complex numbers

I shall now show how the normed vector space $\langle\langle \mathbf{R}^2, +, \cdot \rangle, \|\ \|_2 \rangle$ can be made into a field by defining an appropriate operation of multiplication. The resulting field will contain a *subfield* isomorphic to the field of real numbers, and it will contain an element i such that $i^2 = -1$. It is called the field of *complex numbers*. Historically, the complex numbers were invented precisely to permit the solution of quadratic equations such as $x^2 = -1$, which have no real solution. One might guess that the process of defining new number systems to permit the solution of more and more algebraic equations could continue indefinitely. For example, once we have an **imaginary number** i such that $i^2 = -1$, we might try to define a new kind of number—say a **chimerical number,** j, such that $j^2 = i$. As it turns out, nothing new is needed. In particular, $(1/\sqrt{2})(1 + i)$ is a complex number whose square is i. Indeed, the so-called Fundamental Theorem of Algebra, whose proof will be given later, states that any nonconstant polynomial with complex coefficients has a complex root.

We might think of this as an extraordinary piece of luck, that by defining an enlarged number system to permit the solution of quadratic equations with real coefficients, we already get a system large enough to permit the solution of any polynomial equation with coefficients in the larger system. However, that is not the whole of the story. Complex numbers have also turned out to be extraordinarily useful in analysis, geometry, differential equations, physics, engineering, and elsewhere. Their invention was one of those pieces of great good fortune that convinces men that there must really be an underlying order to the universe, and that mathematics is in fact the discovery of truth rather than an ingenious art of invention. As often happens, the truth lies somewhere to the left of both extremes, whatever that means.

A.   Definition of the complex field

1.   DEFINITION. The **complex product** $(a, b) \times (c, d)$ of two vectors in $\mathbf{R}^2$ is $(ac - bd, ad + bc)$. It is conventional to denote the usual basis vectors $(1, 0)$ and $(0, 1)$ by 1 and i respectively, so that $i \times i = -1$. The **complex number system** is $\langle \mathbf{R}^2, +, \times \rangle$, where $\mathbf{R}^2$ (which we often denote by $\mathbf{C}$ for *complex*) is the set of pairs of real numbers, $+$ is the usual vector addition in $\mathbf{R}^2$, and $\times$ is the complex product.

2.   THEOREM. The system of complex numbers $\langle \mathbf{C}, +, \times \rangle$ is a field whose multiplicative neutral element is $1 = (1, 0)$, whose additive neutral element is $0 = (0, 0)$, and in which $i^2 = -1$. The elements of $\mathbf{C}$ of the form $(r, 0) = r + 0i$ form a subfield of $\mathbf{C}$, which is isomorphic as a field to the field of real numbers.

Proof. We have to verify the field axioms of II A. The group axioms for addition are satisfied because the complex addition is just the addition in the vector space $\mathbf{R}^2$, and addition in a vector space satisfies the group axioms. As to the group axioms for multiplication, associativity and commutativity can be checked by straightforward calculations. $1 = (1, 0)$ is indeed a neutral element because $1 \times (x + iy) = (1, 0) \times (x, y) = (1 \cdot x - 0 \cdot y, 1 \cdot y + 0 \cdot x) = (x, y)$. Likewise, we check that $i^2 = -1$. Constructing the inverse is not completely trivial. Observe that if $x + iy \neq 0$, then $x \neq 0$ or $y \neq 0$, so $x^2 + y^2 \neq 0$. Then it is easy to calculate that $(x + iy)(x/(x^2 + y^2) - iy/(x^2 + y^2)) = 1$, using the fact that $i^2 = -1$. Thus $(x + iy)^{-1} = (x - iy)/(x^2 + y^2)$. Checking the distributive law is also routine, and the nontriviality of $\mathbf{C}$ is trivial.

To see that the set of elements of $\mathbf{C}$ of the form $r + 0i$ form a subfield, we have only to check that 0 and 1 are in this set, as are the sum, product, difference, and quotient of any two of its elements. The other properties such as associativity are inherited from $\mathbf{C}$. Finally, the function $x \mapsto x \cdot 1 + 0i = (x, 0)$ from $\mathbf{R}$ into $\mathbf{C}$ clearly maps $\mathbf{R}$ one-to-one onto this subfield, preserving sums and products, so it is an isomorphism.

3.   DEFINITION. The **complex conjugate** $\bar{z}$ of $z = x + iy$ is $x - iy$. The absolute value of $z = x + iy$ is $|z| = (x^2 + y^2)^{1/2} = ||(x, y)||_2$.

4.   PROPOSITION.

(a) $\overline{z + w} = \bar{z} + \bar{w}$.

(b) $\overline{zw} = \bar{z}\bar{w}$.

(c) $|z|^2 = z\bar{z}$.

(d) $|zw| = |z| \, |w|$.

(e) $|z| = |\bar{z}|$.

(f) $\bar{\bar{z}} = z$.

(g) $|z + w| \leq |z| + |w|$.

Proof. (a) is obvious.

(b) $(x - iy)(a - ib) = (ax - by) - i(xb + ay) = \overline{(ax - by) + i(xb + ay)} =$
$\overline{(x + iy)(a + ib)}$.

(c) $(x + iy)(x - iy) = x^2 - (iy)^2 = x^2 + y^2$.

(d) $|zw|^2 = zw\,\overline{zw} = zw\,\overline{z}\overline{w} = z\overline{z}\,w\overline{w} = |z|^2|w|^2$. Then take square roots, which is permissible since absolute values are always non-negative.

(e) and (f) are obvious.

(g) is simply the triangle inequality for the norm $||(x, y)||_2$.

5.   DEFINITION. If $z = x + iy$ is a complex number, its **complex exponential** $e^z = e^{x+iy}$ is $e^x(\cos y + i \sin y)$. A **complex logarithm** for z is any solution w to the equation $e^w = z$.

6.   THEOREM. (a) $e^{z+w} = e^z e^w$.

(b) $e^0 = 1$, $e^{i\pi} = -1$, $e^{2\pi i} = 1$.

(c) $e^{z+2\pi i} = e^z$.

(d) z has a logarithm $\Leftrightarrow z \neq 0$. If w is a logarithm for z, so is $w + 2n\pi i$ for each integer n.

(e) For any complex number z and any integer n, the equation $r^n = z$ has a solution, r.

Proof. (a) If $w = a + bi$ and $z = x + yi$, then $e^{z+w} = e^{(a+x)+(b+y)i} = e^{a+x}(\cos (b + y) + i \sin (b + y)) = e^a e^x((\cos b \cos y - \sin b \sin y) + i(\sin b \cos y + \sin y \cos b)) = e^a(\cos b + i \sin b)e^x(\cos y + i \sin y) = e^{x+iy}e^{a+ib} = e^z e^w$.

(b) Simply substitute 0, π, and 2π in the sines and cosines.

(c) Use (a) and (b).

(d) 0 can have no logarithm, since $|\cos y + i \sin y|^2 = \cos^2 y + \sin^2 y = 1$, so $|e^{x+iy}| = e^x \neq 0$. On the other hand, if $z \neq 0$, there is an x such that $e^x = |z|$, namely, $x = \log |z|$. We must find a y such that if $z = a + ib$, then $|z| \cos y = a$ and $|z| \sin y = b$. Geometrically, it is obvious that such a y exists, since $z/|z|$

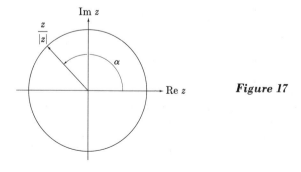

Figure 17

is just a point on the unit circle and if α is the associated angle, we can put $y = \alpha$ (Figure 17). Analytically, we can see that y exists as follows. If $a = 0$, then put $y = \pi/2$ if $b > 0$ and $y = -\pi/2$ if $b < 0$. If $a \neq 0$, then we seek a y such that $\sin y / \cos y = \tan y = b/a$. If $a > 0$, pick such a y between $-\pi/2$ and $\pi/2$. If $a < 0$, pick such a y between $\pi/2$ and $3\pi/2$. Finally, the formula $e^{w+2n\pi i} = e^w$ follows from (c).

(e) If $z = 0$, then put $r = 0$. If $z \neq 0$, let w be any complex logarithm of z. Put $r = e^{w/n}$. Then $r^n = e^{w/n} \cdots e^{w/n} = e^{w/n+\cdots+w/n} = e^w = z$.

7.   DEFINITION. If $x + iy$ is a complex number, its **real part** is $\mathrm{Re}\,(x + iy) = x$ and its **imaginary part** is $\mathrm{Im}\,(x + iy) = y$. Thus $z = \mathrm{Re}\,z + i\,\mathrm{Im}\,z$. Note that $\mathrm{Im}\,z$ is a *real* number.

8.   PROPOSITION. $\mathrm{Re}\,z = (z + \bar{z})/2.\,\mathrm{Im}\,z = (z - \bar{z})/2i.\,\mathrm{Re}\,e^z = e^{\mathrm{Re}\,z}\cos(\mathrm{Im}\,z).$ $\mathrm{Im}\,e^z = e^{\mathrm{Re}\,z}\sin(\mathrm{Im}\,z).$ If y is real, $\cos y = (e^{iy} + e^{-iy})/2 = \mathrm{Re}\,e^{iy},\,\sin y = (e^{iy} - e^{-iy})/2i = \mathrm{Im}\,e^{iy}.\,\mathrm{Re}\,(z + w) = \mathrm{Re}\,z + \mathrm{Re}\,w.\,\mathrm{Im}\,(z + w) = \mathrm{Im}\,z + \mathrm{Im}\,w.$

Proof. Trivial.

Exercises

1. Show that the map $z \mapsto \bar{z}$ is an isomorphism of the complex field onto itself.
2. Using 1, show that if $P(x) = a_n x^n + a_{n-1}x^{n-1} + \cdots + a_1 x + a_0$, where a_n, $a_{n-1}, \ldots, a_0$ are all *real* numbers, and if $P(z) = 0$, then $P(\bar{z}) = 0$.
3. Prove that

$$\tfrac{1}{2} + \sum_{k=1}^{n} \cos kt = \begin{cases} n + \tfrac{1}{2} & \text{if } \sin t/2 = 0 \\ \dfrac{\sin(n + \tfrac{1}{2})t}{2\sin t/2} & \text{if } \sin t/2 \neq 0 \end{cases}$$

   as follows. Note that $\tfrac{1}{2} + \sum_{k=1}^{n}\cos kt = -\tfrac{1}{2} + \mathrm{Re}\sum_{k=0}^{n} e^{ikt}$. Sum the geometric series $\sum_{k=0}^{n} e^{ikt}$ by the usual formula, and find its real part. At some point, you will have to multiply the numerator and denominator of a fraction by $e^{-it/2}$.
4. If z is a complex number, multiplication by z is a map from $\mathbf{R}^2$ to $\mathbf{R}^2$, which is easily seen to be linear: $a + ib \mapsto z \times (a + ib)$. In terms of the usual basis for $\mathbf{R}^2$, namely $1 = (1, 0)$ and $i = (0, 1)$, the matrix representation for this linear map is $\begin{pmatrix} x & -y \\ y & x \end{pmatrix}$, where $z = x + iy$. Call this matrix M_z. *Prove that* $M_z + M_w = M_{z+w}$ and $M_z M_w = M_{zw}$, where the left-hand operations are matrix operations. It is easy to see that the map $\mathbf{C} \rightarrow \mathcal{L}(\mathbf{R}^2, \mathbf{R}^2)$ defined by $x + iy \rightarrow \begin{pmatrix} x & -y \\ y & x \end{pmatrix}$ maps $\mathbf{C}$ one-to-one onto the set of matrices of the form $\begin{pmatrix} a & -b \\ b & a \end{pmatrix}$. Thus, what you will have proved is that this set of 2×2 matrices is a *field* isomorphic to the complex number field, where the field operations on these special matrices are the usual matrix sum and product.

CHAPTER VI
Some applications

One of the main reasons for developing the theory of metric spaces was to formulate some of the arguments that originated in special situations in analysis so that they could conveniently be applied to analysis as a whole. We have seen two examples already—a more or less trivial application of the notion of inner product to proving the triangle inequality for the absolute value function on **C**, and a far less trivial application of the Baire Category Theorem to the question of the existence of nowhere-differential continuous functions. This chapter collects a number of applications of the theory of metric and normed spaces to diverse problems in analysis: uniform convergence, power series, the existence and uniqueness of solutions to differential equations, differentiation, inverse and implicit functions. All are important in themselves, and illustrate how general arguments about metric spaces can be brought to bear on concrete problems in analysis. Most use the notion of *function space*—a metric space whose elements are functions. One argument is used several times: the *Contraction Mapping Principle*.

A. Uniform convergence

1. DEFINITION. Let S be any nonempty set and $\langle M, d \rangle$ any metric space. A sequence of functions $n \mapsto f_n$ from S to M **converges pointwise** to the function $f: S \to M$ if for every $x \in S$, $\lim_{n \to \infty} f_n(x) = f(x)$. That is, for every $x \in S$, and every $\epsilon > 0$, there is an integer N such that $d(f_n(x), f(x)) < \epsilon$ whenever $n > N$. If N can be chosen independently of x, the convergence is **uniform on S**. That is, $n \mapsto f_n$ **converges uniformly to f on S** if for every $\epsilon > 0$, there is an integer N such that $d(f_n(x), f(x)) < \epsilon$ *for all $x \in S$*

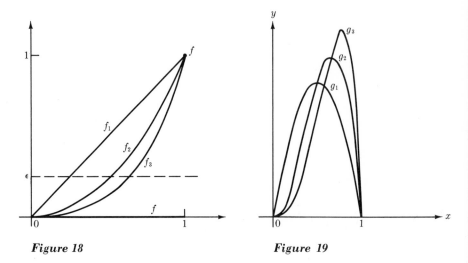

Figure 18   **Figure 19**

whenever $n > N$. Obviously uniform convergence implies pointwise convergence.

2.   EXAMPLES. (a) The sequence $n \mapsto (1/n) \sin nx$ converges uniformly to **0** on **R** since $|(1/n) \sin nx - 0| < \epsilon$ for all $x \in \mathbf{R}$ as soon as $n > 1/\epsilon$.

(b) If $0 \le x < 1$, $\lim\limits_{n \to \infty} x^n = 0$, and if $x = 1$, $\lim\limits_{n \to \infty} x^n = 1$ (compare Exercise II C 3). If $0 \le a < 1$, the sequence of functions $f_n(x) = x^n$ converges uniformly to 0 on $[0, a]$, since if n is large enough so that $a^n < \epsilon$, then $0 \le x^n \le a^n < \epsilon$ whenever $0 \le x \le a$. On the other hand, on the interval $[0, 1]$, the sequence $n \mapsto f_n$ converges *pointwise* to the function f defined by putting $f(x) = 0$ when $0 \le x < 1$ and $f(1) = 1$, but the convergence is *not* uniform. No matter how large n may be, there will always be an $x < 1$ such that $x^n > \frac{1}{2}$, so that $|f_n(x) - f(x)| = |x^n - 0| > \frac{1}{2}$ (Figure 18).

(c)   For $0 \le x \le 1$, put $g_n(x) = (n + 1)(n + 2)x^n(1 - x)$. An argument like that used in Exercise II C 3 readily shows that $\lim\limits_{n \to \infty} g_n(x) = 0$ if $0 \le x < 1$, and $g_n(1) = 0$ for every n. Thus $g_n \to \mathbf{0}$ pointwise on $[0, 1]$. However, it is easy to see that $n \mapsto g_n$ does not converge to **0** uniformly. Indeed, it is easy to verify that on $[0, 1]$, the maximum of g_n occurs when $x = n/(n + 1)$, at which point

$$g_n(x) = (n + 2)(n/(n + 1))^n$$

$$= \frac{(n + 2)}{\left(1 - \dfrac{1}{n + 1}\right)} \left(1 - \frac{1}{n + 1}\right)^{n+1}.$$

As $n \to \infty$,

$$\left(1 - \frac{1}{n + 1}\right)^{n+1} \to 1/e,$$

and

$$\frac{n+2}{1 - \dfrac{1}{n+1}} \to \infty.$$

Thus, $\max_{0 \le x \le 1} g_n(x) \to \infty$ as $n \to \infty$ (Figure 19).

3. THEOREM. Let $\langle M, d \rangle$ and $\langle N, e \rangle$ be metric spaces, and let $n \mapsto f_n$ be a sequence of *continuous* functions from M to N that converges *uniformly* on M to the function f. Then f is continuous. Indeed, if it be assumed only that each f_n is continuous at x, then f is continuous at x as well.

Proof. Let $\epsilon > 0$ be given. Choose N so large that, when $n > N$, $e(f_n(y), f(y)) < \epsilon/3$ for every $y \in M$. Then choose $\delta > 0$ so small that $e(f_{N+1}(y), f_{N+1}(x)) < \epsilon/3$ whenever $d(y, x) < \delta$. Then if $d(y, x) < \delta$, $e(f(y), f(x)) \le e(f(y), f_{N+1}(y)) + e(f_{N+1}(y), f_{N+1}(x)) + e(f_{N+1}(x), f(x)) < \epsilon/3 + \epsilon/3 + \epsilon/3 = \epsilon$.

4. COROLLARY. Let $\langle M, d \rangle$ be a metric space and $\langle V, \| \ \| \rangle$ be a normed linear space. Let $\mathcal{B}(M, V)$ be the vector space of all functions f from M to V such that $| f |_\infty = \text{lub} \{ \|f(x)\| : x \in M \}$ is finite. Let $\mathcal{C}(M, V)$ be the subspace of $\mathcal{B}(M, V)$ consisting of the bounded continuous functions from M to V. Then $\mathcal{C}(M, V)$ is closed in $\langle \mathcal{B}(M, V), | \ |_\infty \rangle$.

Proof. Convergence with respect to the norm $| \ |_\infty$ is just uniform convergence, since $| f - g |_\infty \le \epsilon$ means precisely that $\|f(x) - g(x)\| \le \epsilon$ *for every* $x \in M$. That is why we called $| \ |_\infty$ the *uniform norm*. Thus the corollary says exactly that if $f_n \to f$ uniformly on M and each f_n is continuous, so is f. This is just a restatement of a special case of Theorem 3.

5. COROLLARY. Let $\langle M, d \rangle$ be a metric space, and $\langle V, \| \ \| \rangle$ be a *complete* normed linear space (a *Banach space*). Then the normed linear spaces $\langle \mathcal{B}(M, V), | \ |_\infty \rangle$ and $\langle \mathcal{C}(M, V), | \ |_\infty \rangle$ are complete.

Proof. Since $\mathcal{C}(M, V)$ is closed in $\mathcal{B}(M, V)$, and a closed subspace of a complete space is complete, it is enough to show that $\mathcal{B}(M, V)$ is complete. Let $n \mapsto f_n$ be a Cauchy sequence in $\mathcal{B}(M, V)$. Since, for any $x \in M$, $\|f_n(x) - f_m(x)\| \le | f_n - f_m |_\infty$, $n \mapsto f_n(x)$ is a Cauchy sequence in $\langle V, \| \ \| \rangle$ for each $x \in M$. Since $\langle V, \| \ \| \rangle$ is complete, there is a limit, $f(x) = \lim_{n \to \infty} f_n(x)$. We shall show that f is bounded, so $f \in \mathcal{B}(M, V)$, and then that $\lim_{n \to \infty} | f_n - f |_\infty = 0$. Given any $\epsilon > 0$, choose N so large that when $m, n > N$, $\|f_n(x) - f_m(x)\| \le | f_n - f_m |_\infty < \epsilon$ for all $x \in M$. Then for every $x \in M$ and every $n > N$, $\|f_n(x) - f(x)\| = \lim_{m \to \infty} \|f_n(x) - f_m(x)\| \le \epsilon$. In particular, for any $n > N$, $||f(x)|| \leq ||f_n(x)|| + \epsilon \leq |f_n|_\infty + \epsilon$ for all $x \in M$, so f is bounded. Since $|f_n - f|_\infty \leq \epsilon$ when $n > N$, $\lim_{n \to \infty} |f_n - f|_\infty = 0$.

6.   COROLLARY. Let $\langle M, d \rangle$ be any metric space. The normed linear spaces $\langle \mathcal{C}(M, \mathbf{R}), |\ |_\infty \rangle$ and $\langle \mathcal{C}(M, \mathbf{C}), |\ |_\infty \rangle$ of bounded continuous real- and complex-valued functions on M are complete. $\langle \mathbf{R}^n, ||\ ||_\infty \rangle$ is complete.

Proof. The first assertions are trivial consequences of Corollary 5. To show that $\langle \mathbf{R}^n, ||\ ||_\infty \rangle$ is complete, notice that we can think of n-tuples $(x_1, \ldots, x_n) \in \mathbf{R}^n$ as functions from $\{1, 2, \ldots, n\}$ to $\mathbf{R}$; the n-tuple $(x_1, \ldots, x_n)$ corresponds to the function $k \mapsto x_k$. Then $||(x_1, \ldots, x_n)||_\infty$ is just the uniform norm of the function $k \mapsto x_k$ from $\{1, \ldots, n\}$ to the Banach space $\langle \mathbf{R}, |\ | \rangle$. That is, $\langle \mathbf{R}^n, ||\ ||_\infty \rangle = \langle \mathcal{B}(\{1, 2, \ldots, n\}, \mathbf{R}), |\ |_\infty \rangle$, so the completeness of $\langle \mathbf{R}^n, ||\ ||_\infty \rangle$ is another special case of Corollary 5. (If you wish, you can make $\{1, 2, \ldots, n\}$ into a metric space by giving it the discrete metric.)

7.   REMARKS. (a) The nonuniformity of the convergence of the sequence x^n on $[0, 1]$ follows at once from Theorem 3, since x^n is continuous, but the pointwise limit function is not. The same argument will not suffice, however, to prove that the sequence $(n + 1)(n + 2)x^n(1 - x)$ does not converge uniformly to $\mathbf{0}$ on $[0, 1]$.
(b) Corollary 6 is precisely what is needed to complete the proof of the existence of nowhere-differentiable functions, since in applying the Baire Category Theorem in that case we assumed that the space $\mathcal{C}(\mathbf{R}, \mathbf{R})$ was complete, although the proof was left to an Exercise.

Exercises

1. Find $\lim_{n \to \infty} (1/(1 + x^2))^n$ as a function of x. Does this sequence converge uniformly on $\mathbf{R}$?

*2. Let $g_n(x) = (n + 1)(n + 2)x^n(1 - x)$. Show that for every n, $\int_0^1 g_n(x)\, dx = 1$, so that $\lim_{n \to \infty} \int_0^1 g_n(x)\, dx \neq \int_0^1 \lim_{n \to \infty} g_n(x)\, dx$. Explain what is going on here. Try to formulate and prove a theorem giving sufficient conditions for a sequence of functions $n \mapsto f_n$ converging pointwise to a function f to satisfy $\lim_{n \to \infty} \int_0^1 f_n(x)\, dx = \int_0^1 f(x)\, dx$.

3. Show that the sequence of functions $x^n/n!$ converges uniformly to $\mathbf{0}$ as $n \to \infty$ on every *finite* interval, but *not* on all of $\mathbf{R}$.

B. Absolutely uniform convergence of series.
Tests of convergence.

1. DEFINITION. Let $\langle V, \|\ \|\rangle$ be a normed linear space, and $n \mapsto s_n$ a sequence of vectors in V. The **Nth partial sum** of $n \mapsto s_n$ is $\sum_{n=1}^{N} s_n = s_1 + \cdots + s_N$. If $\lim_{N \to \infty} \sum_{n=1}^{N} s_n = L$, then the **series** $s_1 + s_2 + s_3 + \cdots = \sum_{n=1}^{\infty} s_n$ **converges,** and its **sum** is L. The series $\sum s_n$ is a **Cauchy series** if its partial sums form a Cauchy sequence. It is an **absolutely Cauchy series** if the series $\sum_{n=1}^{\infty} \|s_n\|$ converges to a real number—which is equivalent to the assertion that the partial sums of $\sum_{n=1}^{\infty} \|s_n\|$ form a Cauchy sequence (see Theorem II C 9).

2. PROPOSITION. Every absolutely Cauchy series is a Cauchy series. Hence, in a *complete* normed linear space (a Banach space) every absolutely Cauchy series, and indeed every Cauchy series, converges.

Proof. If $\sum s_n$ is an absolutely Cauchy series, and $M > N$, $\|\sum_{n=1}^{M} s_n - \sum_{n=1}^{N} s_n\| = \|\sum_{n=N+1}^{M} s_n\| \leq \sum_{n=N+1}^{M} \|s_n\|$. This last expression can be made as small as we like by choosing N and M large enough, since the series $\sum \|s_n\|$ is a Cauchy series. The rest follows from the definitions of Cauchy series and convergence. ●

REMARK. This fact is sometimes called the Weierstrass M test.

3. COROLLARY. Let $\langle M, d \rangle$ be a metric space and $\langle V, \|\ \|\rangle$ a *complete* normed linear space (for example, $\mathbf{R}$ or $\mathbf{C}$). Let $n \mapsto f_n$ be a sequence of bounded continuous functions from M to V, such that $\sum_{n=1}^{\infty} |f_n|_\infty$ converges. Then the sequence of partial sums of the series $\sum_{n=1}^{\infty} f_n$ converges uniformly on M to a bounded continuous function.

Proof. We have shown that $\langle \mathcal{C}(M, V), |\ |_\infty \rangle$ is complete. Therefore Proposition 2 applies.

4. REMARK. The notions of convergence of sums defined so far are equivalent to sequential convergence (of partial sums) and depend on the order in which the terms are summed. Generally speaking, the terms of a convergent series cannot be rearranged without changing the sum, although rearrangement of an *absolutely Cauchy* series does not change the sum. More precisely, if $\sigma : \mathbf{N} \to \mathbf{N}$ is one-to-one and onto, the series $\sum_{n=1}^{\infty} a_{\sigma(n)}$ is a *rearrangement* of the series $\sum_{n=1}^{\infty} a_n$. If $\sum a_n$ is a convergent absolutely Cauchy series, it can be shown (see Exercises 4 and 5) that $\sum_{n=1}^{\infty} a_{\sigma(n)} = \sum_{n=1}^{\infty} a_n$. If $\sum a_n$ is not an absolutely Cauchy series, $\sum a_{\sigma(n)}$ may converge to a different sum, or even diverge (see Exercise 6).

For example, the Alternating Series Test 13 will show that the series $1 - \frac{1}{2} + \frac{1}{3} - \frac{1}{4} + \frac{1}{5} - \frac{1}{6} + \cdots$ converges to a limit $L > 0$ (in fact,

$L = \log 2$), although the series is *not* absolutely Cauchy. If we rearrange the series by taking one positive term, then two negative terms, then one positive term, then two negative terms, and so on, we get $1 - \frac{1}{2} - \frac{1}{4} + \frac{1}{3} - \frac{1}{6} - \frac{1}{8} + \frac{1}{5} - \frac{1}{10} - \frac{1}{12} + \cdots = (1 - \frac{1}{2}) - \frac{1}{4} + (\frac{1}{3} - \frac{1}{6}) - \frac{1}{8} + (\frac{1}{5} - \frac{1}{10}) - \frac{1}{12} + \cdots = \frac{1}{2} - \frac{1}{4} + \frac{1}{6} - \frac{1}{8} + \frac{1}{10} - \frac{1}{12} + \cdots = \frac{1}{2}(1 - \frac{1}{2} + \frac{1}{3} - \frac{1}{4} + \frac{1}{5} - \frac{1}{6} + \cdots) = \frac{1}{2}L$.

There is an important notion of *unordered* or *unconditional convergence* of series that is independent of the order in which the terms are taken. Let S be any *countable* set and $\langle V, \| \ \| \rangle$ be a normed space. If f is a function from S to V and $E = \{x_1, \ldots, x_n\}$ is any *finite* subset of S, we define $\sum_{x \in E} f(x)$ in the obvious way to be $f(x_1) + \cdots + f(x_n)$. Then we say that L is the *unordered* or *unconditional sum* of f if for every $\epsilon > 0$, there is a *finite* set $E(\epsilon) \subseteq S$ with the following property: if H is any finite subset of S that contains $E(\epsilon)$, then $\|L - \sum_{x \in H} f(x)\| < \epsilon$. In particular, it is not hard to see that if $S = \mathbf{N}$, a function $f: S \to V$ has an unordered sum whenever $\sum_{n=1}^{\infty} f(n)$ is a convergent absolutely Cauchy series (Exercise 4). Conversely, if V is *finite-dimensional*, it can be shown that every unconditionally convergent series is an absolutely Cauchy series, but if V is an *infinite-dimensional* normed space, it will generally contain unconditionally convergent series that fail to converge absolutely. For example, let V be the space of bounded continuous functions on the *open* interval $]0, \pi[= \{x \in \mathbf{R}: 0 < x < \pi\}$, with the norm $\|f\|_2 = \left(\int_0^\pi (f(x))^2 \, dx \right)^{1/2}$. Then it is shown in the theory of Fourier series that $\sum_{n=1}^{\infty} (1/(2n - 1)) \sin (2n - 1)x$ converges in $\langle V, \| \ \|_2 \rangle$ to the constant function $\pi/4$. In fact, the convergence is unconditional, but the series is *not* absolutely Cauchy.

5.   THEOREM (THE COMPARISON TEST). Let $\sum b_n$ be a Cauchy series of nonnegative real numbers, and $\sum x_n$ a series of vectors in the normed linear space $\langle V, \| \ \| \rangle$ such that $\|x_n\| \leq b_n$ for every $n \geq N$ (where N is some fixed integer). Then $\sum x_n$ is an absolutely Cauchy series.

Proof. Let $\epsilon > 0$ be given. Choose $M \geq N$ so large that $\sum_{k=n+1}^{m} b_k < \epsilon$ whenever $m > n > M$. Then if $m > n > M$, $\sum_{k=n+1}^{m} \|x_k\| \leq \sum_{k=n+1}^{m} b_k < \epsilon$.

6.   COROLLARY. If $\sum b_n$ is a *non-Cauchy* series of nonnegative real numbers, and $\sum x_n$ is a series of vectors such that $\|x_n\| \geq b_n$ for every $n \geq N$, then $\sum x_n$ is *not* an absolutely Cauchy series. In particular, if $x_n \in \mathbf{R}$ and $x_n \geq b_n \geq 0$ for every n, then $\sum x_n$ diverges if $\sum b_n$ diverges (does not converge).

Proof. If $\sum x_n$ were an absolutely Cauchy series, $\sum \|x_n\|$ would be a Cauchy series. Then $\sum b_n$ would be a Cauchy series by the Comparison Test. ●

The Comparison Test provides one of the most useful techniques for verifying the convergence of series, but to use it, it is essential to have some standard comparison series available. What follows is a list of some.

7. THE GEOMETRIC SERIES. If $-1 < r < 1$, the geometric series $1 + r + r^2 + r^3 + \cdots$ converges to $1/(1 - r)$.

Proof. Put $s_n = 1 + r + r^2 + \cdots + r^{n-1}$. Then $rs_n - s_n = r^n - 1$, or $s_n = (1 - r^n)/(1 - r)$. We know that $\lim_{n \to \infty} r^n = 0$ (see Exercise II C 3). Therefore, $\lim_{n \to \infty} s_n = 1/(1 - r)$ provided that $|r| < 1$.

8. COROLLARY (THE RATIO TEST). Let $\sum x_n$ be a series of positive real numbers. Suppose $\lim_{n \to \infty} x_{n+1}/x_n = r$. Then

(a)   if $r < 1$, $\sum x_n$ converges,
(b)   if $r > 1$, $\sum x_n$ diverges,
(c)   if $r = 1$, the test doesn't work.

Proof. (a) If $r < 1$, then for some N, $x_{n+1}/x_n < s = (1 + r)/2$ whenever $n > N$. Then

$$x_{n+k} = x_n \left(\frac{x_{n+1}}{x_n} \right) \cdots \left(\frac{x_{n+k}}{x_{n+k-1}} \right) < x_n s^k = (x_n s^{-n})s^{n+k}.$$

Now use the Comparison Test to compare the series $\sum x_n$ with the geometric series $(x_n s^{-n})(1 + s + s^2 + s^3 + \cdots)$.
(b) If $r > 1$, then for some N, $n > N \Rightarrow x_{n+1}/x_n > s > 1$. Then if $n > N$ and $k \geq 0$, $x_{n+k} \geq x_n s^k \to +\infty$ as $k \to \infty$. Clearly, then, $\sum x_n$ diverges.
(c) We shall shortly see that $1 + \frac{1}{2} + \frac{1}{3} + \frac{1}{4} + \cdots$ diverges, and $1 + (1/2^2) + (1/3^2) + (1/4^2) + \cdots$ converges. In both cases the limiting ratio is 1.

9. THEOREM. Let $n \mapsto s_n$ be a sequence of real numbers that is monotonically increasing and bounded above: $s_n \leq s_{n+1} \leq B$ for every $n \in \mathbf{N}$. Then $\lim_{n \to \infty} s_n = \text{lub } \{s_n : n \in \mathbf{N}\}$.

Proof. Let $L = \text{lub } \{s_n : n \in \mathbf{N}\}$. According to Theorem II C 1, for every $\epsilon > 0$, there is an $N \in \mathbf{N}$ such that $L - \epsilon < s_N \leq L$. Then if $n \geq N$, $L - \epsilon < s_N \leq s_n \leq L$.

10. COROLLARY. Let $\sum a_n$ be a series of nonnegative real numbers whose partial sums form a bounded set. Then the series converges to the least upper bound of the set of its partial sums.

Proof. Let $s_n = \sum_{k=1}^n a_k$. Then $n \mapsto s_n$ is a bounded monotonic sequence.

11. THE INTEGRAL TEST. Let $f \colon \mathbf{R^+} \cup \{0\} \to \mathbf{R^+} \cup \{0\}$ be a Riemann-integrable function such that $x < y \Rightarrow f(y) \leq f(x)$. Then $\sum_{n=1}^{\infty} f(n)$ converges $\Leftrightarrow$ $\int_0^\infty f(t)\, dt = \lim_{n \to \infty} \int_0^n f(t)\, dt$ exists. If $\int_0^\infty f(t)\, dt$ exists, $0 \leq \int_0^\infty f(t)\, dt -$ $\sum_{n=1}^\infty f(n) \leq f(0)$. [*Note:* Riemann-integrable means that $\int_0^n f(t)\, dt$ exists for each $n \in \mathbf{N}$.]

Proof. For each $n \in \mathbf{N}$, $f(n-1) = \int_{n-1}^n f(n-1)\, dt \geq \int_{n-1}^n f(t)\, dt \geq$ $\int_{n-1}^n f(n)\, dt = f(n)$. Thus $\sum_{k=0}^{n-1} f(k) \geq \sum_{k=1}^n \int_{k-1}^k f(t)\, dt = \int_0^n f(t)\, dt \geq$ $\sum_{k=1}^n f(k)$. It follows that $f(0) \geq f(0) - f(n) = \sum_{k=0}^{n-1} f(k) - \sum_{k=1}^n f(k) \geq$ $\int_0^n f(t)\, dt - \sum_{k=1}^n f(k) \geq 0$. If $\lim_{n \to \infty} \int_0^n f(t)\, dt$ exists, then because $f(t) \geq 0$ for all t, $\int_0^\infty f(t)\, dt \geq \int_0^n f(t)\, dt \geq \sum_{k=1}^n f(k)$ for all n. Corollary 10 then shows that $\sum_{k=1}^\infty f(k)$ exists, and the inequalities above show that

$$f(0) \geq \lim_{n \to \infty} \int_0^n f(t)\, dt - \lim_{n \to \infty} \sum_{k=1}^n f(k) = \int_0^\infty f(t)\, dt - \sum_{k=1}^\infty f(k) \geq 0.$$

Conversely, if $\sum_{k=1}^\infty f(k)$ exists, then

$$f(0) + \sum_{k=1}^\infty f(k) \geq f(0) + \sum_{k=1}^{n-1} f(k) = \sum_{k=0}^{n-1} f(k) \geq \int_0^n f(t)\, dt,$$

so that $\lim_{n \to \infty} \int_0^n f(t)\, dt$ exists by Theorem 9. ●

12. COROLLARY. $\sum_{n=1}^\infty n^{-p}$ converges if $p > 1$ and diverges if $p \leq 1$.

Proof. If $p \neq 1$, $\int_1^n t^{-p}\, dt = (-p+1)^{-1}(n^{-p+1} - 1)$. If $p > 1$, $n^{-p+1} \to 0$ as $n \to \infty$. If $p < 1$, $n^{-p+1} \to +\infty$ as $n \to \infty$. $\int_1^n t^{-1}\, dt = \log n - \log 1 \to +\infty$ as $n \to \infty$. Now apply the Integral Test (which works in the same way if the lower limit of integration is 1 instead of 0).

13. ALTERNATING SERIES TEST. Suppose that for each n, $a_n \geq a_{n+1} \geq 0$ and that $\lim_{n \to \infty} a_n = 0$. Then $a_1 - a_2 + a_3 - a_4 + \cdots$ converges to a limit L, and for every n, $|L - \sum_{k=1}^n (-1)^{k+1} a_k| \leq a_{n+1}$.

Proof. Let $s_n = \sum_{k=1}^n (-1)^{k+1} a_k$. Then for every $n \in \mathbf{N}$, $s_{2n} \leq s_{2n} + (a_{2n+1} - a_{2n+2}) = s_{2n+2} = s_{2n+3} - a_{2n+3} \leq s_{2n+3} \leq s_{2n+3} + (a_{2n+2} - a_{2n+3}) = s_{2n+1}$. By induction, it follows that $s_2 \leq s_4 \leq s_6 \leq \cdots \leq s_{2n} \leq s_{2n+1} \leq \cdots \leq s_5 \leq s_3 \leq s_1$. Thus the sequence $n \mapsto s_{2n}$ is monotonically increasing and bounded above by s_1. It converges to its least upper bound L. Also the sequence $n \mapsto s_{2n+1}$ is monotonically decreasing and bounded below by s_2. It converges to its greatest lower bound G. But $L = \lim_{n \to \infty} s_{2n} =$

$\lim_{n \to \infty} (s_{2n+1} - a_{2n+1}) = \lim_{n \to \infty} s_{2n+1} - \lim_{n \to \infty} a_{2n+1} = G - 0 = G$. This shows that $\lim_{n \to \infty} s_n = L = G$. Moreover, $s_{2n} \leq L = G \leq s_{2n+1} \leq s_{2n-1}$ for each $n \in \mathbf{N}$, so $0 \leq L - s_{2n} \leq s_{2n+1} - s_{2n} = a_{2n+1}$ and $0 \leq s_{2n-1} - L \leq s_{2n-1} - s_{2n} = a_{2n} = a_{(2n-1)+1}$. ●

Most of the preceding theorems not only allow us to recognize when a series converges, but also provide estimates of the error committed in approximating the limit of a series by one of its partial sums.

Exercises

1. Show that
$$\sum_{n=2}^{\infty} \frac{1}{n \log n}$$
diverges, but
$$\sum_{n=2}^{\infty} \frac{1}{n(\log n)^2}$$
converges.

2. Prove that $\sum_{n=1}^{\infty} nr^{n-1}$ converges (by using the Ratio Test) if $|r| < 1$.

3. Show that the series $\sum_{n=1}^{\infty} (1/n^2) \sin nx$ converges uniformly for all $x \in \mathbf{R}$.

*4. (a) Starting with the definitions, prove that if $\sum x_n$ is a convergent absolutely Cauchy series in the normed space $\langle V, \| \ \| \rangle$, then $\sum_{n=1}^{\infty} x_n$ is the unordered sum of the function $n \mapsto x_n$ from $\mathbf{N}$ into V. *Hint:* Let $E(\epsilon) = \{n \in \mathbf{N}: 1 \leq n \leq N\}$ for an appropriate $N \in \mathbf{N}$.

(b) Conversely, show that if the function $f: \mathbf{N} \to \mathbf{R}$ has an unordered sum L, then $\sum f(n)$ is an absolutely Cauchy series whose limit is L. *Hint:* Show that if $\sum |f(n)|$ diverges, then the sum of the positive terms of the series $\sum f(n)$ diverges or else the sum of the negative terms of the series $\sum f(n)$ diverges.

*5. Let S be a countable set, $\langle V, \| \ \| \rangle$ a normed space, and $f: S \to V$ a function that has an unordered sum L. Show that if $\sigma: \mathbf{N} \to S$ is one-to-one and onto, then $\sum_{n=1}^{\infty} f(\sigma(n)) = L$. Interpret this fact as a theorem about the rearrangement of absolutely Cauchy series (use 4).

*6. Let $\sum a_n$ be a convergent series of real numbers that is *not* an absolutely Cauchy series. Show that if L is *any* real number, there is a rearrangement of $\sum a_n$ that converges to L. Show also that $\sum a_n$ can be rearranged so as to diverge to $-\infty$, $+\infty$, or in an oscillatory manner. *Hint:* First sum up just enough positive terms of $\sum a_n$ to attain a partial sum $> L$, then add just enough negative terms to attain a partial sum $< L$, then just enough positive terms to attain a partial sum $> L$, and so on.

C. The radius of convergence of a power series

1. DEFINITION. A **power series about z_0** is a series of the form $a_0 + a_1(z - z_0) + a_2(z - z_0)^2 + a_3(z - z_0)^3 + \cdots$. In general, the coefficients $a_0, a_1, a_2, \ldots$ and z and z_0 can be complex numbers, although they may some-

times be restricted to being real. It would even make sense to let the coefficients be vectors, while z and z_0 are scalars. The series **converges at z** if $\sum_{k=0}^{\infty} a_k(z - z_0)^k$ has a limit.

If S is a subset of $\mathbf{C}$, we can think of a power series as a series of functions in $\mathcal{C}(S, \mathbf{C})$. We shall be particularly interested in those sets S on which a power series converges uniformly.

2.   THEOREM. Suppose that $\{|a_n(z_1 - z_0)^n|\}$ is a bounded set of real numbers or, in particular, that the power series $\sum a_n(z - z_0)^n$ converges at z_1. Then the series converges uniformly on the closed disk $B_r^c(z_0) \subseteq \mathbf{C}$ for any non-negative real number $r < |z_1 - z_0|$, as do the series $\sum na_n(z - z_0)^{n-1}$ and $\sum (1/(n + 1))a_n(z - z_0)^{n+1}$.

Proof. It is clear that convergence of the series $\sum a_n(z - z_0)^n$ implies boundedness of its terms. Suppose that for every n, $|a_n(z_1 - z_0)^n| \leq B$. Then, for every n, $|a_n| \leq B|z_1 - z_0|^{-n}$. Thus if $|z - z_0| \leq r$,

$$|a_n(z - z_0)^n| \leq \frac{B}{|z_1 - z_0|^n} r^n = Bs^n,$$

where

$$s = \frac{r}{|z_1 - z_0|} < 1.$$

Now let V be the space of bounded continuous complex-valued functions on $B_r^c(z_0)$ with the norm $|f|_\infty = \text{lub } \{|f(z)| : z \in B_r^c(z_0)\}$. For each n, the function $z \mapsto a_n(z - z_0)^n$ is in V. We have just shown that $|a_n(z - z_0)^n|_\infty \leq Bs^n$. Therefore we can compare with the geometric series $B(1 + s + s^2 + s^3 + \cdots)$ to show that $\sum a_n(z - z_0)^n$ converges in the Banach space $\langle V, | \ |_\infty \rangle$ to a continuous complex-valued function on $B_r^c(z_0)$. Likewise, the Ratio Test shows that the series $B(1 + 2s + 3s^2 + \cdots)$ and $B(s + s^2/2 + s^3/3 + s^4/4 + \cdots)$ converge. Using these as comparison series, the Comparison Test establishes the convergence of $\sum na_n(z - z_0)^{n-1}$ and $\sum (n + 1)^{-1}a_n(z - z_0)^{n+1}$.

3.   DEFINITION. Let $\sum a_n(z - z_0)^n$ be a power series, and $S = \{s \in \mathbf{R} : \sum a_n s^n$ converges$\}$. Clearly $0 \in S$. If S is bounded above, the **radius of convergence** R of the series is lub S. Otherwise $R = +\infty$.

4.   THEOREM. Let R be the radius of convergence of the power series $\sum a_n(z - z_0)^n$.

(a)   The series converges pointwise on $B_R(z_0)$.

(b)   For each r such that $0 \leq r < R$, the series converges uniformly on $B_r^c(z_0)$.

(c)   If $|z_1 - z_0| > R$, the series does not converge at z_1.

(d)   The series $\sum a_n(z - z_0)^n$ has the same radius of convergence as the series $\sum_{n=1}^{\infty} na_n(z - z_0)^{n-1}$ and $\sum_{n=0}^{\infty} (n + 1)^{-1}a_n(z - z_0)^{n+1}$ obtained by differentiating it and integrating it term by term.

Proof. (b) If $0 \leq r < R$, there is, by the definition of R, a number s such that $r < s < R$ and $\sum a_n s^n = \sum a_n (z_1 - z_0)^n$ converges, where $z_1 = z_0 + s$. Apply Theorem 2.

(a) If $z \in B_R(z_0)$, $|z - z_0| = r < R$, and therefore $\sum a_n (z - z_0)^n$ converges by (b).

(c) If $|z_1 - z_0| > R$, and $\sum a_n (z_1 - z_0)^n$ converged, then we could put $s = \frac{1}{2}(R + |z_1 - z_0|)$, and s would be in $S = \{s \in \mathbf{R} \colon \sum a_n s^n \text{ converges}\}$ by Theorem 2. But then R would not be an upper bound for S since $s > R$, contradicting the definition of R.

(d) Let R_1 and R_2 be respectively the radii of convergence of the series $\sum_{n=1}^{\infty} n a_n (z - z_0)^{n-1}$ and $\sum_{n=0}^{\infty} (n+1)^{-1} a_n (z - z_0)^{n+1}$. Suppose $0 \leq t < R$. Choose s such that $t < s < R$. According to (a), $\sum a_n s^n$ converges, and according to Theorem 2 with $z_1 = z_0 + t$, $\sum n a_n t^{n-1}$ and $\sum (n+1)^{-1} a_n t^{n+1}$ converge as well. This shows that $t \leq R_1$ and $t \leq R_2$. It follows that $R \leq R_1$ and $R \leq R_2$. But $\sum_{n=0}^{\infty} a_n (z - z_0)^n$ is obtained by term-by-term differentiation of the series $\sum_{n=0}^{\infty} (n+1)^{-1} a_n (z - z_0)^{n+1}$. Applying what we have just proved to this latter series, we see that $R_2 \leq R$. Similarly, $R_1 \leq R$. ●

5. REMARKS. (a) If $|z_1 - z_0| = R$, the series $\sum a_n (z_1 - z_0)^n$ may or may not converge. For example, one can verify by the Ratio Test that the series $\sum_{n=1}^{\infty} z^n / n^2$ has radius of convergence 1. Therefore, so does the series $\sum_{n=1}^{\infty} z^n / n = z \sum_{n=1}^{\infty} n(z^{n-1}/n^2)$. The first series converges when $z = 1$, but the second does not. However, the series $\sum_{n=1}^{\infty} z^n / n$ does converge (by the Alternating Series Test) when $z = -1$.

(b) Suppose the series $\sum a_n (z - z_0)^n$ has only real coefficients and that z_0 is real. Then the *circle of convergence* $B_R(z_0)$ will intersect the real axis in an interval centered at z_0 of diameter $2R$. That is, the radius of convergence can be determined entirely by looking at what the series does as a real-valued series of functions of a real variable.

(c) The study of complex power series can illuminate rather mysterious phenomena in the theory of real power series. For example, the function $(1 + x^2)^{-1} = 1 - x^2 + x^4 - x^6 + x^8 - \cdots$ is well-behaved as a function of the real variable x on all of $\mathbf{R}$. Therefore, it is rather mysterious why its power series expansion fails to converge when $|x| \geq 1$. It is not clear what goes wrong when $|x| = 1$ if we look only at real values of x. However, if we think of the series $1 - z^2 + z^4 - z^6 + \cdots$ as representing the function $(1 + z^2)^{-1}$ of a *complex* variable when $|z| < 1$, we see what goes wrong when $|z| = 1$— namely, that at $z = i$ or $z = -i$, the denominator of the fraction $1/(1 + z^2)$ vanishes. These ideas will be developed in greater detail later on.

*Exercises

Determine the radii of convergence of the following power series. (Use the Ratio test.)

(a) $1 + \dfrac{z}{1!} + \dfrac{z^2}{2!} + \dfrac{z^3}{3!} + \dfrac{z^4}{4!} + \cdots .$

(b) $z + \dfrac{z^2}{2} + \dfrac{z^3}{3} + \dfrac{z^4}{4} + \cdots .$

(c) $1 + \frac{1}{3}x + \dfrac{\frac{1}{3}(\frac{1}{3} - 1)}{2!}\, x^2 + \dfrac{\frac{1}{3}(\frac{1}{3} - 1)(\frac{1}{3} - 2)}{3!}\, x^3$

$$+ \dfrac{\frac{1}{3}(\frac{1}{3} - 1)(\frac{1}{3} - 2)(\frac{1}{3} - 3)}{4!}\, x^4 + \cdots .$$

D.   The contraction mapping theorem

Let us consider a technique for solving nonlinear equations such as $x^2 = 2$. First we perform some algebraic sleight of hand on the equation

$$x^2 = 2 \Leftrightarrow x = \frac{2}{x} \Leftrightarrow x + x = \frac{2}{x} + x \Leftrightarrow x = \tfrac{1}{2}\left(x + \frac{2}{x}\right).$$

We now have the equation in the form $x = f(x)$. If we think of f as a mapping from **R** to **R**, we see that what we want is a point $(\sqrt{2})$ that is *left fixed* by the mapping f.

1.   DEFINITION.   Let S be a set, and f a function from S to S. A *fixed point* of f is any $x \in S$ such that $x = f(x)$.   ●

Here is an idea how to find the fixed point: start with some number x_0 as a first guess. In general, x_0 will not be a fixed point, but we may hope that $x_1 = f(x_0)$ will be a better guess. Repeat the procedure, to get $x_2 = f(x_1)$, $\ldots , x_{n+1} = f(x_n)$. With luck, this procedure will converge to a fixed point.

If we use $x_0 = 1$ and $f(x) = \tfrac{1}{2}(x + (2/x))$, we get

$x_0 = 1 = 1.000000,$

$x_1 = 3/2 = 1.500000,$

$x_2 = 17/12 = 1.4166666 \ldots ,$

$x_3 = 577/408 = 1.4142156 \ldots ,$

while $\sqrt{2} = 1.414214 \ldots .$

In this case, at least, the method works with remarkable efficiency. We can see why it works (although not why it is so efficient) by the following observations. First, if $x_n \geq 1$, so is $x_{n+1} = \tfrac{1}{2}(x_n + (2/x_n))$. Indeed if $x_n \geq 2$, then $x_{n+1} \geq \tfrac{1}{2}x_n \geq 1$. If $1 \leq x_n < 2$, then $x_{n+1} \geq \tfrac{1}{2}x_n + (1/x_n) > \tfrac{1}{2} + \tfrac{1}{2} = 1$, since $1/x_n > \tfrac{1}{2}$. Thus f maps the set $M = \{x \in \mathbf{R} : x \geq 1\}$ into itself. Now $f'(x) = \tfrac{1}{2}(1 - (2/x^2))$. If $x \geq 1$, $0 \leq 2/x^2 \leq 2$, so $-1 \leq 1 - (2/x^2) \leq 1$. Hence $|f'(x)| \leq \tfrac{1}{2}$. Since $\sqrt{2} \geq 1$, we can apply the Mean Value Theorem: $|f(x) - \sqrt{2}| = |f(x) - f(\sqrt{2})| = |f'(\eta)(x - \sqrt{2})|$ for some η between x and $\sqrt{2}$, and therefore $|f(x) - \sqrt{2}| = |f'(\eta)|\,|x - \sqrt{2}| \leq \tfrac{1}{2}|x - \sqrt{2}|$. That is, if our approximation x_n is in error by the amount $|x_n - \sqrt{2}|$, the error in the next approximation, namely $|f(x_n) - \sqrt{2}| = |x_{n+1} - \sqrt{2}|$, will be at most half as much. $|x_n - \sqrt{2}| \leq 2^{-n}|x_0 - \sqrt{2}|$; $\lim_{n \to \infty} x_n = \sqrt{2}$.

These ideas can be made to work in complete metric spaces in general. The following definition will help.

2. DEFINITION. Let $\langle M, d \rangle$ be a metric space. A function $f: M \to M$ is a **contraction** with **Lipschitz constant** L if L is a constant such that $0 \leq L < 1$ and that for every $x, y \in M$, $d(f(x), f(y)) \leq Ld(x, y)$. (More generally, if f is any mapping from the metric space $\langle M, d \rangle$ to the metric space $\langle N, \delta \rangle$ such that $\delta(f(x), f(y)) \leq Ld(x, y)$, then f **satisfies a Lipschitz condition with constant** L, even if $L \geq 1$.)

3. THE CONTRACTION MAPPING THEOREM (BANACH FIXED POINT THEOREM). A contraction of a complete metric space has a unique fixed point.

In more detail, if f is a contraction of the complete metric space $\langle M, d \rangle$ with Lipschitz constant L, if x_0 is any point of M and $x_{n+1} = f(x_n)$ for $n = 0, 1, 2, \ldots$, then the sequence $n \mapsto x_n$ converges to the unique fixed point x^* of f, and $d(x^*, x_0) \leq d(x_1, x_0)/(1 - L)$.

Proof. For every $n \geq 1$, $d(x_{n+1}, x_n) = d(f(x_n), f(x_{n-1})) \leq Ld(x_n, x_{n-1})$. Thus, by induction, $d(x_{n+1}, x_n) \leq Ld(x_n, x_{n-1}) \leq L^2d(x_{n-1}, x_{n-2}) \leq \cdots \leq L^nd(x_1, x_0)$. If $n > m$,

$$d(x_n, x_m) \leq d(x_n, x_{n-1}) + d(x_{n-1}, x_{n-2}) + \cdots + d(x_{m+1}, x_m)$$
$$\leq (L^{n-1} + L^{n-2} + \cdots + L^m)d(x_1, x_0)$$
$$= L^m((1 - L^{n-m})/(1 - L))d(x_1, x_0) \leq \frac{L^m}{1 - L} d(x_1, x_0).$$

If N is so large that $(L^N/(1 - L))d(x_1, x_0) < \epsilon$, $d(x_n, x_m) < \epsilon$ whenever $n, m > N$. This shows that $n \mapsto x_n$ is a Cauchy sequence. Since M is complete, $n \mapsto x_n$ has a limit x^*. Since f is a contraction, it is clearly continuous. Thus $x^* = \lim_{n \to \infty} x_{n+1} = \lim_{n \to \infty} f(x_n) = f(\lim_{n \to \infty} x_n) = f(x^*)$, so x^* is a fixed point of f. Moreover, we know that $d(x_n, x_0) \leq ((1 - L^n)/(1 - L))d(x_1, x_0) \leq (1/(1 - L))d(x_1, x_0)$, so, taking the limit as $n \to \infty$, $d(x^*, x_0) \leq (1/(1 - L))d(x_1, x_0)$. Finally, if y is any point in M such that $y = f(y)$, then $d(y, x^*) = d(f(y), f(x^*)) \leq Ld(y, x^*)$, or $(1 - L)d(y, x^*) \leq 0$. Since $1 - L > 0$ and $d(y, x^*) \geq 0$, this is possible only if $y = x^*$. That is, the fixed point of f is unique. ●

The following corollary shows how the Contraction Mapping Theorem may conveniently be applied to the solution of nonlinear equations.

4. COROLLARY. Let I be a closed (possibly infinite) interval in **R**, and f a

differentiable function from I to I such that $|f'(x)| \le L < 1$ for every $x \in I$. Then the equation $x = f(x)$ has a unique solution in I.

Proof. f is a contraction of I, since $|f(x) - f(y)| = |f'(\zeta)(x - y)| = |f'(\zeta)| \, |x - y| \le L|x - y|$ for some ζ between x and y, by the Mean Value Theorem. Any solution to the equation $x = f(x)$ is a fixed point of f. $\bullet$

Next, we shall apply the Contraction Mapping Theorem to the calculation of the inverse of a bounded linear operator. First, however, we must check the fact that the space of bounded linear operators on a Banach space is complete.

5.   PROPOSITION. If $\langle V, \| \ \| \rangle$ is a Banach space, so is the space $\langle B(V, V), | \ | \rangle$ of bounded linear operators from V to V.

Proof. Let $n \mapsto L_n$ be a Cauchy sequence of bounded linear operators. For each $x \in V$, the sequence $n \mapsto L_n(x)$ is a Cauchy sequence in V, since $\|L_n(x) - L_m(x)\| = \|(L_n - L_m)(x)\| \le |L_n - L_m| \, \|x\|$. Since V is a Banach space, it has a limit $L(x)$. Then $L(x + y) = \lim_{n \to \infty} L_n(x + y) = \lim_{n \to \infty} (L_n(x) + L_n(y)) = L(x) + L(y)$, and similarly $L(rx) = rL(x)$ for each $r \in \mathbf{R}$. Moreover, $\|L(x)\| = \lim_{n \to \infty} \|L_n(x)\| = \lim_{n \to \infty} \|(L_m + (L_n - L_m))(x)\|$, and if m is so large that $|L_n - L_m| < 1$ when $n > m$, this will be less than $[|L_m| + 1]\|x\|$. Thus $L \in B(V, V)$. Finally, $\|L(x) - L_m(x)\| = \lim_{n \to \infty} \|L_n(x) - L_m(x)\| \le \lim_{n \to \infty} |L_n - L_m| \, \|x\| \le \epsilon\|x\|$ if m is so large that $|L_n - L_m| \le \epsilon$ when $n > m$. Thus $|L - L_m| \le \epsilon$, so $\lim_{m \to \infty} L_m = L$. Alternatively, we could observe that the sequence $n \mapsto L_n$ is a sequence of continuous functions on V converging uniformly on the unit ball $B_1(0)$ in V. Therefore the limit L is continuous on $B_1(0)$. Since L is linear, it is continuous everywhere. We then prove as above that $\lim_{n \to \infty} L_n = L$ in $B(V, V)$.

6.   THEOREM. Let L be a map in $B(V, V)$ that has a bounded linear inverse in $B(V, V)$. Then for every $M \in B(V, V)$ such that $|L - M| < 1/|L^{-1}|$, M has a bounded linear inverse M^{-1} in $B(V, V)$, and $|L^{-1} - M^{-1}| \le (|L - M| \, |L^{-1}|^2)/(1 - |L - M| \, |L^{-1}|)$.

Proof. Put $A = M - L$. Then $N = L^{-1} + B$ is an inverse for $M \Leftrightarrow MN = NM = I$. We can write the equation $MN = I$ as $(L + A)(L^{-1} + B) = I$, or $LL^{-1} + LB + AL^{-1} + AB = I$, or $LB = -AB - AL^{-1}$, or $B = -L^{-1}AB - L^{-1}AL^{-1}$. If we put $f(B) = -L^{-1}AB - L^{-1}AL^{-1}$, this equation has the form $B = f(B)$. Since $|f(B) - f(B')| = |-L^{-1}A(B - B')| \le |L^{-1}| \, |A| \, |B - B'|$, and $|L^{-1}| \, |A| = |L^{-1}| \, |L - M| < 1$, f is a contraction of $B(V, V)$. It has a unique fixed point B. Then $(L + A)(L^{-1} + B) = MN = I$. Similarly, the equation $N'M = I$ can be expressed as $B' = -B'AL^{-1} - L^{-1}AL^{-1}$ if we put $N' = L^{-1} + B'$. By the same argument, it has a solution as well. Now $N' = N'I = N'(MN) = (N'M)N = IN = N$. Therefore $N = N'$ is the unique inverse of M. Moreover, the Contraction Mapping Theorem tells us that

$$| M^{-1} - L^{-1} | = | N - L^{-1} | = | B |$$

$$= | B - 0 | \leq \frac{1}{1 - |A||L^{-1}|} | f(0) - 0 |$$

$$= \frac{| L^{-1}AL^{-1} |}{1 - |A||L^{-1}|} \leq \frac{| L^{-1} |^2 |A|}{1 - |A||L^{-1}|}. \quad \bullet$$

7. COROLLARY. The set of elements in $B(V, V)$ that have an inverse is open in $B(V, V)$, and the function $L \to L^{-1}$ is continuous on that set.

Proof. We have shown that if L has an inverse, so does every point in the ball of radius $1/| L^{-1} |$ about L in $B(V, V)$, so the set of invertible elements in $B(V, V)$ is open. Since $(| L - M || L^{-1} |^2)/(1 - | L - M || L^{-1} |)$ can be made as small as we please by making $| L - M |$ small, and since this number is a bound for $| L^{-1} - M^{-1} |$, the inverse is continuous wherever it exists.

8. COROLLARY. The inverse of an $n \times n$ matrix over $\mathbf{R}$ is a continuous function on the set of $n \times n$ matrices whose determinant is not 0.

(*Remark.* That the set of invertible $n \times n$ matrices is open follows from the facts that its complement is the inverse image of 0 under the determinant function, and that det is clearly continuous.)

Exercises

1. Observe that $(a + x)^2 = a^2 + y \Leftrightarrow a^2 + 2ax + x^2 = a^2 + y \Leftrightarrow 2ax + x^2 = y \Leftrightarrow x = (y - x^2)/2a$. Show that if $2|y| \leq a^2$, the function $f(x) = (y - x^2)/2a$ maps the interval $[-|a|/2, |a|/2]$ into itself, and that f is a contraction of this interval. Use this method to compute $\sqrt{2}$ to two or three decimal places by taking a to be a fairly close approximation to $\sqrt{2}$, and taking $y = 2 - a^2$.

2. Apply the method of Theorem 6 to the special case that $L = L^{-1} = I$, so that the relevant equation becomes $B = -AB - A$. Show that if we start with $B_0 = 0$, and put $B_{n+1} = -AB_n - A$, that in general $B_n = -A + A^2 - A^3 + \cdots + (-A)^n$. Show that this can be interpreted as an application of the geometric series $I - A + A^2 - A^3 + \cdots$ to compute $(I + A)^{-1}$ when $| A | < 1$.

E. Existence and uniqueness theorems for ordinary differential equations

A **differential equation** is one such as $f''(x) = \sin f'(x) + 3$ or

$$\frac{\partial^2 f}{\partial x^2}(x, y) + \frac{\partial^2 f}{\partial y^2}(x, y) = 0$$

relating a function and some of its derivatives. An **ordinary differential equation** (ODE) is one involving differentiation with respect to only one variable. The second example above exhibits a **partial differential equation,** involving partial derivatives.

Consider a differential equation $\Phi(x, f(x), f'(x), \ldots, f^{(n)}(x)) = 0$ **of the nth order.** A function f which is n-times differentiable on the interval $[a, b]$ is a **solution** to this equation if the equation $\Phi(x, f(x), f'(x), \ldots, f^{(n)}(x)) = 0$ is true for every $x \in [a, b]$. In general, a differential equation need not have any solution. For example, if we look for a *real-valued* solution to the ODE $[f'(x)]^2 = -1 - (f(x))^2$, we cannot find one, because the left-hand expression would have to be nonnegative, while the right-hand expression is negative. It is also possible that there be a solution on some intervals, but not on others. This more interesting case, of more frequent occurrence in practice, is illustrated by the equation $f'(x) = 1 + [f(x)]^2$. Suppose f is a solution. Then $f'(x)/(1 + [f(x)]^2) = 1$. If f is a solution on the interval $[a, b]$, we find that $\int_a^x (f'(t)/(1 + [f(t)]^2)) \, dt = x - a$. The left-hand integral can be evaluated. It is $\arctan f(x) - \arctan f(a)$. Thus $\arctan f(x) = (x - a) + \arctan f(a)$, or $f(x) = \tan (\arctan f(a) + x - a)$. Note that if f is any solution, it must be differentiable, and therefore continuous. Then it follows from the differential equation that $f'(x)$ is continuous as well, since $1 + [f(x)]^2$ is continuous. This justifies the integration. We have now found solutions to our ODE; in fact, we have found all of them. Note, however, that the function $\tan (x + c)$ is not defined on any interval of length greater than π, since the tangent becomes infinite on any interval of length $> \pi$. Therefore, the ODE $f'(x) = 1 + [f(x)]^2$ can have no solution on any interval of length $> \pi$.

On the other hand, since every function of the form $\tan (x - c)$ is a solution to the DE on some interval, it is clear that the DE $f' = 1 + f^2$ has infinitely many distinct solutions. This should be expected on physical grounds. That is, if we look at the way in which a differential equation arises in physics, it is immediately apparent that we should expect there to be infinitely many solutions. Consider, for example, a large ball (such as the earth) with a hole drilled through the middle (from the north pole to the south pole) (Figure 20). Let x be the coordinate measuring distance from the center of the ball along the axis. It can be shown that the force of gravity on a particle at the point with coordinate x is approximately $-kx$, where k is some constant depending on the density of the ball and the mass of the particle (provided certain relatively small physical effects are ignored). According to Newton's law of motion, Force = Mass $\times$ Acceleration, if a particle is at a distance $x(t)$ from the center of the ball at time t, $-kx(t) = mx''(t)$, where m is the mass of the particle. Clearly, however, this differential equation does not contain all the information needed to determine the motion of the particle. We also must know where the particle starts out (say at time 0) and how fast it is going when it begins its journey. The differ-

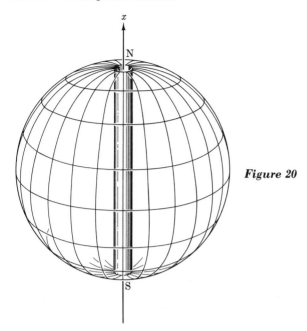

Figure 20

ential equation, *together with this* **initial data,** determines the motion of the particle. For example, if we give the initial value $f(a)$ at $x = a$ of a solution to the DE $f'(x) = 1 + [f(x)]^2$, we thereby pick out a unique solution $f(x) = \tan (x - a + f(a))$.

Physical considerations also suggest that we consider not just differential equations of the form $\Phi(x, f(x), \ldots, f^{(n)}(x)) = 0$, where f is a real- or complex-valued function, but that we allow f to take its values in a vector space (over **R** or **C**). For example, the law of motion mentioned above can be expressed as a differential equation $mx''(t) = F(t, x(t), x'(t))$. That is, the mass m of a particle times its acceleration (the second derivative with respect to time of the *vector* in $\mathbf{R}^3$ that gives the position of the particle) is equal to the (vector-valued) force acting on the particle. In this equation, the force acting on the particle is allowed to depend on time, the position of the particle, and its velocity. Such an equation describes more or less correctly, for example, the motion of an electron through the magnetic field produced by an alternating current in a wire.

A vector-valued differential equation $\Phi(x, f(x), f'(x), \ldots) = \mathbf{0}$ can be interpreted as a system of simultaneous differential equations in the components $f_1, \ldots, f_n$ of the vector-valued function f:

$$\Phi_1(x, f_1(x), \ldots, f_n(x), f'_1(x), \ldots, f'_n(x), \ldots) = 0$$
$$\vdots$$
$$\Phi_m(x_1 f_1(x), \ldots, f_n(x), f'_1(x), \ldots, f'_n(x), \ldots) = 0.$$

This interpretation is useful, for it allows us to reduce nth-order equations to systems of n first-order equations by a trick. The trick will be made clear by two examples.

EXAMPLE. The differential equation $f'''(x) + \sin f''(x) + 3xf'(x) + [f(x)]^2 + e^x = 0$ with initial conditions $f(0) = 0$, $f'(0) = 1$, $f''(0) = 3$ is equivalent to the system of simultaneous first-order equations

$$f'(x) = g(x),$$
$$g'(x) = h(x),$$
$$h'(x) = -[\sin h(x) + 3xg(x) + [f(x)]^2 + e^x],$$

with initial conditions $f(0) = 0$, $g(0) = 1$, $h(0) = 3$.

EXAMPLE. The system of differential equations

$$f''(x) + f'(x) + g'(x) = 0,$$
$$g'(x) + 3g(x) = 0,$$

with initial conditions $g(0) = 1$, $f(0) = 1$, $f'(0) = 0$ can first be transformed into the system

$$f''(x) + f'(x) - 3g(x) = 0,$$
$$g'(x) + 3g(x) = 0,$$

and then into the system

$$f'(x) = h(x),$$
$$h'(x) = -h(x) + 3g(x),$$
$$g'(x) = -3g(x),$$

with initial conditions $f(0) = 1$, $g(0) = 1$, $h(0) = 0$.

By means of this trick, most questions about nth-order ODE's, or systems of ODE's can be transformed into questions about systems of first-order ODE's, or equivalently, into questions about vector-valued ODE's. This is true in particular of questions of existence and uniqueness for the initial value problem, which we now formulate for *first-order vector-valued* ODE's.

The general first-order ODE has the form $\Phi(x, f(x), f'(x)) = \mathbf{0}$. As I have already remarked, there is no guarantee that in this generality the equation will have any solution. Accordingly, we shall deal with a more restricted class of equations, of the form $f'(x) = \Phi(x, f(x))$, in which the derivative can be expressed as a function of x and $f(x)$. This equation has a natural physical interpretation. To simplify the illustration, let us consider the *time-independent case* $f'(x) = \Phi(f(x))$. We can interpret Φ as a function that assigns to each point y in a region in space (say the interior of a pipe as shown in Figure 21) a *velocity vector* (representing, say, the velocity of a particle of fluid passing through that point during a steady flow in the pipe).

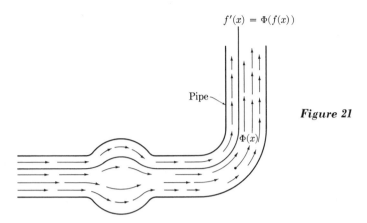

$$f'(x) = \Phi(f(x))$$

Pipe

$\Phi(x)$

Figure 21

A *solution* to the differential equation describes the motion of a particle moving in such a way that when it passes through the point y, its velocity is $\Phi(y)$. The vector $\Phi(y)$ is thus tangent to the solution curve $y = f(x)$ at the time x when the particle passes through the point y. In the *time-dependent case*, the *velocity field* changes with time as well as from point to point.

Let us now attempt a solution to the ODE $f'(x) = \Phi(x, f(x))$ with initial condition $f(x_0) = y_0$. If $\Phi(x, y)$ were independent of y, the solution to the DE $f'(x) = \Phi(x)$ could be obtained trivially by integrating both sides of the equation: $f(x) = y_0 + \int_{x_0}^{x} \Phi(t)\, dt$. In general, if f is a solution to the ODE $f'(x) = \Phi(x, f(x))$ on the interval $[x_0, x_1]$ with initial condition $f(x_0) = y_0$, then f must first of all be differentiable on that interval. If $\Phi(x, y)$ is continuous, then f' must also be continuous. Therefore, we can conclude by integrating both sides of the equation that $f(x) = y_0 + \int_{x_0}^{x} \Phi(t, f(t))\, dt$. Conversely, if Φ is continuous, and f is a continuous solution to this **integral equation**, then first of all $f(x_0) = y_0 + \int_{x_0}^{x_0} \Phi(t, f(t))\, dt = y_0$ and secondly $f'(x) = \dfrac{d}{dx} \int_{x_0}^{x} \Phi(t, f(t))\, dt = \Phi(x, f(x))$. Thus the *integral equation* $f(x) = y_0 + \int_{x_0}^{x} \Phi(t, f(t))\, dt$ is equivalent to the *differential equation* $f'(x) = \Phi(x, f(x))$ *with initial condition* $f(x_0) = y_0$.

Some details require further attention. Later we shall define a vector-valued integral, which will allow us to integrate any continuous function with values in a Banach space. Once we have such an integral, we shall be able to apply the technique to be developed presently to differential equations with values in any Banach space. For the present, however, let us define the integral $\int_a^b f(x)\, dx$ of a vector-valued function f (with values in $\mathbf{R}^n$) by

☛ $\displaystyle \int_a^b (f_1(x), \ldots, f_n(x))\, dx = \left(\int_a^b f_1(x)\, dx, \ldots, \int_a^b f_n(x)\, dx \right).$

That is, integrate each component separately. As our norm on $\mathbf{R}^n$, let us take the uniform norm $||(x_1, \ldots, x_n)||_\infty = \max\limits_{1 \le k \le n} |x_k|$. Then the following proposition generalizes a well-known fact in the one-dimensional case.

1.   PROPOSITION. *If f is a continuous function from $[a, b]$ to $\langle \mathbf{R}^n, || \; ||_\infty \rangle$, then $\left\| \int_a^b f(x) \, dx \right\|_\infty \le \int_a^b ||f(x)||_\infty \, dx$.*

Proof. By definition of $|| \; ||_\infty$, for each $k = 1, \ldots, n$, $||f(x)||_\infty \ge |f_k(x)|$. Therefore $\int_a^b ||f(x)||_\infty \, dx \ge \int_a^b |f_k(x)| \, dx \ge \left| \int_a^b f_k(x) \, dx \right|$. Since this is true for each k, $\int_a^b ||f(x)||_\infty \, dx \ge \max\limits_{1 \le k \le n} \left| \int_a^b f_k(x) \, dx \right| = \left\| \int_a^b f(x) \, dx \right\|_\infty$.   ●

Similarly, we can interpret $f'(x)$ as $(f_1'(x), \ldots, f_n'(x))$.

Now observe that our integral equation has the form $f = Tf$, where Tf is the function defined by $Tf(x) = y_0 + \int_{x_0}^x \Phi(t, f(t)) \, dt$. What we want, then, is a function left fixed by the operator T. This situation is set up for an application of the Contraction Mapping Theorem. Let $\langle B, \mathbf{|} \; \mathbf{|}_\infty \rangle$ be the Banach space of (bounded) continuous functions from $[x_0, x_1]$ to the space $\langle \mathbf{R}^n, || \; ||_\infty \rangle$ (compare Corollary A 5). The norm is $\mathbf{|} f \mathbf{|}_\infty = \text{lub } \{ ||f(x)||_\infty : x_0 \le x \le x_1 \}$. We would like T to be a contraction of B. That is, we want $\mathbf{|} Tf - Tg \mathbf{|}_\infty$ to be $\le L \mathbf{|} f - g \mathbf{|}_\infty$ for some constant $L < 1$:

$$\mathbf{|} Tf - Tg \mathbf{|}_\infty = \text{lub} \left\{ \left\| \int_{x_0}^x [\Phi(t, f(t)) - \Phi(t, g(t))] \, dt \right\|_\infty : x_0 \le x \le x_1 \right\}$$

$$\le \text{lub} \left\{ \int_{x_0}^x ||\Phi(t, f(t)) - \Phi(t, g(t))||_\infty \, dt : x_0 \le x \le x_1 \right\}$$

$$\le \int_{x_0}^{x_1} ||\Phi(t, f(t)) - \Phi(t, g(t))||_\infty \, dt.$$

Now suppose that $\Phi(x, y)$ satisfies a Lipschitz condition with respect to y uniformly in x: $||\Phi(x, y) - \Phi(x, y')||_\infty \le K||y - y'||_\infty$ for all $x \in [x_0, x_1]$. Then the last integral will be bounded by

$$\int_{x_0}^{x_1} K||f(t) - g(t)||_\infty \, dt \le K \int_{x_0}^{x_1} \mathbf{|} f - g \mathbf{|}_\infty \, dt = K(x_1 - x_0) \mathbf{|} f - g \mathbf{|}_\infty.$$

Therefore T will be a contraction if we choose x_1 close enough to x_0 so that $K(x_1 - x_0) < 1$. In this case, our integral equation $f = Tf$ will have a *unique* solution on the interval $[x_0, x_1]$ according to the Contraction Mapping Theorem.

The condition that Φ be a Lipschitz function with respect to y is essential. Without it, the DE $f' = \Phi(x, f)$, with initial condition $f(x_0) = y_0$ may have more than one solution. For example, let $\Phi(y) = 3y^{2/3}$. Then Φ is a continuous function from $\mathbf{R}$ to $\mathbf{R}$. The differential equation $f'(x) = \Phi(f(x)) = 3[f(x)]^{2/3}$ has at least two solutions, both of which take the value 0 when $x = 0$. They are $f(x) \equiv 0$ and $f(x) = x^3$. $\Phi(y)$ does not satisfy a Lipschitz condition with respect to y even at 0. In fact, the ratio

$$\left| \frac{\Phi(y) - \Phi(0)}{y - 0} \right| = \left| \frac{3y^{2/3}}{y} \right| = |3y^{-1/3}|$$

is not bounded as $y \to 0$.

The argument above already has established the existence and uniqueness of solutions to the initial value problem $f'(x) = \Phi(x, f(x))$, $f(x_0) = y_0$ on a *sufficiently* small interval $[x_0, x_1]$ provided that $\Phi(x, y)$ satisfies a Lipschitz condition with respect to y. The remainder of this section will be devoted to improving this result by using a more subtle analysis of the behavior of the successive approximations $Tf_0 = f_1$, $Tf_1 = f_2$, $Tf_2 = f_3$, ..., which occur in the proof of this special case of the Contraction Mapping Theorem.

2.   LEMMA. Let Φ be a continuous function from $[x_0, x_1] \times \mathbf{R}^n$ to $\mathbf{R}^n$ that satisfies the *Lipschitz condition:* there is a constant L such that for every $x \in [x_0, x_1]$, and for each $y, y' \in \mathbf{R}^n$, $||\Phi(x, y) - \Phi(x, y')||_\infty \le L||y - y'||_\infty$. Let

$$Tf(x) = y_0 + \int_{x_0}^{x} \Phi(t, f(t)) \, dt$$

for each continuous function f from $[x_0, x_1]$ to $\mathbf{R}^n$. Suppose that when $x \in [x_0, x_1]$, $||f(x) - g(x)||_\infty \le M(x - x_0)^k$. Then if $x \in [x_0, x_1]$, $||Tf(x) - Tg(x)||_\infty \le LM(x - x_0)^{k+1}/(k + 1)$.

Proof.

$$||Tf(x) - Tg(x)||_\infty = \left\| \left(y_0 + \int_{x_0}^{x} \Phi(t, f(t)) \, dt \right) - \left(y_0 + \int_{x_0}^{x} \Phi(t, g(t)) \, dt \right) \right\|_\infty$$

$$= \left\| \int_{x_0}^{x} [\Phi(t, f(t)) - \Phi(t, g(t))] \, dt \right\|_\infty \le \int_{x_0}^{x} ||\Phi(t, f(t)) - \Phi(t, g(t))||_\infty \, dt$$

$$\le L \int_{x_0}^{x} ||f(t) - g(t)||_\infty \, dt \le LM \int_{x_0}^{x} (t - x_0)^k \, dt$$

$$= \frac{LM(x - x_0)^{k+1}}{k + 1}.$$

3.   GLOBAL PICARD EXISTENCE AND UNIQUENESS THEOREM FOR ODE'S. Let Φ be a continuous function from $[x_0, x_1] \times \mathbf{R}^n$ to $\mathbf{R}^n$ that satisfies the *Lipschitz condition:* $||\Phi(x, y) - \Phi(x, y')||_\infty \le L||y - y'||_\infty$ for every $x \in [x_0, x_1]$ and all $y, y' \in \mathbf{R}^n$. Then the initial value problem $f'(x) = \Phi(x, f(x))$, $f(x_0) = y_0$ has a unique solution $f \in \mathcal{C}([x_0, x_1], \mathbf{R}^n)$.

Proof. Let T be the *integral operator* from $\mathcal{C}([x_0, x_1], \mathbf{R}^n)$ to itself defined by $Tf(x) = y_0 + \int_{x_0}^{x} \Phi(t, f(t)) \, dt$. We know that $f = Tf$ if, and only if, f is a continuously differentiable solution to the initial value problem. Let g_0 be *any* continuous function from $[x_0, x_1]$ to $\mathbf{R}^n$. For example, g_0 might be the constant function y_0. Then put $g_1 = Tg_0$, $g_2 = Tg_1$, ..., $g_{n+1} = Tg_n$. Put $M = |g_1 - g_0|_\infty$. Then by Lemma 2,

$$||g_2(x) - g_1(x)||_\infty = ||Tg_1(x) - Tg_0(x)||_\infty \le LM \frac{(x - x_0)}{1!},$$

$$||g_3(x) - g_2(x)||_\infty \le \frac{L^2 M}{1!} \frac{(x - x_0)^2}{2} = \frac{L^2 M (x - x_0)^2}{2!},$$

and in general, by induction,

$$||g_{n+1}(x) - g_n(x)||_\infty \le \frac{M(L(x - x_0))^n}{n!}.$$

In particular,

$$|g_{n+1} - g_n|_\infty \le \frac{M[L(x_1 - x_0)]^2}{n!}.$$

Now consider the **telescoping series** $g_0 + (g_1 - g_0) + (g_2 - g_1) + (g_3 - g_2) + \cdots$, whose nth partial sum is g_n. The norm of the nth term is

$$|g_n - g_{n-1}|_\infty \le \frac{M[L(x_1 - x_0)]^{n-1}}{(n-1)!}.$$

Since the series

$$M\left[1 + \frac{L(x_1 - y_0)}{1!} + \frac{(L(x_1 - x_0))^2}{2!} + \cdots\right]$$

converges to $Me^{L(x_1 - x_0)}$, the telescoping series converges in the Banach space $\mathcal{C}([x_0, x_1], \mathbf{R}^n)$ by the Comparison Test B 5. That is, the partial sums g_n converge to a continuous function f. As in the proof of the Contraction Mapping Theorem, $f = \lim_{n\to\infty} g_{n+1} = \lim_{n\to\infty} Tg_n = Tf$. Moreover, the fixed point f of T is unique, since if $f = Tf$ and $h = Th$, then several applications of Lemma 2 show that $||f(x) - h(x)||_\infty = ||Tf(x) - Th(x)||_\infty \le L(x - x_0)|f - h|_\infty \Rightarrow$ $||f(x) - h(x)||_\infty \le \frac{L^2(x - x_0)^2}{2!} |f - h|_\infty \Rightarrow \cdots \Rightarrow ||f(x) - h(x)||_\infty \le$ $(L^n(x - x_0)^n/n!) |f - h|_\infty \to 0$ as $n \to \infty$. Thus $f = h$.

4.   REMARKS. (a) It was assumed in the proof that T is continuous (when I wrote $\lim_{n\to\infty} Tg_n = Tf$). This follows from the relation $|Tf - Tg|_\infty \le$ $L(x_1 - x_0)|f - g|_\infty$, which is a consequence of Lemma 2.

(b)   The solution f is automatically continuously differentiable, since $f'(x) = \Phi(x, f(x))$, and the latter function is continuous.

(c)   The Picard method can actually be used to calculate numerically approximate solutions to an ODE. It is clear from the proof that a good choice of the initial approximation will produce rapid convergence. The proof moreover provides error estimates, a fact that is exploited in the following theorem.

(d) The preceding theorem was formulated for initial data given at the left end point of the interval $[x_0, x_1]$ or, in physical terms, for data given at time x_0, with the result to be followed forward in time. There is no reason why the initial data cannot be given at the right end point. In fact, this case can be reduced to the former case by a trick. To deal with the equation $f'(x) = \Phi(x, f(x))$ and the initial condition $f(x_1) = y_0$ on the interval $[x_0, x_1]$, make a change of variable $x = x_1 - t$. If we put $g(t) = f(x) = f(x_1 - t)$, then g satisfies the differential equation $g'(t) = -\Phi(x_1 - t, g(t)) = \Psi(t, g(t))$ with initial condition $g(0) = y_0$ on the interval $[0, x_1 - x_0]$ if, and only if, f satisfies the ODE $f'(x) = \Phi(x, f(x))$ with initial condition $f(x_1) = y_0$ on the interval $[x_0, x_1]$.

Likewise, we can give initial data at a point x_2 interior to $[x_0, x_1]$. Existence and uniqueness of solutions to Df's under these conditions can be proved by piecing together one solution on $[x_0, x_2]$ with initial data at the right end point with another on $[x_2, x_1]$ with initial data at the left end point.

(e) Theorem 3 can also be used to prove existence and uniqueness of solutions on an infinite interval. We simply use the observation that there is a unique function f from $[x_0, +\infty[= \{x \in \mathbf{R} : x \geq x_0\}$ to $\mathbf{R}^n$ such that $f(x_0) = y_0$ and $f'(x) = \Phi(x, f(x))$ if there is a unique function having these properties on every interval $[x_0, x_1]$ for $x_1 \geq x_0$. In fact, if we have one such function f on $[x_0, x_1]$ and another g on $[x_0, x_2]$, where $x_2 > x_1$, then the restriction of g to $[x_0, x_1]$ must be f (by uniqueness—$g(x_0) = y_0$ and g satisfies the same DE as f on $[x_0, x_1]$). We can then put together all the functions obtained in this way on finite intervals to get a function defined on all of $[x_0, +\infty[$.

5.   THEOREM (CONTINUOUS DEPENDENCE ON INITIAL DATA). Let Φ satisfy the conditions of Theorem 3, and let f_1 and f_2 be solutions to the differential equation $f'(x) = \Phi(x, f(x))$ on the interval $[x_0, x_1]$. Then for each $x \in [x_0, x_1]$, $\|f_1(x) - f_2(x)\|_\infty \leq \|f_1(x_0) - f_2(x_0)\|_\infty e^{L(x-x_0)}$.

Proof. Let $Tf(x) = f_1(x_0) + \int_{x_0}^x \Phi(t, f(t))\,dt$ and $Uf(x) = f_2(x_0) + \int_{x_0}^x \Phi(t, f(t))\,dt$. We know that if g_0 and h_0 are the constant functions $f_1(x_0)$ and $f_2(x_0)$ respectively, and if $g_{n+1} = Tg_n$ and $h_{n+1} = Uh_n$, then $\lim_{n \to \infty} g_n = f_1 = Tf$ and $\lim_{n \to \infty} h_n = f_2 = Uf_2$. To prove the theorem, it will be enough to show that for every n, $\|g_n(x) - h_n(x)\|_\infty \leq \|f_1(x_0) - f_2(x_0)\|_\infty e^{L(x-x_0)}$ and then take the limit as $n \to \infty$. Since $e^{L(x-x_0)} \geq 1$ when $x_0 \leq x \leq x_1$, this assertion is true when $n = 0$. For a proof by induction, assume that $\|g_n(x) - h_n(x)\|_\infty \leq \|g_n(x_0) - h_n(x_0)\|_\infty e^{L(x-x_0)}$. Then

$$\|g_{n+1}(x) - h_{n+1}(x)\|_\infty = \|Tg_n(x) - Uh_n(x)\|_\infty = \|f_1(x_0) - f_2(x_0)$$

$$+ \int_{x_0}^x [\Phi(t, g_n(t)) - \Phi(t, h_n(t))]\,dt\|_\infty \leq \|f_1(x_0) - f_2(x_0)\|_\infty$$

$$+ \int_{x_0}^x \|\Phi(t, g_n(t)) - \Phi(t, h_n(t))\|_\infty\,dt \leq \|f_1(x_0) - f_2(x_0)\|_\infty$$

$$+ \int_{x_0}^{x} L||f_1(x_0) - f_2(x_0)||_{\infty} e^{L(t-x_0)} \, dt = ||f_1(x_0) - f_2(x_0)||_{\infty}$$

$$+ ||f_1(x_0) - f_2(x_0)||_{\infty}[e^{L(x-x_0)} - 1] = ||f_1(x_0) - f_2(x_0)||_{\infty} e^{L(x-x_0)}. \quad \bullet$$

Even in the case of such a simple function as $\Phi(x, y) = 1 + y^2$ it will not generally be true that $||\Phi(x, y) - \Phi(x, y')||_{\infty} \leq L||y - y'||_{\infty}$ for a constant L that works for all y, $y' \in \mathbf{R}^n$. Therefore, although this function $\Phi(x, y) = 1 + y^2$ is defined for all x, we cannot apply the Global Existence Theorem to conclude that there is a solution valid for all x. In fact, as we have seen, there can be no solution to the DE $y' = 1 + y^2$ on an interval of length $> \pi$. The next theorem proves that solutions will exist *close enough* to the initial point, provided that Φ satisfies a Lipschitz condition in some region of the form $[x_0, x_1] \times B_r^c(y_0)$ in $[x_0, x_1] \times \mathbf{R}^n$. Note that $B_r^c(y_0) = \{y \in \mathbf{R}^n : ||y - y_0||_{\infty} \leq r\}$. In this respect it gives *local* information in contrast to the *global* information afforded by Theorem 3.

6. LOCAL PICARD EXISTENCE AND UNIQUENESS THEOREM FOR ODE's. Let Φ be a continuous function from $[x_0, x_1] \times B_r^c(y_0)$ to $\mathbf{R}^n$ ($B_r^c(y_0) = \{y \in \mathbf{R}^n : ||y - y_0||_{\infty} \leq r\}$) that satisfies the *Lipschitz condition:* $||\Phi(x, y) - \Phi(x, y')||_{\infty} \leq L||y - y'||_{\infty}$ for every $x \in [x_0, x_1]$ and all y, $y' \in B_r^c(y_0)$. Suppose further that for every $\langle x, y \rangle \in [x_0, x_1] \times B_r^c(y_0)$, $||\Phi(x, y)||_{\infty} \leq M$. Let ϵ be the smaller of the numbers $x_1 - x_0$ and r/M. Then there is a unique continuously differentiable function f from $[x_0, x_0 + \epsilon]$ to $B_r^c(x_0)$ such that $f'(x) = \Phi(x, f(x))$ and $f'(x_0) = y_0$.

Proof. The proof is the same as that of the Global Picard Theorem, except that we no longer can be sure that the Lipschitz condition is satisfied for all values of y. To make the proof work, we have only to show that each of the successive approximations, $g_0(x) \equiv y_0, \ldots, g_{n+1}(x) = y_0 + \int_{x_0}^{x} \Phi(t, g_n(t)) \, dt$, stays within the ball $B_r^c(x_0)$, where the Lipschitz condition holds, so long as $x \in [x_0, x_0 + \epsilon]$. It is clear that for *every* x, $g_0(x) \in B_r^c(x_0)$. Therefore we can complete a proof by induction if we show that $g_{n+1}(x)$ maps $[x_0, x_0 + \epsilon]$ into $B_r^c(x_0)$ on the assumption that $g_n(x)$ does. But if $x \in [x_0, x_0 + \epsilon]$,

$$||g_{n+1}(x) - y_0||_{\infty} = \left|\left|y_0 + \int_{x_0}^{x} \Phi(t, g_n(t)) \, dt - y_0\right|\right|_{\infty} = \left|\left|\int_{x_0}^{x} \Phi(t, g_n(t)) \, dt\right|\right|_{\infty}$$

$$\leq \int_{x_0}^{x} ||\Phi(t, g_n(t))||_{\infty} \, dt \leq \int_{x_0}^{x} M \, dt = M(x - x_0) \leq M\epsilon \leq r. \quad \bullet$$

Remark. Theorem 6 shows that there is only one function f from $[x_0, x_0 + \epsilon]$ into $B_r^c(y_0)$ that satisfies the DE $f'(x) = \Phi(x, f(x))$ with initial condition $f(x_0) = y_0$. If Φ is actually defined on all of $[x_0, x_0 + \epsilon] \times \mathbf{R}^n$, it might seem possible a priori that there be some other function $g : [x_0, x_0 + \epsilon] \to \mathbf{R}^n$ that satisfies the same differential equation and the same initial condition, but does not stay inside $B_r^c(y_0)$. However, this cannot happen, for if $g([x_0, x_0 + \epsilon])$

were not included in $B_r^c(y_0)$, $\{x \in [x_0, x_0 + \epsilon]: g(x) \notin B_r^c(y_0)\}$ would be non-empty and bounded below by x_0. If B were the greatest lower bound of this set, then for every $\delta > 0$ there would have to be an $x \in [x_0, x_0 + \epsilon]$ such that $B \leq x \leq B + \delta$ and $||g(x) - y_0||_\infty > r$ (Theorem II C 1). Because g is continuous, it would follow that $||g(B) - y_0||_\infty \geq r$. On the other hand, if $x_0 \leq x \leq B$, $g(x) \in B_r^c(y_0)$. The proof of Theorem 6 shows then that $r \leq ||g(B) - y_0|| \leq M(B - x_0)$. It follows that $B - x_0 \geq r/M \geq \epsilon$. This, together with the fact that $B \in [x_0, x_0 + \epsilon]$ implies that $B = x_0 + \epsilon$. But then there is in fact no element x of $[x_0, x_0 + \epsilon]$ such that $||g(x) - y_0||_\infty > r$, contradicting our assumption that g does not stay in $B_r^c(y_0)$.

Exercises

This group of exercises consists of two parts. The first is a conventional set of exercises on the theorems we have just proved. The second is a series of exercises intended to instruct you on some tricks for finding explicit formulaic solutions to some of the simple ODE's for which tricks will do any good.

Part I

*1. Apply the Picard process of successive approximations to the initial value problem $f'(x) = f(x), f(0) = 1$ and see what you get. Do enough steps (at least 3 or 4) to get a clear picture of what the successive approximations look like.

*2. Do the same for the initial value problem $f''(x) + f(x) = 0$, $f(0) = 0$, $f'(0) = 1$. Remember first to convert this second-order DE into a system of two simultaneous first-order DE's. Make at least 5 or 6 steps, and compare your results with the exact solution $f(x) = \sin x$.

3. Apply the Picard process to the DE $y'(t) = [y(t)]^{17}$, $y(0) = 0$. What can you say about the solution?

4. (a) Show that $\Phi(y) = \sin y$ satisfies a Lipschitz condition with Lipschitz constant 1.

   (b) Solve the initial value problem $f'(x) = \sin f(x)$, $f(0) = 0$ explicitly. (Try a few Picard approximations to see what happens.)

   (c) Show that for any constant A, the initial value problem $f'(x) = \sin f(x)$, $f(0) = A$ has a unique solution f_A on the interval $[0, \pi]$.

   (d) Put $g(A) = f_A(\pi)$. Prove that g is a continuous function of A.

   (e) Show that $|f_1(x)| \leq e^x$ for $0 \leq x \leq \pi$. Thus, even though you cannot find an explicit formula for f_1, you can still get useful information about it.

*5. Find an explicit formula for the solution to the integral equation $f(x) = 1 + \int_0^x tf(t) \, dt$. *Hint:* Use the method employed in the introduction to section E, which is further elaborated in Part II of these exercises.

*6. Prove that the integral equation $f(x) - \frac{1}{2} \int_0^1 x \cos(xy)f(y) \, dy = e^x$ has a unique solution f in $C([0, 1], \mathbf{R})$. (This is an instance of **Volterra's equation of the second kind,** also known as **Fredholm's equation.**) *Hint:* Use the same method as in the proof of Picard's theorem, applied to the operator T defined by

$$(Tf)(x) = e^x + \frac{1}{2} \int_0^1 x \cos(xy)f(y) \, dy.$$

Part II. Exercises on tricks for solving DE's

If a differential equation can be put in the form $\Phi(f(x))f'(x) = \Psi(x)$ (briefly $\Phi(y)y' = \Psi(x)$) in which the *variables are separated*, it may be possible to find the solutions simply by integrating both sides: $\int_{x_0}^{x} \Phi(f(t))f'(t)\, dt = \int_{x_0}^{x} \Psi(t)\, dt$. But $\int_{x_0}^{x} \Phi(f(t))f'(t)\, dt = \int_{f(x_0)}^{f(x)} \Phi(y)\, dy$. It may be possible to calculate this integral explicitly and solve for $f(x)$ in terms of $f(x_0)$ and the known function $\int_{x_0}^{x} \Psi(t)\, dt$.

*1. Solve these differential equations by the method of **separation of variables:**
   (a) $f'(x) = f(x)$,
   (b) $f'(x) = 2xf(x)$,
   (c) $f'(x) = (1 - (f(x))^2)^{1/2}$,
   (d) $xf'(x) = 1/f(x)$.

2. Show that the solution to the **first-order linear differential equation** $f'(x) = P(x)f(x)$ with initial conditions $f(x_0) = y_0$ is

$$y_0 \exp\left(\int_{x_0}^{x} P(t)\, dt\right) = f(x).\quad \bullet$$

We say that a function $\Phi(x, y)$ is *homogeneous in x and y of degree 0* if for each constant $c \neq 0$, $\Phi(cx, cy) = \Phi(x, y)$. For example, $\Phi(x, y) = x/y$ is homogeneous of degree 0. When Φ is homogeneous of degree 0, the following trick will separate the variables in the DE $f' = \Phi(x, f)$. Put $f(x) = g(x)x$ or $g(x) = (f(x))/x$. Then $f'(x) = xg'(x) + g(x)$ and $\Phi(x, f(x)) = \Phi(1, g(x))$. The DE then becomes $xg'(x) + g(x) = \Phi(1, g(x))$, or

$$\frac{g'(x)}{\Phi(1, g(x)) - g(x)} = \frac{1}{x}.$$

*3. Solve these DE's:
   (a) $y' = (x - y)/(x + y)$,
   (b) $xy' - y = e^{y/x}$ (a little manipulation is called for here).

Exact DE's

*4. Let E be a function from $[x_0, x_1] \times [y_0, y_1]$ into **R** that is continuously differentiable, and f a function from $[x_0, x_1]$ into $[y_0, y_1]$ that is continuously differentiable. Show that $E(x, f(x))$ is a constant function on $[x_0, x_1]$ if, and only if, f satisfies the differential equation

$$\frac{\partial E}{\partial x}(x, f(x)) + \frac{\partial E}{\partial y}(x, f(x))f'(x) = 0.$$

*5. Let P and Q be continuously differentiable functions from $[x_0, x_1] \times [y_0, y_1]$ into **R**. Show that in order for there to exist a function E from $[x_0, x_1] \times [y_0, y_1]$ into **R** such that $P = \partial E/\partial x$ and $Q = \partial E/\partial y$, it is necessary and sufficient that $\partial P/\partial y = \partial Q/\partial x$.

Hint. If P, Q, and E are continuously differentiable functions such that $P = \partial E/\partial x$ and $Q = \partial E/\partial y$, the fact that $\partial P/\partial y \equiv \partial Q/\partial x$ is an elementary theorem about partial derivatives. Conversely, suppose $\partial P/\partial y \equiv \partial Q/\partial x$. Define functions

$$E_1(x, y) = \int_{x_0}^x P(t, y_0)\, dt + \int_{y_0}^y Q(x, s)\, ds \quad \text{and} \quad E_2(x, y) = \int_{y_0}^y Q(x_0, s)\, ds + \int_{x_0}^x$$

$P(t, y)\, dt$. Show that

$$E_1(x, y) - E_2(x, y) = \int_{x_0}^x \int_{y_0}^y \left(\frac{\partial Q}{\partial x}(t, s) - \frac{\partial P}{\partial y}(t, s) \right) ds\, dt = 0.$$

Set $E = E_1 = E_2$. Then use the Fundamental Theorem of Calculus to show that $P = \partial E_2 / \partial x$, $Q = \partial E_1 / \partial y$. ●

A differential equation is called **exact** if it is of the form $\dfrac{\partial E}{\partial x} + \dfrac{\partial E}{\partial y} f'(x) = 0$.
Exercise 4 shows that the solutions to an exact DE all describe curves on which E is constant. That is, they are *solution curves* of the equation $E(x, f(x)) = C$. Which solution we get depends on C. Exercise 5 tells us how to recognize which DE's are exact. For example, the DE $x + yy' = 0$ is exact. To find the function E, note that $\partial E / \partial x = x$, so $E(x, y) = (x^2/2) + \phi(y)$. But $d\phi/dy = \partial E / \partial y = y$, so $\phi(y) = (y^2/2) + C$. Thus all solutions of the DE $x + yy' = 0$ are curves on which $E(x, y) = (x^2 + y^2)/2$ is constant. That is, they are arcs of circles centered at $(0, 0)$ in the x-y-plane.

*6. Solve these exact DE's:
  (a) $xe^{xy}y' + ye^{xy} = 0$,
  (b) $xy' + 2x + y = 0$,
  (c) $(xy + 1)y' + y^2/2 = 0$. ●

Integrating factors
If the DE $P(x, f(x)) + Q(x, f(x))f'(x) = 0$ is not exact, it may be possible to find an **integrating factor** $R(x, y)$ such that $RP + RQf' = 0$ is an exact DE. For example, $f'(x) + (f(x)/x) = 0$ is not an exact DE, but multiplication of this DE by the integrating factor $R(x, y) = x$ makes it exact.

*7. Find integrating factors for the following DE's, and solve the resulting exact equations:
  (a) $f'(x) + xf(x) = 0$,
  (b) $(2x + 2x/f(x))f'(x) + f(x) + 2 = 0$.

F. Linear differential equations

We shall now apply several of the preceding developments to the class of linear ODE's. They are of interest because of their frequent occurrence in applied mathematics, and because their theoretical development is especially satisfying.

1. DEFINITION. A **(homogeneous) linear ordinary differential equation** is one of the form $f'(x) = \Phi(x, f(x))$, where Φ is a function from $[x_0, x_1] \times \mathbf{R}^n$ to $\mathbf{R}^n$, and for each fixed x, the function $y \mapsto \Phi(x, y)$ is a *linear map* from $\mathbf{R}^n$ to $\mathbf{R}^n$.

2. REMARKS. (a) For each fixed x, the linear map $y \mapsto \Phi(x, y)$ can be given

in terms of the components of y by a matrix (whose entries are functions of x). That is, Φ has the form

$$\Phi(x, y_1, \ldots, y_n) = (c_{11}(x)y_1 + \cdots + c_{1n}(x)y_n, \ldots, c_{n1}(x)y_1 + \cdots + c_{nn}(x)y_n).$$

In particular, $\Phi(x, e_j) = (c_{1j}(x), \ldots, c_{nj}(x))$, where e_j is the vector $(0, \ldots, 1, \ldots, 0)$ with the 1 in the jth place. Thus, if Φ is continuous as a function of x for each *fixed* y, all the coefficients $c_{ij}(x)$ are continuous functions of x. It is then apparent that Φ is continuous as a function of x and y together.

(b)   As a linear map from $\langle \mathbf{R}^n, \| \ \|_\infty \rangle$ into itself, the function $y \mapsto \Phi(x, y)$ has a bound, $|\Phi(x, \cdot)|$. In general, if L is a linear map from $\langle \mathbf{R}^n, \| \ \|_\infty \rangle$ into itself whose matrix representation is

$$\begin{pmatrix} a_{11} & \cdots & a_{1n} \\ \cdots & \cdots & \cdots \\ a_{n1} & \cdots & a_{nn} \end{pmatrix}$$

in terms of the usual basis for $\mathbf{R}^n$, then we see that

$$L(y) = (a_{11}y_1 + \cdots + a_{1n}y_n, \ldots, a_{n1}y_1 + \cdots + a_{n1}y_n).$$

The jth component of $L(y)$ is $a_{j1}y_1 + \cdots + a_{jn}y_n$, which is bounded by

$$|a_{j1}|\,|y_1| + \cdots + |a_{jn}|\,|y_n| \le |a_{j1}|\,\|y\|_\infty + \cdots + |a_{jn}|\,\|y\|_\infty$$
$$= [|a_{j1}| + \cdots + |a_{jn}|]\|y\|_\infty.$$

Therefore $\|L(y)\|_\infty$, which is the maximum of the absolute values of its components, is bounded by $\max\limits_{1 \le j \le n} [|a_{j1}| + \cdots + |a_{jn}|]\|y\|_\infty$. This shows that $|L| \le \max\limits_{1 \le j \le n} [|a_{j1}| + \cdots + |a_{jn}|]$. It is not hard to see that $|L|$ is actually equal to this expression. In fact, if j is chosen so that $|a_{j1}| + \cdots + |a_{jn}|$ is maximized, we can choose $y_i = \pm 1$ for $i = 1, \ldots, n$ so that $a_{ji}y_i = |a_{ji}|$. Then the jth component of $L(y)$ is exactly $|a_{j1}| + \cdots + |a_{jn}|$ and $\|y\|_\infty = 1$. This shows that $|L| \ge |a_{j1}| + \cdots + |a_{jn}|$, and completes the proof.

We see then that $|\Phi(x, \cdot)| = \max\limits_{1 \le j \le n} [|c_{j1}(x)| + \cdots + |c_{jn}(x)|]$ is a continuous function of x. Therefore, on a finite closed interval $[x_0, x_1]$, it is bounded above by a constant K. Then for any $x \in [x_0, x_1]$ and any $y, y' \in \mathbf{R}^n$, $\|\Phi(x, y) - \Phi(x, y')\|_\infty = \|\Phi(x, y - y')\|_\infty \le |\Phi(x, \cdot)|\,\|y - y'\|_\infty \le K\|y - y'\|_\infty$, so Φ satisfies a Lipschitz condition with the same constant K for every $x \in [x_0, x_1]$. To sum up: *If Φ is continuous as a function of x for each fixed y, and linear as a function of y for each fixed x, then it is continuous with respect to both variables together, and satisfies a Lipschitz condition with respect to y uniformly in x.*

This puts us in position to apply the Global Picard Theorem. In fact, we can do much more. The results are summarized below.

3. BASIC THEOREM ON HOMOGENEOUS LINEAR ODE's. Let Φ be a function from $[x_0, x_1] \times \mathbf{R}^n$ to $\mathbf{R}^n$ such that for each fixed $y \in \mathbf{R}^n$, $\Phi(x, y)$ is a continuous function of x, and for each fixed x, $y \mapsto \Phi(x, y)$ is a linear map.

(a) For each $x^* \in [x_0, x_1]$ and each $y^* \in \mathbf{R}^n$, there is a unique differentiable function f from $[x_0, x_1]$ to $\mathbf{R}^n$ such that $f'(x) = \Phi(x, f(x))$ and $f(x^*) = y^*$. f is actually continuously differentiable.

(b) The collection of solutions to the differential equation $f'(x) = \Phi(x, f(x))$ is a linear subspace of the vector space of all continuously differentiable functions from $[x_0, x_1]$ to $\mathbf{R}^n$. Its dimension is n.

(c) If $f_1, \ldots, f_k$ are solutions to the DE $f'(x) = \Phi(x, f(x))$ such that the vectors $f_1(x), \ldots, f_k(x)$ are linearly independent at *any one* point $x \in [x_0, x_1]$, then these vectors are linearly independent at *every* point in $[x_0, x_1]$. If $k = n$, $f_1, \ldots, f_k$ are then a basis for the space of solutions.

Proof. (a) In view of Remark 2(b), we can apply the Global Picard Theorem E 3 to our DE to conclude that there is a unique solution satisfying the initial condition $f(x^*) = y^*$. Take note of Remarks (b) and (d) following the proof of the Global Picard Theorem.

(b) Consider the map $D - \Phi$ defined by $(D - \Phi)f(x) = f'(x) - \Phi(x, f(x))$ from the space of all continuously differentiable functions on $[x_0, x_1]$ to the space of continuous functions on that interval. Because the differentiation operator D and the map $y \mapsto \Phi(x, y)$ are both linear, $D - \Phi$ is a linear map. Its kernel (Definition III D 4, Theorem III D 5) is a linear subspace of the space of continuously differentiable functions. But the kernel of $D - \Phi$ is just the set of solutions to our DE.

Now take any point x^* in $[x_0, x_1]$, and let $b_1, \ldots, b_n$ be a basis for $\mathbf{R}^n$. According to part (a) of the theorem, there are solutions $f_1, \ldots, f_n$ to the DE such that $f_1(x^*) = b_1, \ldots, f_n(x^*) = b_n$. I claim that $f_1, \ldots, f_n$ are a basis for the space of solutions. First of all, they are linearly independent, for if $c_1 f_1 + \cdots + c_n f_n$ is the zero function, then in particular, $c_1 b_1 + \cdots + c_n b_n = c_1 f_1(x^*) + \cdots + c_n f_n(x^*) = \mathbf{0}$, and then $c_1 = \cdots = c_n = 0$ because $b_1, \ldots, b_n$ are independent. If g is any solution to our DE, then $g(x^*) \in \mathbf{R}^n$, so there are scalars $c_1, \ldots, c_n$ such that $g(x^*) = c_1 b_1 + \cdots + c_n b_n = c_1 f_1(x^*) + \cdots + c_n f_n(x^*)$ (because $b_1, \ldots, b_n$ span $\mathbf{R}^n$). But then $c_1 f_1 + \cdots + c_n f_n$ is a linear combination of $f_1, \ldots, f_n$ that takes the same value at x^* as does g. According to part (a), there is only one solution to our DE with this property, so $g = c_1 f_1 + \cdots + c_n f_n$. Thus $f_1, \ldots, f_n$ span the space of solutions.

(c) The proof of (c) is part of the proof of (b). In fact, if $f_1, \ldots, f_k$ are independent at x, then $c_1 f_1 + \cdots + c_k f_k = \mathbf{0} \Rightarrow c_1 f_1(x) + \cdots + c_1 f_k(x) = \mathbf{0} \Rightarrow c_1 = \cdots = c_k = 0$. If $k = n$, then $f_1(x), \ldots, f_n(x)$ are a basis for $\mathbf{R}^n$, and we have just shown that this implies that $f_1, \ldots, f_n$ are a basis for the space of solutions to our DE. ●

Now let us consider a related but more general class of equations.

4. DEFINITION. An **inhomogeneous linear ODE** is one of the form $f'(x) = \Phi(x, f(x)) + \Psi(x)$, where for each fixed x, the function $y \mapsto \Phi(x, y)$ is a linear map from $\mathbf{R}^n$ into itself. The function Ψ is called the **inhomogeneous part**. The **related homogeneous DE** is the equation $f'(x) = \Phi(x, f(x))$, corresponding to the case $\Psi = \mathbf{0}$. ●

There is a simple relation between the solutions to an inhomogeneous linear DE and the related homogeneous equation, which I shall not dignify by the name of theorem. If f_1 and f_2 are solutions to the inhomogeneous DE, $f' = \Phi(x, f) + \Psi$, their difference is a solution to the related homogeneous DE:

$$(f_1 - f_2)' = f_1' - f_2' = \Phi(x, f_1) + \Psi - \Phi(x_1 f_2) - \Psi = \Phi(x, f_1 - f_2).$$

Conversely, if f is a solution to the inhomogeneous DE and h is a solution to the related homogeneous DE, then $f + h$ solves the inhomogeneous DE as well: $(f + h)' = f' + h' = \Phi(x, f) + \Psi + \Phi(x, h) = \Phi(x, f + h) + \Psi$. In sum, having found any one solution to an inhomogeneous linear ODE, all the others can be constructed by adding solutions to the related homogeneous DE. It is a remarkable fact that the one **particular solution** to the inhomogeneous DE we need can be constructed out of solutions to the related homogeneous DE as well. The following discussion is intended to motivate the construction that will accomplish this miracle.

Suppose we have a stagnant pond in which a colony of the bacterium *Toxica toxica* (Kripke) is growing. Imagine that this pond is being fed by a putrid stream, which at time t is supplying new *T. toxica* at the rate $\Psi(t)$. Suppose these bacteria increase at such a rate that if $f(t)$ of them are present at time t, their rate of increase will be $(1/10)f(t)$ per hour. (Our unit of time will be the hour.) If the stream didn't flow, the size of the colony would be governed by the differential equation $f' = (1/10)f$, whose solution is $f(t) = f(0)e^{(1/10)t}$, until it grew so large that the decreased supply of nutrients and the accumulation of metabolic wastes caused its growth rate to slow. (If the growth rate failed to slow, the colony would attain a mass equal to that of the earth in at most one or two months.)

Now let us take into account the effects of the stream. Suppose at time 0, $f(0)$ bacteria are present. At time t, they will have $f(0)e^{(1/10)t}$ descendants. During a short time interval Δ, approximately $\Psi(0)\Delta$ new bacteria will enter the colony. At time t, they will have approximately $\Psi(0)e^{(1/10)t}\Delta$ descendants. During the time interval from $\Delta \le t \le 2\Delta$, there will be about $\Psi(\Delta)\Delta$ newcomers, whose descendants, having grown for $t - \Delta$ hours at time t, will number $\Psi(\Delta)e^{(1/10)(t-\Delta)}\Delta$. The next batch of $\Psi(2\Delta)\Delta$, arriving at times between $t = 2\Delta$ and $t = 3\Delta$, will have about $\Psi(2\Delta)e^{(1/10)(t-2\Delta)}\Delta$ descendants at time t. The approximate total at time t will thus be

$$f(0)e^{(1/10)t} + \sum_{k=0}^{(t/\Delta)-1} \Psi(k\Delta)e^{(1/10)(t-k\Delta)}\Delta.$$

If we shrink the time interval Δ, the sum becomes an integral,

$\int_0^t \Psi(s)e^{(1/10)\,(t-s)}\,ds$. We can analyze the integrand $\Psi(s)e^{(1/10)\,(t-s)}$ as follows: it represents the number of descendants at time t of $\Psi(s)$ bacteria introduced at time s. That is, the total number of bacteria in the pond is governed by the DE $f'(t) = (1/10)f(t) + \Psi(t)$. *The integrand* $\Psi(s)e^{(1/10)\,(t-s)}$ *is the solution to the related homogeneous DE* $f' = (1/10)f$, *which takes the initial value* $\Psi(s)$ *at time* s.

The function that assigns to each pair $\langle s, t \rangle$ the linear map $y \mapsto e^{(1/10)\,(t-s)}y$ is called the *Green's function* for the differential equation.

Let us guess that a similar formula works in the vector-valued case. In the simple case of the equation $y' = (1/10)y + \Psi$, the space of solutions to the related homogeneous DE is one-dimensional, with the single function $f \colon t \mapsto e^{(1/10)t}$ as a basis. The integrand is obtained by multiplying this basic solution by a certain coefficient function $e^{-(1/10)s}\Psi(s)$. The coefficient function itself depends linearly on Ψ. In the general case, $f'(x) = \Phi(x, f(x)) + \Psi(x)$, and we might guess that if the functions $f_1, \ldots, f_n$ are a basis for the vector space of solutions to the related homogeneous equation, $f'(x) = \Phi(x, f(x))$, that the integrand will have the form $c_1(s)f_1(t) + \cdots + c_n(s)f_n(t)$. We might expect that the coefficient functions $c_1, \ldots, c_n$ will depend linearly on $\Psi \colon c_i(s) = G_i(s)(\Psi(s))$, where for each s, $G_i(s)$ is a linear functional $G_i(s)\colon$ $\mathbf{R}^n \to \mathbf{R}$. We can write the expression $G_i(s)(y)$ less clumsily as $G_i(s)\cdot y$. (See III E 6.)

In the preceding discussion of the bacterial equation, the integrand $\langle s, t \rangle \mapsto e^{-(1/10)s}\Psi(s)e^{(1/10)t}$ had the property that when $t = s$, its value is $\Psi(s)$. In general, then, we might guess that the coefficients $c_1, \ldots, c_n$ should be chosen so that $c_1(s)f_1(t) + \cdots + c_n(s)f_n(t)$ takes the value $\Psi(s)$ when $t = s$. Finally, we want the coefficient functions $c_1(s), \ldots, c_n(s)$ to be continuous. The following lemma says that it is possible to accomplish all these things.

5.   LEMMA. Let Φ be a function from $[x_0, x_1] \times \mathbf{R}^n$ to $\mathbf{R}^n$ such that for each fixed $y \in \mathbf{R}^n$, $\Phi(x, y)$ is a continuous function of x, and for each fixed $x \in [x_0, x_1]$, $\Phi(x, y)$ is a linear function of y. Let $f_1, \ldots, f_n$ be a basis for the space of solutions to the DE $f'(x) = \Phi(x, f(x))$ on the interval $[x_0, x_1]$. Then there are continuous functions $G_1, \ldots, G_n$ from $[x_0, x_1]$ into the space $\mathcal{L}(\mathbf{R}^n, \mathbf{R})$ of linear functionals on $\mathbf{R}^n$ with the following property. For each $y \in \mathbf{R}^n$ and each $s \in [x_0, x_1]$, $y = (G_1(s)\cdot y)f_1(s) + \cdots + (G_n(s)\cdot y)f_n(s)$.

Proof. Observe that for each s, the map $(g_1, \ldots, g_n) \mapsto g_1 f_1(s) + \cdots + g_n f_n(s)$ is a linear map $L(s)$ from $\mathbf{R}^n$ to $\mathbf{R}^n$. Because $f_1, \ldots, f_n$ are continuous functions of s, L is a continuous function from $[x_0, x_1]$ to $B(\mathbf{R}^n, \mathbf{R}^n)$. Indeed,

$$\|(L(s) - L(t))\cdot(g_1, \ldots, g_n)\|_\infty$$
$$= \|g_1(f_1(s) - f_1(t)) + \cdots + g_n(f_n(s) - f_n(t))\|_\infty$$
$$\leq \|g\|_\infty[\|f_1(s) - f_1(t)\|_\infty + \cdots + \|f_n(s) - f_n(t)\|_\infty],$$

so $|L(s) - L(t)| \leq \|f_1(s) - f_1(t)\|_\infty + \cdots + \|f_n(s) - f_n(t)\|_\infty$. Now $L(s)$ is actually an isomorphism for each s, because its range, which contains $f_1(s), \ldots, f_n(s)$, contains a basis for $\mathbf{R}^n$ and is therefore n-dimensional (compare Theorem III D 5(d)). Therefore, for each s, $L(s)$ has an inverse $G(s): \mathbf{R}^n \to \mathbf{R}^n$ (Theorem III E 8(c)). But it was proved in Section D that the inverse of a linear map is a continuous function of the map (Corollary D 7). Thus G is a continuous function. By its definition, $G(s) \cdot y = (G_1(s) \cdot y, \ldots, G_n(s) \cdot y)$ is a vector $(g_1, \ldots, g_n)$ such that $y = L(s) \cdot (g_1, \ldots, g_n) = g_1 f_1(s) + \cdots + g_n f_n(s)$. $\bullet$

6. DEFINITION. We call the function that sends $\langle s, x \rangle$ into the linear map $y \mapsto (G_1(s) \cdot y) f_1(x) + \cdots + (G_n(s) \cdot y) f_n(x)$ the **Green's function** of the inhomogeneous linear ODE $f' = \Phi(x, f) + \Psi$. $\bullet$

In the bacterial example studied above, the Green's function sends $\langle s, x \rangle$ into the linear map from $\mathbf{R}$ to $\mathbf{R}$ given by $y \mapsto (e^{-(1/10)s} \cdot y) e^{(1/10)x}$.

7. THEOREM. Let Φ satisfy the conditions of Lemma 5. If Ψ is any continuous function from $[x_0, x_1]$ to $\mathbf{R}^n$, the solution to the ODE $f'(x) = \Phi(x, f(x)) + \Psi(x)$ that satisfies the initial condition $f(x_0) = 0$ is given in terms of the Green's function by the formula

$$f(x) = \int_{x_0}^{x} \{[G_1(s) \cdot \Psi(s)] f_1(x) + \cdots + [G_n(s) \cdot \Psi(s)] f_n(x)\} \, ds.$$

$$= \left(\int_{x_0}^{x} G_1(s) \cdot \Psi(s) \, ds \right) f_1(x) + \cdots + \left(\int_{x_0}^{x} G_n(s) \cdot \Psi(s) \, ds \right) f_n(x).$$

Proof. It is clear that the formula defines a function f such that $f(x_0) = 0$, since $f(x_0) = \int_{x_0}^{x_0} (\text{stuff}) = 0$. We must check that f satisfies our differential equation. To do this, let us define the coefficient $A_j(x) = \int_{x_0}^{x} [G_j(s) \cdot \Psi(s)] \, ds$, so that the expression for f can be rewritten $f(x) = A_1(x) f_1(x) + \cdots + A_n(x) f_n(x)$. Then $f'(x) = [A_1'(x) f_1(x) + \cdots + A_n'(x) f_n(x)] + \{A_1(x) f_1'(x) + \cdots + A_n(x) f_n'(x)\}$. The first term, in square brackets, involves the coefficient $A_j'(x) = G_j(x) \cdot \Psi(x)$. By the definition of G, $[G_1(x) \cdot \Psi(x)] f_1(x) + \cdots + [G_n(x) \cdot \Psi(x)] f_n(x) = \Psi(x)$. In the second term, in braces, we have $f_j'(x) = \Phi(x, f_j(x))$, since f_j is a solution to the homogeneous DE. Thus the second term is $A_1 \Phi(x, f_1) + \cdots + A_n \Phi(x, f_n) = \Phi(x, A_1 f_1 + \cdots + A_n f_n) = \Phi(x, f)$. $\bullet$

Remark. A common name for this technique for finding solutions to an inhomogeneous linear ODE is **variation of parameters.**

EXAMPLE. Let us find the solution to the DE $f''(x) - 3f'(x) + 2f(x) = x$ with the initial conditions $f(0) = 1$, $f'(0) = 2$.

Step 1. Convert the equation to a system of first-order equations:
$$f'(x) = g(x),$$
$$g'(x) = 3g(x) - 2f(x) + x,$$
with initial condition $f(0) = 1$, $g(0) = 2$.

Step 2. Find all solutions to the related homogeneous equation $f''(x) - 3f'(x) + 2f(x) = 0$, or equivalently:
$$f'(x) = g(x),$$
$$g'(x) = 3g(x) - 2f(x).$$
Using the techniques developed in the Exercises, we can see that $f_1 \colon x \mapsto e^x$ and $f_2 \colon x \mapsto e^{2x}$ are solutions. Since these two functions are obviously linearly independent, they form a basis for the space of solutions (Theorem 3(c)). Corresponding to them are functions $g_1 = f_1' \colon x \mapsto e^x$ and $g_2 = f_2' \colon x \to 2e^{2x}$. Thus the vector-valued functions (f_1, g_1) and (f_2, g_2) are a basis for the space of solutions to the vector form of the equation.

Step 3. Find the Green's function of the equation, using the technique of Theorem 7. The function $L(s) \colon (y_1, y_2) \mapsto y_1(f_1(s), g_1(s)) + y_2(f_2(s), g_2(s)) = (y_1 e^s + y_2 e^{2s}, y_1 e^s + 2y_2 e^{2s})$ has the matrix
$$\begin{pmatrix} e^s & e^{2s} \\ e^s & 2e^{2s} \end{pmatrix}.$$

Let $\begin{pmatrix} a & b \\ c & d \end{pmatrix}$ be the inverse of this matrix. Then
$$\begin{pmatrix} e^s & e^{2s} \\ e^s & 2e^{2s} \end{pmatrix} \begin{pmatrix} a & b \\ c & d \end{pmatrix} = \begin{pmatrix} 1 & 0 \\ 0 & 1 \end{pmatrix}.$$
To find a, b, c, and d, solve the two systems of linear equations
$$\begin{pmatrix} e^s & e^{2s} \\ e^s & 2e^{2s} \end{pmatrix} \begin{pmatrix} a \\ c \end{pmatrix} = \begin{pmatrix} 1 \\ 0 \end{pmatrix}$$
and
$$\begin{pmatrix} e^s & e^{2s} \\ e^s & 2e^{2s} \end{pmatrix} \begin{pmatrix} b \\ d \end{pmatrix} = \begin{pmatrix} 0 \\ 1 \end{pmatrix}.$$
This can be accomplished in a simple way by row operations using the technique of Section III B. The computations are shown in Table 12. The inverse matrix is $\begin{pmatrix} 2e^{-s} & -e^{-s} \\ -e^{-2s} & e^{-2s} \end{pmatrix}$, as can readily be checked.

Thus the Green's function assigns to each s, t the map $(y_1, y_2) \to (2e^{-s}y_1 - e^{-s}y_2)(e^t, e^t) + (-e^{-2s}y_1 + e^{-2s}y_2)(e^{2t}, 2e^{2t})$.

Step 4. Find a *particular solution* to the inhomogeneous DE using the tech-

---

Table 12

$$\begin{pmatrix} e^s & e^{2s} & 1 & 0 \\ e^s & 2e^{2s} & 0 & 1 \end{pmatrix} \rightarrow \begin{pmatrix} e^s & e^{2s} & 1 & 0 \\ 0 & e^{2s} & -1 & 1 \end{pmatrix} \rightarrow$$

$$\begin{pmatrix} e^s & 0 & 2 & -1 \\ 0 & e^{2s} & -1 & 1 \end{pmatrix} \rightarrow \begin{pmatrix} 1 & 0 & 2e^{-s} & -e^{-s} \\ 0 & 1 & -e^{-2s} & e^{-2s} \end{pmatrix}$$

---

nique of Theorem 7. The *inhomogeneous part* is the function $\Psi: t \mapsto (0, t)$. The coefficients are

$$A_1(x) = \int_0^x (2e^{-s} \cdot 0 - e^{-s} \cdot s) \, ds = (se^{-s} + e^{-s})\Big|_0^x = xe^{-x} + e^{-x} - 1$$

and

$$A_2(x) = \int_0^x (-e^{-2s} \cdot 0 + e^{-2s} \cdot s) \, ds = (-\tfrac{1}{2}se^{-2s} - \tfrac{1}{4}e^{-2s})\Big|_0^x$$

$$= \tfrac{1}{4} - \tfrac{1}{4}e^{-2x} - \tfrac{1}{2}xe^{-2x}.$$

The *particular solution* is

$$p(x) = (xe^{-x} + e^{-x} - 1)e^x + (\tfrac{1}{4} - \tfrac{1}{4}e^{-2x} - \tfrac{1}{2}xe^{-2x})e^{2x}$$

$$= x + 1 - e^x - \tfrac{1}{2}x - \tfrac{1}{4} + \tfrac{1}{4}e^{2x} = \tfrac{1}{2}x + \tfrac{3}{4} - e^x + \tfrac{1}{4}e^{2x}.$$

It satisfies the initial conditions $p(0) = p'(0) = 0$.

Step 5. Add a solution h to the related homogeneous equation that satisfies the initial conditions $h(0) = 1$, $h'(0) = 2$. The Green's function gives us the required coefficients: $h(x) = (2e^{-0} \cdot 1 - e^{-0} \cdot 2)e^x + (-e^{-2 \cdot 0} \cdot 1 + e^{-2 \cdot 0} \cdot 2)e^{2x} = e^{2x}$. The final solution $f(x) = p(x) + h(x) = \tfrac{1}{2}x + \tfrac{3}{4} - e^x + \tfrac{5}{4}e^{2x}$.

Remark. This solution involves an unnecessary step, because it is not necessary to find the Green's function in order to find the coefficients c_1 and c_2 whose integrals are A_1 and A_2. In fact c_1 and c_2 satisfy the system of equations $\begin{pmatrix} e^s & e^{2s} \\ e^s & 2e^{2s} \end{pmatrix}\begin{pmatrix} c_1 \\ c_2 \end{pmatrix} = \begin{pmatrix} 0 \\ s \end{pmatrix}$. The solution to this system is worked out in Table 13. Thus $A_1(x) = \int_0^x - se^{-s} \, ds$ and $A_2(x) = \int_0^x se^{-2s} \, ds$. ●

Table 13

$$\begin{pmatrix} e^s & e^{2s} & 0 \\ e^s & 2e^{2s} & s \end{pmatrix} \rightarrow \begin{pmatrix} e^s & e^{2s} & 0 \\ 0 & e^{2s} & s \end{pmatrix} \rightarrow$$

$$\begin{pmatrix} e^s & 0 & -s \\ 0 & e^{2s} & s \end{pmatrix} \rightarrow \begin{pmatrix} 1 & 0 & -se^{-s} \\ 0 & 1 & se^{-2s} \end{pmatrix}$$

Exercises

*1. The **general first-order inhomogeneous linear ODE** has the form $f'(x) = \Phi(x)f(x) + \Psi(x)$ (where Φ is a function from $[x_0, x_1]$ to **R**). Show that the solution that takes the initial value y_0 at x_0 can be given explicitly in terms of Φ and Ψ. *Hint:* Use Exercise II 2 of Section E, the remarks following Definition F 4, and Theorem F 7.

*2. We have seen that the general solution to the DE $f''(x) + f(x) = 0$ is $f(x) = A \sin x + B \cos x$. Use this information to find all the solutions to the inhomogeneous DE $f''(x) + f(x) = \sin x$. *Hint:* If we use the standard trick to put our DE in the form

$$f'(x) = g(x),$$
$$g'(x) = -f(x) + \sin x,$$

we see that the general solution to the related homogeneous equation is

$$(f(x), g(x)) = (A \sin x + B \cos x, A \cos x - B \sin x)$$
$$= A(\sin x, \cos x) + B(\cos x, -\sin x).$$

According to Theorem 7, we must choose functions $A(s)$ and $B(s)$ such that

$$A(s) \sin(s) + B(s) \cos(s) = 0,$$
$$A(s) \cos(s) - B(s) \sin(s) = \sin(s).$$

Then a particular solution to the inhomogeneous DE will have the form $\left(\int_0^x A(s)\, ds \right) \sin x + \left(\int_0^x B(s)\, ds \right) \cos x$.

3. Define the linear map $D - \lambda I$ from the space of all infinitely many times differentiable functions into itself by $(D - \lambda I)f(x) = f'(x) - \lambda f(x)$.
   (a) Show that $(D - \lambda I)x^n e^{\lambda x} = nx^{n-1}e^{\lambda x}$.
   (b) Deduce from (a) that if P is a polynomial of degree $<n$ in x, then $(D - \lambda I)^n P(x)e^{\lambda x} = 0$.
   *(c) Find all the solutions to the differential equation $(D - \lambda I)^n f = 0$.

4. The operators $D - \lambda I$ can be manipulated like ordinary polynomials in D. Thus $(D - \lambda I)(D - \mu I)$ is the same as the operator $D^2 - (\lambda + \mu)D + \lambda\mu I$, which sends f into $f'' - (\lambda + \mu)f' + \lambda\mu f$. Using this remark, show that if λ is a root of the polynomial equation $x^n + c_{n-1}x^{n-1} + \cdots + c_1 x + c_0 = 0$, then the operator $D^n + c_{n-1}D^{n-1} + \cdots + c_1 D + c_0 I$ can be factored as $T(D - \lambda I)$, where T is some operator of the form $D^{n-1} + b_{n-2}D^{n-2} + \cdots + b_0 I$. Conclude that $e^{\lambda x}$ is a solution to the differential equation

$$f^{(n)} + c_{n-1}f^{(n-1)} + \cdots + c_1 f' + c_0 f = (D^n + c_{n-1}D^{n-1} + \cdots + c_1 D + c_0 I)f = 0.$$

More generally, if $x^n + c_{n-1}x^{n-1} + \cdots + c_1 x + c_0$ factors as $(x - \lambda_1)^{e_1} \cdots (x - \lambda_k)^{e_k}$ (with $e_1 + \cdots + e_k = n$) show that $P(x)e^{\lambda_j}$ is a solution to this DE for each polynomial P in x of degree $<e_j$, and each $j = 1, \ldots, k$.

Remark. It is actually true that the general solution to an nth order linear ODE with *constant coefficients* has the form $P_1(x)e^{\lambda_1 x} + \cdots + P_k(x)e^{\lambda_k x}$, where P_j is a polynomial in x of degree $<e_j$. Exercise 4 shows only that *every* such function is a solution. The proof of the converse requires a more elaborate algebraic argument than I care to put in these exercises.

5. Define the operator A_λ by $A_\lambda f(x) = \int_0^x e^{\lambda(x-s)} f(s)\, ds$. Show that $(D - \lambda I)A_\lambda = I$, so that A_λ is a *right inverse* for the operator $D - \lambda I$. Conclude that the function $A_{\lambda_n} A_{\lambda_{n-1}} \cdots A_{\lambda_1} g$ is a solution to the DE $(D - \lambda_1 I) \cdots (D - \lambda_n I)f = g$.

*6. Use the method of Exercises 4 and 5 to find all solutions to the DE $f'' - 3f' + 2f = 1$.

G. Differentiation of functions of several variables. The inverse and implicit function theorems.

Let f be a function from **R** to **R** which has a derivative at x_0. Then there is a linear function $L(h) = Dh$ from **R** to **R** which is a good approximation to $f(x_0 + h) - f(x_0)$ in the sense that

$$|f(x_0 + h) - f(x_0) - Dh|/|h| \to 0 \quad \text{as } h \to 0.$$

Just take $D = f'(x_0)$. Then, by the definition of f',

$$|f(x_0 + h) - f(x_0) - Dh|/|h| = |(f(x_0 + h) - f(x_0))/h - D| \to 0 \quad \text{as } h \to 0.$$

Conversely, if there is any good linear approximation in this sense, the relation $|(f(x_0 + h) - f(x_0))/h - D| \to 0$ as $h \to 0$ shows precisely that f is differentiable at x_0 and $f'(x_0) = D$. That is, we can think of the derivative of f at x_0 as giving a good linear approximation $L(h)$ to the function $f(x_0 + h) - f(x_0)$ ("good" in the sense that the error goes to 0 faster than h does).

1.   DEFINITION. Let U be an open set in $\mathbf{R}^n$, and f a function from U to $\mathbf{R}^m$. We say that f is **differentiable** at $x_0 \in U$ if there is a linear map $L: \mathbf{R}^n \to \mathbf{R}^m$ such that

$$||f(x_0 + h) - f(x_0) - L(h)||_\infty / ||h||_\infty \to 0 \quad \text{as } h \to 0$$

in $\mathbf{R}^n$. In this case, we call the **linear map** L the **Frechet differential** of f at x_0, and write $L = df_{x_0}$.

2.   REMARK. A function can have at most one Frechet differential. In fact, if L_1 and L_2 are both linear maps having the property that $||f(x_0 + h) - f(x_0) - L(h)||_\infty / ||h||_\infty \to 0$ as $h \to 0$, then for any nonzero $h \in \mathbf{R}^n$, we have

$$||L_1(h) - L_2(h)||_\infty = \frac{1}{|t|} ||L_1(th) - L_2(th)||_\infty$$

$$\leq \frac{1}{|t|} [||f(x_0 + th) - f(x_0) - L_2(th)||_\infty + ||f(x_0 + th) - f(x_0) - L_1(th)||_\infty]$$

$$= ||h|| \left[\frac{||f(x_0 + th) - f(x_0) - L_2(th)||_\infty}{||th||_\infty} \right.$$

$$\left. + \frac{||f(x_0 + th) - f(x_0) - L_1(th)||_\infty}{||th||_\infty} \right] \to 0$$

as $t \to 0$. (Use the triangle inequality to prove this inequality.)

EXAMPLE. Consider the map $f: (x, y) \mapsto x^2 + y^2$ from $\mathbf{R}^2$ to $\mathbf{R}$. We have $f(x_0 + \alpha, y_0 + \beta) - f(x_0, y_0) = (x_0 + \alpha)^2 - x_0^2 + (y_0 + \beta)^2 - y_0^2 = \alpha(2x_0) + \beta(2y_0) + \alpha^2 + \beta^2$. The map $L: (\alpha, \beta) \mapsto 2x_0\alpha + 2y_0\beta$ is linear. The error $\alpha^2 + \beta^2 = f(x_0 + \alpha, y_0 + \beta) - f(x_0, y_0) - L(\alpha, \beta)$ has the property that $(\alpha^2 + \beta^2)/||(\alpha, \beta)||_\infty \leq |\alpha| + |\beta| \to 0$ as $(\alpha, \beta) \to (0, 0)$. Thus f is Frechet-differentiable at (x_0, y_0) and its Frechet differential is the linear map $(\alpha, \beta) \mapsto 2x_0\alpha + 2y_0\beta$. Note that

$$2x_0 = \frac{\partial f}{\partial x}(x_0, y_0), \qquad 2y_0 = \frac{\partial f}{\partial y}(x_0, y_0).$$

3. THEOREM.

(a) If f is differentiable at x_0, it is continuous at x_0.

(b) The map $f \mapsto df_{x_0}$ from the space of functions on U which are differentiable at x_0 into $\mathcal{L}(\mathbf{R}^n, \mathbf{R}^m)$ is linear.

(c) If L is a linear map from $\mathbf{R}^n$ into $\mathbf{R}^m$, $dL_{x_0} = L$.

Proof. (a) Suppose f is differentiable at x_0. Then as $h \to 0$, $||f(x_0 + h) - f(x_0) - df_{x_0}(h)||_\infty \to 0$ as $h \to 0$ even faster than $||h||_\infty \to 0$. Since df_{x_0} is a linear map, $df_{x_0}(h) \to 0$ as $h \to 0$. Therefore, $||f(x_0 + h) - f(x_0)||_\infty \to 0$ as $h \to 0$.

(b) Suppose f and g are differentiable at x_0. Then $||(Af + Bg)(x_0 + h) - (Af + Bg)(x_0) - (A\, df_{x_0}(h) + B\, dg_{x_0}(h))||_\infty \leq |A|\, ||f(x_0 + h) - f(x_0) - df_{x_0}(h)||_\infty + |B|\, ||g(x_0 + h) - g(x_0) - dg_{x_0}(h)||_\infty$, and the latter expression goes to 0 faster than $||h||_\infty$ as $h \to 0$. This shows that $d(Af + Bg) = A\, df + B\, dg$.

(c) $L(x_0 + h) - L(x_0) = L(h)$. That is, L is a very good linear approximation to itself when L is linear.

4. THEOREM. If f is differentiable at x_0, and f maps a neighborhood of x_0 in $\mathbf{R}^n$ into $\mathbf{R}^m$, then for each $h \in \mathbf{R}^n$,

$$df_{x_0}(h) = \lim_{t \to 0} \frac{f(x_0 + th) - f(x_0)}{t} = \left[\frac{d}{ds} f(x_0 + sh)\right]_{s=0}.$$

Proof. By the definition of df_{x_0}, if $h \neq 0$,

$$||(f_0 + th) - f(x_0))/t - df_{x_0}(h)||_\infty = ||(f(x_0 + th) - f(x_0) - df_{x_0}(th))/t||_\infty$$
$$= ||h||_\infty[||f(x_0 + th) - f(x_0) - df_{x_0}(th)||_\infty]/||th||_\infty \to 0$$

as $t \to 0$.

5. DEFINITION. $\lim_{t \to 0} (f(x_0 + th) - f(x_0))/t$ is the **derivative of f along the vector h at x_0**. We may denote it by the symbol

$$\frac{\partial f}{\partial h}(x_0). \quad \bullet$$

☛ $\partial f/\partial h$ is to be interpreted as a directional derivative, that is, along the line through x_0 in the direction of h. Theorem 4 shows that a differentiable function has a directional derivative in every direction, and that its directional derivative depends linearly on the direction. This fact allows us to calculate the differential of a function. In fact, let $e_j = (0, \ldots, 0, 1, 0, \ldots, 0)$ (1 in the jth place) be one of the usual basis vectors for $\mathbf{R}^n$. Then $df_{x_0}(e_j) = \dfrac{\partial f}{\partial e_j}(x_0)$ is just the jth partial derivative $\dfrac{df}{\partial x_j}(x_0)$ of f at x_0. If $f = (f_1, \ldots, f_m)$, then

$$\frac{\partial f}{\partial x_j}(x_0) = \left(\frac{\partial f_1}{\partial x_j}(x_0), \ldots, \frac{\partial f_m}{\partial x_j}(x_0) \right).$$

This shows that the matrix representation of the linear map df is

$$\begin{pmatrix} \dfrac{\partial f_1}{\partial x_1} & \cdots & \dfrac{\partial f_1}{\partial x_n} \\ \cdots\cdots\cdots \\ \cdots\cdots\cdots \\ \cdots\cdots\cdots \\ \dfrac{\partial f_m}{\partial x_1} & \cdots & \dfrac{\partial f_m}{\partial x_n} \end{pmatrix},$$

which is the **Jacobian matrix.**

☛ *Warning!* As Exercise 1 will show, a function of two or more variables may have a directional derivative along every vector without even being continuous, much less differentiable.

If f is a function from an open subset of $\mathbf{R}^n$ into $\mathbf{R}$, then df_{x_0} is a *linear functional* on $\mathbf{R}^n$. As shown in Exercise III E 4, there is a unique vector $\nabla f(x_0)$ in $\mathbf{R}^n$ such that $df_{x_0}(y) = [y, \nabla f(x_0)]$. We call $\nabla f(x_0)$ the **gradient of f at x_0.**

6. MEAN VALUE PROPERTIES. Let f be a function from an open set $U \subseteq \mathbf{R}^n$ into $\mathbf{R}^m$ that has a differential at each point of U. Suppose that U is **convex** in the sense that if y' and y are in U, so is the **line segment joining them,** $\{y + t(y' - y) : 0 \le t \le 1\}$.

(a) If $m = 1$, there is a point ζ on the line segment joining y and y' such that
$$f(y') - f(y) = df_\zeta(y' - y) = [\nabla f(\zeta), y' - y].$$
(b) For every m, $\|f(y') - f(y)\|_\infty \le B\|y' - y\|_\infty$, if B is a bound for $\{|df_x| : x \in U\}$.

Proof. (a) Consider the function $g(t) = f(y + t(y' - y))$ for $0 \le t \le 1$. Theorem 4 shows that $g'(t) = df_{y+t(y'-y)}(y' - y) = [y' - y, \nabla f(y + t(y' - y))]$. Therefore, by the ordinary mean value theorem for functions from $\mathbf{R}$ to $\mathbf{R}$, $f(y') - f(y) = g(1) - g(0) = (1 - 0)g'(\eta) = [y' - y, \nabla f(y + \eta(y' - y))]$ for some η between 0 and 1. We can then take $\zeta = y + \eta(y' - y)$.

(b) If B is a bound for $|df_x|$ for $x \in U$, then by part (a) applied to each component of f, $|f_j(y') - f_j(y)| = |df_\zeta^j(y - y')| \le |df_\zeta^j|\,\|y' - y\|_\infty \le B\|y' - y\|_\infty$, where df_ζ^j is the linear map that sends y into the jth component of $df_\xi(y)$. Clearly $|\, df_\xi^j\, | \leq |\, df_\xi\, |$, since df_ξ^j is just one component of df_ξ. Since this inequality is true for every component of f, $||f(y') - f(y)||_\infty \leq B||y' - y||_\infty$. ●

For functions of one variable, the chain rule states that the derivative of a composite function is a product of derivatives: if $h = f \circ g$, $h'(x) = f'(g(x)) \cdot g'(x)$. Jacobian matrices of functions of several variables correspond to derivatives of functions of one variable. The corresponding *chain rule formula* is: the Jacobian matrix of the composite function $f \circ g$ is the product of the Jacobian matrices of f and g. A precise statement follows.

7. THE CHAIN RULE. Let U be an open subset of $\mathbf{R}^m$ and V be an open subset of $\mathbf{R}^k$. Let $f: U \to V$ and $g: V \to \mathbf{R}^n$. Suppose that f is differentiable at x and g is differentiable at $f(x)$. Then $g \circ f$ is differentiable at x, and $d(g \circ f)_x = (dg_{f(x)}) \circ (df_x)$. If M is the Jacobian matrix representing $dg_{f(x)}$ and L is the Jacobian matrix representing df_x, then the Jacobian matrix representing $d(g \circ f)_x$ is the product ML. In particular, if $n = 1$,

$$\frac{\partial g \circ f}{\partial x_j}(x_0) = \sum_{i=1}^{k} \frac{\partial g}{\partial y_i}(f(x_0)) \cdot \frac{\partial f_i}{\partial x_j}(x_0).$$

Proof. We have to show that the composite linear map $dg_{f(x)} \circ df_x$ is a good approximation to the error function $y \mapsto g \circ f(x + y) - g \circ f(x)$. To simplify the notation, define functions $\delta: V \to \mathbf{R}^n$ and $\epsilon: U \to \mathbf{R}^k$ by the formulas

$$\delta(y) = g(f(x) + y) - g(f(x)) - dg_{f(x)}(y),$$

$$\epsilon(z) = f(x + z) - f(x) - df_x(z).$$

The hypotheses that f is differentiable at x and that g is differentiable at $f(x)$ can be expressed by the formulas

$$\lim_{z \to 0} \frac{||\epsilon(z)||_\infty}{||z||_\infty} = 0, \qquad \lim_{y \to 0} \frac{||\delta(y)||_\infty}{||y||_\infty} = 0.$$

Put $\rho(0) = 0$, $\rho(y) = (1/||y||_\infty)\delta(y)$. Then ρ is continuous at 0 and $\delta(y) = ||y||_\infty \rho(y)$. Using the functions δ, ϵ, and ρ, and the fact that $dg_{f(x)}$ and df_x are linear functions, we can calculate that

$$g \circ f(x + z) - g \circ f(x) - dg_{f(x)} \circ df_x(z)$$
$$= g(f(x + z) - f(x) + f(x)) - g(f(x)) - dg_{f(x)}(f(x + z) - f(x))$$
$$+ dg_{f(x)}(f(x + z) - f(x) - df_x(z)) = \delta(f(x + z) - f(x)) + dg_{f(x)}(\epsilon(z))$$
$$= ||f(x + z) - f(x)||_\infty \rho(f(x + z) - f(x)) + dg_{f(x)}(\epsilon(z)).$$

Thus

$$||g \circ f(x + z) - g \circ f(x) - dg_{f(x)} \circ df_x(z)||_\infty / ||z||_\infty$$
$$= (||f(x + z) - f(x)||_\infty / ||z||_\infty) ||\rho(f(x + z) - f(x))||$$
$$+ ||dg_{f(x)}((1/||z||_\infty)\epsilon(z))||.$$

Since $dg_{f(x)}$ is a linear map from $\mathbf{R}^k$ to $\mathbf{R}^n$, it is continuous (see IV A 3(f)). Thus

$$\lim_{z \to 0} dg_{f(x)}((1/||z||_\infty)\epsilon(z)) = dg_{f(x)}(\lim_{z \to 0}((1/||z||_\infty)\epsilon(z))) = dg_{f(x)}(\mathbf{0}) = \mathbf{0}.$$

Moreover,

$$||f(x + z) - f(x)||_\infty/||z||_\infty \leq (||df_x(z)||_\infty + ||\epsilon(z)||_\infty)/||z||_\infty$$
$$= ||df_x((1/||z||_\infty)z)||_\infty + ||\epsilon(z)||_\infty/||z||_\infty.$$

Since df_x is a bounded linear map (IV A 3(f)), $||(1/||z||_\infty)z||_\infty = 1$, and

$$\lim_{z \to 0} ||\epsilon(z)||_\infty/||z||_\infty = 0,$$

it follows that $||f(x + z) - f(x)||_\infty/||z||_\infty$ is bounded in a neighborhood of $\mathbf{0}$ as a function of z. But as $z \to \mathbf{0}$, $f(x + z) - f(x) \to \mathbf{0}$, so that $\rho(f(x + z) - f(x)) \to \mathbf{0}$, and

$$||\rho(f(x + z) - f(x))||_\infty||f(x + z) - f(x)||_\infty/||z||_\infty \to 0$$

as well. These facts demonstrate that

$$\lim_{z \to 0} ||g \circ f(x + z) - g \circ f(x) - dg_{f(x)} \circ df_x(z)||_\infty/||z||_\infty = 0,$$

which shows, by definition, that $g \circ f$ is differentiable at x and that $d(g \circ f)_x = dg_{f(x)} \circ df_x$.

The matrix of the composite linear map $dg_{f(x)} \circ df_x$ is just the matrix product ML (see section III E). Finally, in case $n = 1$, the matrices in question are

$$M = \left(\frac{\partial g}{\partial y_1}, \ldots, \frac{\partial g}{\partial y_k}\right), \qquad L = \begin{pmatrix} \dfrac{\partial f_1}{\partial x_1} & \cdots & \dfrac{\partial f_1}{\partial x_m} \\ \cdots\cdots\cdots \\ \cdots\cdots\cdots \\ \cdots\cdots\cdots \\ \dfrac{\partial f_k}{\partial x_1} & \cdots & \dfrac{\partial f_k}{\partial x_m} \end{pmatrix}$$

The jth entry of the product, according to the product rule for matrices (III E 6) is $\dfrac{\partial g}{\partial y_1}\dfrac{\partial f_1}{\partial x_j} + \cdots + \dfrac{\partial g}{\partial y_k}\dfrac{\partial f_k}{\partial x_j}$, which is $\dfrac{\partial g \circ f}{\partial x_j}$. ●

Figure 22 gives a dramatic (?) interpretation of the chain rule for the case $m = 3$, $k = 2$, $n = 1$. The factories represent functions. One puts glop into the muckle works f and gets out muckle; f: glop → muckle. g turns muckle into widgets; g: muckle → widgets. Suppose that each ton of high-grade muckle yields $\partial g/\partial y_1 = 1000$ widgets, and each ton of low-grade muckle yields $\partial g/\partial y_2 = 800$ widgets. Suppose that each barrel of anthracite glop yields $\partial f_1/\partial x_2 = 1/50$ ton of high-grade muckle and $\partial f_2/\partial x_2 = 1/40$ ton of low-grade muckle. Then for each barrel of anthracite glop, we should get

$$\frac{\partial g}{\partial y_1}\frac{\partial f_1}{\partial x_2} + \frac{\partial g}{\partial y_2}\frac{\partial f_2}{\partial x_2} = 1000 \times (1/50) + 800(1/40) = 20 + 20 = 40 \text{ widgets.}$$

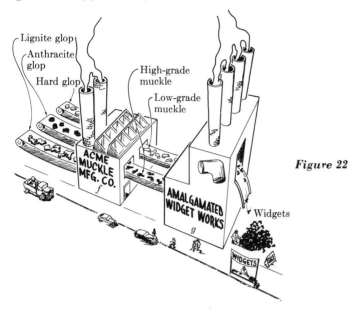

Figure 22

If a barrel of lignite glop yields $\partial f_1/\partial x_3 = 1/100$ ton of high-grade muckle and $\partial f_2/\partial x_3 = 1/20$ ton of low-grade muckle, each barrel of lignite glop should yield

$$\frac{\partial g}{\partial y_1}\frac{\partial f_1}{\partial x_3} + \frac{\partial g}{\partial y_2}\frac{\partial f_2}{\partial x_3} = 1000 \times (1/100) + 800 \times (1/20)$$

$$= 10 + 40 = 50 \text{ widgets.}$$

If you wish, you can think of x_1 as the total amount of hard glop brought to the muckle works, $f_2(x_1, x_2, x_3)$ as the total amount of low-grade muckle produced from x_1 barrels of hard glop, x_2 barrels of anthracite glop, and x_3 barrels of lignite glop. $g(y_1, y_2)$ is the total number of widgets produced from y_1 tons of high-grade muckle and y_2 tons of low-grade muckle. Then the formula $\partial f_1/\partial x_2 = (f_1(x_1, x_2 + \delta, x_3) - f_1(x_1, x_2, x_3))/\delta = 1/50$ simply expresses the fact that if x_1 barrels of hard glop, x_2 barrels of anthracite glop, and x_3 barrels of lignite glop have been received, and then δ new barrels of anthracite glop come in, the muckle works will produce $\delta/50$ new tons of high-grade muckle. Et cetera.

We can give a less fanciful geometric interpretation to the gradient vector of a function h (Figure 23). The figure shows a contour map of two neighboring mountain peaks. Let $h(x, y)$ be the height of the land at the point whose map coordinates are (x, y), measured in feet above sea level. For example, the top of Cockatoo Peak is at a height of 10,027 feet, about 200 feet higher than the top of Elfenegg. Let us imagine that an airplane flies over this terrain at a height of 20,000 feet above sea level at a speed of 400 feet per second, equipped

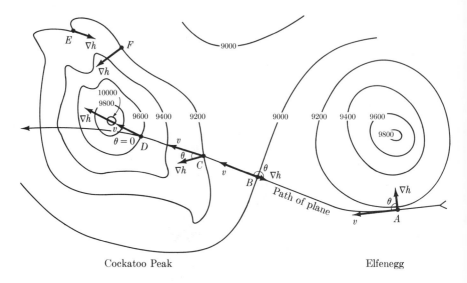

Cockatoo Peak   Elfenegg

Figure 23

with a radar device that enables the crew to measure their height above the ground, and consequently to measure the height of the ground above sea level. Let $H(t)$ be the height of the ground immediately below the plane at time t, and let $v(t)$ be the velocity of the plane at time t. Since the speed is always 400 feet per second, $||v(t)||_2 = 400$. According to the definition of the gradient vector, if the plane passes over the point $P(t)$ at time t, $H'(t) = [\nabla h(P(t)), v(t)] = ||\nabla h||_2 ||v||_2 \cos \theta = 400||\nabla h||_2 \cos \theta$, where θ is the angle between the vectors $\nabla h(P(t))$ and $v(t)$ (See Theorem VI G 4 and formula III C 3).

As the plane passes over point A at time t_1, it is flying above one of the contour lines (level curves) of Elfenegg. Thus $[\nabla h(A), v(t_1)] = 400 \cos \theta_1 ||\nabla h(A)||_2 = H'(t_1) = 0$, because h is *constant* along level lines. This shows that $\nabla h(A)$ is *perpendicular* to $v(t_1)$. Then $\nabla h(A)$ *is perpendicular to the level curve through* A. Clearly $400 \cos \theta ||\nabla h||$ will be largest when $\theta = 0$, so that the plane is flying parallel to ∇h. This is the case when the plane flies over point D at time t_4. Of course $H'(t)$ is largest when the plane is flying toward the upslope. Thus ∇h *points uphill*. Finally, if the plane flies toward the upslope as it does from point C to point D, then $H'(t)$ is proportional simply to $||\nabla h||_2$. This shows that $||\nabla h||_2$ *measures the steepness of the terrain*. In sum, $\nabla h(P)$ *is perpendicular to the level curve through* P, points uphill (in the direction of increasing values of h) *and its length measures the steepness of the terrain*.

Of course $\nabla h(P)$ does not have to point toward the top of the hill, since it depends only on the terrain very near P. This is shown by $\nabla h(E)$, which actually points toward the top of Elfenegg rather than toward the top of Cockatoo Peak, and by $\nabla h(F)$ and $\nabla h(C)$, which point away from both peaks.

8. DEFINITION. Let f be a function mapping an open subset U of $\mathbf{R}^n$ into $\mathbf{R}^m$. f is **continuously differentiable at y** if f is differentiable in a neighborhood of y and the function $x \mapsto df_x$ is continuous at y as a function from a neighborhood of y into the space $B(\mathbf{R}^n, \mathbf{R}^m)$ of bounded linear maps from $\mathbf{R}^n$ into $\mathbf{R}^m$.

9. INVERSE FUNCTION THEOREM. Let U be an open subset of $\mathbf{R}^n$ and f be a function from U to $\mathbf{R}^n$ that is continuously differentiable at the point $v \in U$. Suppose that the linear map $L = df_v$ has an inverse. Then there are open neighborhoods V of v $(V \subseteq U)$ and W of $w = f(v)$ and an *inverse function* $g: W \rightarrow V$ such that:

(a) f maps V one-to-one onto W and g maps W one-to-one onto V;
(b) $g \circ f(x) \equiv x$ on V and $f \circ g(y) \equiv y$ on W;
(c) for each $x \in V$, f is differentiable at x, the linear map df_x has an inverse, g is differentiable at the point $y = f(x)$, and $dg_y = (df_x)^{-1}$;
(d) in particular, g is continuously differentiable at v and $dg_v = L^{-1}$.

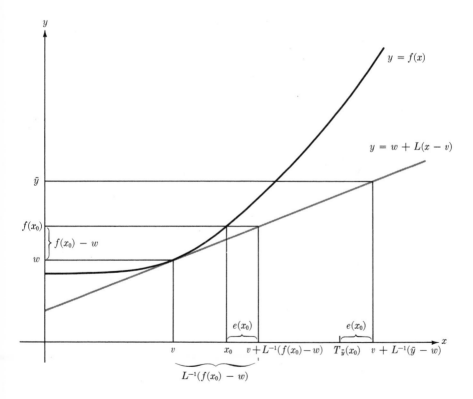

Figure 24. Illustration of the proof of the Inverse Function Theorem for $n = 1$.

Proof. If f were actually linear, so that $f = L$, we could solve the equation $f(x) = y$ as follows. Subtract $f(v) = w$ from both sides: $f(x) - f(v) = y - w$. Then apply L^{-1}:

$$x - v = L^{-1} \circ L(x - v) = L^{-1}(L(x) - L(v)) = L^{-1}(f(x) - f(v)) = L^{-1}(y - w),$$

so that $x = v + L^{-1}(y - w)$. Since f is only approximately linear near v in the general case, the formula $x = v + L^{-1}(y - w)$ yields only an approximate solution to the equation $y = f(x)$. The error can be measured by putting $f(x)$ for y, and seeing how close $v + L^{-1}(f(x) - w)$ is to x. We define the error function $e \colon U \to \mathbf{R}^n$ by $e(x) = v + L^{-1}(f(x) - w) - x$ (see Figure 24). Then a straightforward calculation shows that

$$y = f(x) \Leftrightarrow x = v + L^{-1}(y - w) - e(x).$$

That is, in the general case, the solution x to the equation $y = f(x)$ is given approximately by the formula $x \approx v + L^{-1}(y - w)$ (which works exactly when f is linear), and the error in the approximation is $e(x)$. Now define $T_y(x)$ to be $v + L^{-1}(y - w) - e(x)$. The equation $y = f(x)$ can be reformulated as $x = T_y(x)$. Thus we seek an x that is left fixed by T_y. This is a setup for the Contraction Mapping Theorem. There are many details to be checked, so the remainder of the proof will be broken down into a series of lemmas.

10. LEMMA. There is an $r \in \mathbf{R}^+$ so small that

(a)　e is differentiable on $B^c_{2r}(v) = \{x \in \mathbf{R}^n \colon ||x - v||_\infty \leq 2r\}$,

(b)　$|\, de_x \,| \leq \frac{1}{2}$ for each $x \in B^c_{2r}(v)$,

(c)　for each x, $x' \in B^c_{2r}(v)$, $||e(x) - e(x')||_\infty \leq \frac{1}{2}||x - x'||_\infty$.

Proof. By hypothesis, f is differentiable on a neighborhood of v, so there is a ball $B_s(v)$ contained in U on which f is differentiable. Let I be the identity map on $\mathbf{R}^n$, $I \colon x \mapsto x$. Using the Chain Rule and the fact that L^{-1} is linear, so that $dL^{-1} = L^{-1}$, we conclude that if $x \in B_s(v)$,

$$de_x = d(v + L^{-1} \circ (f - w) - I)_x = (dv + dL^{-1} \circ d(f - w) - dI)_x$$
$$= 0 + L^{-1} \circ (df_x - 0) - I = L^{-1} \circ df_x - I.$$

In particular, e is differentiable at each $x \in B_s(v)$, and $de_v = L^{-1} \circ df_v - I = L^{-1} \circ L - I = 0$. Since the function $x \mapsto df_x$ is continuous, there is an $r \in \mathbf{R}^+$ such that $2r < s$ and that $||x - v||_\infty \leq 2r \Rightarrow |\, df_x - df_v \,| \leq (1/2) \,|\, L^{-1} \,|^{-1} \Rightarrow$ $|\, de_x \,| = |\, L^{-1} \circ df_x - I \,| = |\, L^{-1} \circ (df_x - df_v) + L^{-1} \circ df_v - I \,| = $ $|\, L^{-1} \circ (df_x - df_v) \,| \leq |\, L^{-1} \,|\,|\, df_x - df_v \,| \leq 1/2$ (see IV A 3(f)). This proves (a) and (b). (c) follows from (b) and Theorem 6(b) (Mean Value Property). ●

　　Now put $W = \{y \in \mathbf{R}^n \colon ||L^{-1}(y - w)||_\infty < r\}$. Because the map $y \mapsto L^{-1}(y - w)$ is continuous, W is open (Theorem IV D 3). Moreover, $w \in W$.

11. LEMMA. For each $y \in W$, T_y is a contraction of the complete metric space $B^c_{2r}(w)$.

Proof. $\langle \mathbf{R}^n, \|\ \|_\infty \rangle$ is complete (Corollary VI A 6) and $B_{2r}^c(v)$ is closed in $\langle \mathbf{R}^n, \|\ \|_\infty \rangle$. Theorem IV C 10 shows that $B_{2r}^c(v)$ is complete. It remains to show that $T_y : B_{2r}^c(v) \to B_{2r}^c(v)$ when $y \in W$ and that T_y is a contraction. Note that $e(v) = 0$ and that $y \in W \Leftrightarrow \|L^{-1}(y - w)\|_\infty < r$. Using Lemma 10(c), we see that $x \in B_{2r}^c(v) \Rightarrow$

$$\|T_y(x) - v\|_\infty = \|L^{-1}(y - w) - e(x)\|_\infty = \|L^{-1}(y - w) - (e(x) - e(v))\|_\infty$$
$$\leq \|L^{-1}(y - w)\|_\infty + \|e(x) - e(v)\|_\infty < r + (1/2)\|x - v\|_\infty$$
$$\leq r + (1/2)(2r) = 2r$$

$\Rightarrow T_y(x) \in B_{2r}^c(v)$. Moreover, $T_y(x) - T_y(x') = e(x') - e(x)$, so the fact that T_y is a contraction also follows from 10(c). $\bullet$

The Contraction Mapping Theorem shows that for each $y \in W$, there is a unique $x \in B_{2r}^c(v)$ such that $x = T_y(x)$ (which means that $y = f(x)$). Let this fixed point of T_y be $g(y)$ for each $y \in W$. Then the proof of Lemma 11 actually shows that $\|g(y) - v\|_\infty < 2r$, so that $g : W \to B_{2r}(v)$. Put $V = B_{2r}(v) \cap f^{-1}(W)$. Because f is continuous (indeed differentiable) on $B_{2r}(v)$, V is open (Theorem IV D 3), and $v \in V$.

We have just shown that for each $y \in W$ there is an $x \in V$ such that $f(x) = y$, and that this x is unique. This shows that f maps V one-to-one onto W. By definition, $g = f^{-1} : W \to V$, so that g maps W one-to-one onto V, $g \circ f(x) \equiv x$ on V and $f \circ g(y) \equiv y$ on W. We have proved parts (a) and (b) of the Theorem.

12. LEMMA. *If* $y, y' \in W$, *then* $\|g(y) - g(y')\|_\infty \leq 2|L^{-1}|\,\|y - y'\|_\infty$.

Proof. Put $x = g(y)$ and $x' = g(y')$. By the definition of g, $x = T_y(x)$ and $x' = T_{y'}(x')$. Thus

$$\|x - x'\|_\infty = \|T_y(x) - T_{y'}(x')\|_\infty$$
$$= \|v + L^{-1}(y - w) - e(x) - (v + L^{-1}(y' - w) - e(x'))\|_\infty$$
$$= \|L^{-1}(y - y') - (e(x) - e(x'))\|_\infty \leq \|L^{-1}(y - y')\|_\infty$$
$$+ \|e(x) - e(x')\|_\infty \leq |L^{-1}|\,\|y - y'\|_\infty + (1/2)\|x - x'\|_\infty$$

by the definition of $|L^{-1}|$ (IV A 3(f)) and Lemma 10(c). Now subtract $\frac{1}{2}\|x - x'\|_\infty$ from both sides of this inequality. The result is $\frac{1}{2}\|g(y) - g(y')\|_\infty = \frac{1}{2}\|x - x'\|_\infty \leq |L^{-1}|\,\|y - y'\|_\infty$. $\bullet$

13. LEMMA. *If* $y \in W$, *then* g *is differentiable at* y, dg_y *is invertible, and* $dg_y = (df_x)^{-1}$, *where* $x = g(y)$.

Proof. If $y \in W$, put $x = g(y)$. Then $x \in V \subseteq B_{2r}(v)$. In particular, from the way r was chosen in the proof of Lemma 10, $|df_x - df_v| = |df_x - L| \leq (1/2)|L^{-1}|^{-1}$. According to Theorem VI D 6, $M = df_x$ is invertible. By the definition of df_x, the function ρ, defined by

$\rho(x) = 0, \rho(x') = ||x - x'||_\infty^{-1}(f(x') - f(x) - df_x(x' - x))$ when $x' \in V - \{x\}$,

is continuous at x. If $y' \in W$ and $x' = g(y')$, then

$$||g(y') - g(y) - M^{-1}(y' - y)||_\infty = ||x' - x - M^{-1}(f(x') - f(x))||_\infty$$

$$= ||M^{-1}(M(x' - x) - f(x') + f(x))||_\infty$$

$$\leq |\, M^{-1} \,|\, ||f(x') - f(x) - M(x' - x)||_\infty$$

$$= |\, M^{-1} \,|\, ||x' - x||_\infty ||\rho(x')||_\infty$$

$$= |\, M^{-1} \,|\, ||g(y') - g(y)||_\infty ||\rho(g(y'))||_\infty$$

$$\leq 2\,|\, M^{-1} \,|\,|\, L^{-1} \,|\, ||y' - y||_\infty ||\rho \circ g(y')||_\infty$$

(see Lemma 12). Therefore

$$||y' - y||_\infty^{-1} ||g(y') - g(y) - M^{-1}(y' - y)||_\infty \leq 2\,|\, M^{-1} \,|\,|\, L^{-1} \,|\, ||\rho \circ g(y')||_\infty,$$

which goes to 0 as $y' \to y$, since $\rho \circ g$ is continuous at y and $\rho \circ g(y) = \rho(x) = 0$. This shows (Definition 1) that g is differentiable at y and that $dg_y = M^{-1} = (df_x)^{-1}$. Clearly M^{-1} is invertible, since $(M^{-1})^{-1} = M$. ●

Lemma 13 proves part (c) of the Theorem. That $dg_v = L^{-1}$ follows from (c) and the formula $L = df_x$. Finally, the function $M \mapsto M^{-1}$ is continuous (Corollary VI D 7), the function $x \mapsto df_x$ is continuous at v, and therefore the composite function $x \mapsto (df_x)^{-1}$ is continuous at v. This proves part (d). ●

14. REMARKS. (a) If $U = \mathbf{R}^n$, there need not be an inverse for f defined on all of $\mathbf{R}^n$. For example, let $f: x \mapsto \sin x$. Then f is defined on all of $\mathbf{R}^1$, and df_0 is the linear map $y \mapsto f'(0)y = y$, which is surely invertible. Nonetheless, f is not one-to-one on all of $\mathbf{R}$. It is only the *restriction* of f to a neighborhood of 0 that is one-to-one.

(b) If df_v is not invertible, f need not be one-to-one in any neighborhood of v. An example is the function $x \mapsto x^2$, which is not one-to-one in any neighborhood of $0 \in \mathbf{R}$.

(c) Even if f is one-to-one, the inverse function g cannot be differentiable at w if df_v is not invertible. An example is the function $f: x \mapsto x^3$ from $\mathbf{R}$ to $\mathbf{R}$. Although f maps $\mathbf{R}$ one-to-one onto $\mathbf{R}$ and is continuously differentiable everywhere, the inverse function $y \mapsto y^{1/3}$ is not differentiable at 0. In general, if $g = f^{-1}$, f is differentiable at v and g is differentiable at $w = f(v)$, then $I = dI_v = d(g \circ f)_v = dg_w \circ df_v$, so that $dg_w = (df_v)^{-1}$. This shows that df_v must be invertible if g is to be differentiable at w. (See Theorem III E 8(b).)

(d) The Inverse Function Theorem gives one of the best examples of a situation in which one *must* distinguish between the functions defined by a single formula on two different domains. For example, the formula $y = \sin x$ defines one inverse $x = (\text{sine}_1)^{-1}(y)$ of the function $\text{sine}_1: [-\pi/2, \pi/2] \to [-1, 1]$ and an entirely different inverse of the function $\text{sine}_2: [\pi/2, 3\pi/2] \to [-1, 1]$. There is no way to put $(\text{sine}_1)^{-1}$ and $(\text{sine}_2)^{-1}$ together into a single function from $[-1, 1]$ to $[-\pi/2, 3\pi/2]$. This is why we took such pains in section I B to distinguish a formula from the function or functions it defines.

With the Inverse Function Theorem now in hand, we can quickly deal with the question of functions defined implicitly. Consider the equation $f(x, y) = 0$. It may be possible to solve this equation for y as a function of x. This problem includes the problem of finding the inverse function, which is the problem of solving the equation $f(x) - y = 0$ for x as a function of y. In the case of the Inverse Function Theorem, only the local existence of an inverse function was proved, and for good reason (see Remarks 14). We can expect something similar in the general case.

We can see what should be expected as follows. Suppose $f(x, y) = \phi(x) + L(y)$, where L is a linear function that has an inverse. In the present case, where x and y vary over $\mathbf{R}$, this means just that $L(y) = Ay$, where $A \neq 0$. Then we can solve for y as follows: $f(x, y) = 0 \Leftrightarrow y = -L^{-1}(\phi(x))$. In the general case, there will only be a good linear approximation to work with.

A notational device will help us. If $x \in \mathbf{R}^n$ and $y \in \mathbf{R}^m$, let $(x, y) \in \mathbf{R}^{n+m}$ be the vector $(x_1, \ldots, x_n, y_1, \ldots, y_m)$.

15. IMPLICIT FUNCTION THEOREM. Let U be an open neighborhood of 0 in $\mathbf{R}^{n+m}$, and f a function from U to $\mathbf{R}^m$. Suppose that f is continuously differentiable at $\mathbf{0}$, that $f(\mathbf{0}) = \mathbf{0}$, and that the linear map $y \mapsto df_0(\mathbf{0}, y)$ from $\mathbf{R}^m$ into $\mathbf{R}^m$ is invertible. Then there are an open neighborhood $\mathfrak{N}$ of $\mathbf{0}$ in $\mathbf{R}^n$, an open neighborhood $\mathfrak{M}$ of $\mathbf{0}$ in $\mathbf{R}^m$, and a unique continuous function $\phi \colon \mathfrak{N} \to \mathfrak{M}$ such that

(a) ϕ is continuously differentiable at $\mathbf{0}$,
(b) $\phi(\mathbf{0}) = \mathbf{0}$,
(c) $f(x, \phi(x)) = \mathbf{0}$ on $\mathfrak{N}$.

Proof. We apply the Inverse Function Theorem. Let $g \colon U \to \mathbf{R}^{n+m}$ be defined by $g(x, y) = (x, f(x, y))$. I claim that if f is differentiable at (v, w), g is too, and $dg_{(v,w)}(x, y) = (x, df_{(v,w)}(x, y))$. That is, $dg_{(v,w)}$ has the matrix representation with respect to the usual bases:

$$
\begin{pmatrix}
1 & \cdots & 0 & 0 & \cdots & 0 \\
0 & \cdots & 0 & 0 & \cdots & 0 \\
\multicolumn{6}{c}{\cdots\cdots\cdots\cdots\cdots\cdots} \\
0 & \cdots & 1 & 0 & \cdots & 0 \\
\dfrac{\partial f_1}{\partial x_1} & \cdots & \dfrac{\partial f_1}{\partial x_n} & \dfrac{\partial f_1}{\partial y_1} & \cdots & \dfrac{\partial f_1}{\partial y_m} \\
\multicolumn{6}{c}{\cdots\cdots\cdots\cdots\cdots\cdots} \\
\dfrac{\partial f_m}{\partial x_1} & \cdots & \dfrac{\partial f_m}{\partial x_n} & \dfrac{\partial f_m}{\partial y_1} & \cdots & \dfrac{\partial f_m}{\partial y_m}
\end{pmatrix}
$$

In fact,

$$g(x, y) - g(v, w) - (x - v, df_{(v,w)}(x - v, y - w))$$
$$= (x, f(x, y)) - (v, f(v, w)) - (x - v, df_{(v,w)}(x - v, y - w))$$
$$= (0, f(x, y) - f(v, w) - df_{(v,w)}(x - v, y - w)).$$

Thus

$$||(x - v, y - w)||_\infty^{-1}||g(x, y) - g(v, w) - (x - v, df_{(v,w)}(x - v, y - w))||_\infty$$
$$= ||(x - v, y - w)||_\infty^{-1}||f(x, y) - f(v, w) - df_{(v,w)}(x - v, y - w)||_\infty,$$

which goes to 0 as $(x, y) \to (v, w)$ by the definition of $df_{(v,w)}$.

It is apparent from this formula that g is continuously differentiable at $\mathbf{0}$. We know that $g(\mathbf{0}) = \mathbf{0}$. Finally, $dg_\mathbf{0}$ is invertible because it maps $\mathbf{R}^{n+m}$ onto $\mathbf{R}^{n+m}$ (see Theorem III E 8(c)). To see this, take any $(v, w) \in \mathbf{R}^{n+m}$. Because the map $y \mapsto df_\mathbf{0}(0, y)$ is invertible (and hence onto) we can choose $z \in \mathbf{R}^m$ such that $df_\mathbf{0}(0, z) = w - df_\mathbf{0}(v, 0)$. Then $dg_\mathbf{0}(v, z) = (v, df_\mathbf{0}(v, z)) = (v, df_\mathbf{0}(v, 0) + df_\mathbf{0}(0, z)) = (v, w)$. Therefore we can apply the Inverse Function Theorem to g to conclude that there exist a neighborhood V of $\mathbf{0}$ in $\mathbf{R}^{n+m}$, another neighborhood W in $\mathbf{R}^{n+m}$, and a differentiable function $h\colon W \to V$ such that h is continuously differentiable at $\mathbf{0}$, that $h \circ g(x, y) \equiv (x, y)$ on V, and that $g \circ h(x, y) \equiv (x, y)$ on W. Let $h(x, y) = (h_1(x, y), \ldots, h_{n+m}(x, y))$, and put $\psi(x) = (h_1(x, 0), \ldots, h_n(x, 0))$, $\phi(x) = (h_{n+1}(x, 0), \ldots, h_{n+m}(x, 0))$. Then if $(x, 0) \in W$, $(x, 0) = g \circ h(x, 0) = g(\psi(x), \phi(x)) = (\psi(x), f(\psi(x), \phi(x)))$. From this it follows first that $x = \psi(x)$, then that $0 = f(x, \phi(x))$ by comparing components. Since V is open and $\mathbf{0} \in V$, we can choose a ball of radius $r > 0$, $B_r(\mathbf{0}) = \{(x, y) \in \mathbf{R}^{n+m}: ||(x, y)||_\infty < r\}$, which is included in V. Let $\mathfrak{N} = \{x \in \mathbf{R}^n: (x, 0) \in W\} \cap \{x \in \mathbf{R}^n: (x, \phi(x)) \in B_r(\mathbf{0})\}$ and let $\mathfrak{M} = \{y \in \mathbf{R}^m: ||y||_\infty < r\}$. Then $\{x \in \mathbf{R}^n: (x, \phi(x)) \in B_r(\mathbf{0})\}$ is open because $B_r(\mathbf{0})$ is open and $x \mapsto (x, \phi(x))$ is continuous, and it follows easily that $\mathfrak{N}$ is open. $0 \in \mathfrak{N}$, $\mathfrak{M}$ is open, $0 \in \mathfrak{M}$, and by the definition of $\mathfrak{N}$, $\phi\colon \mathfrak{N} \to \mathfrak{M}$. Moreover, if $\chi\colon \mathfrak{N} \to \mathfrak{M}$ is any function such that $f(x, \chi(x)) \equiv 0$ on $\mathfrak{N}$, then $x \in \mathfrak{N} \Rightarrow (x, 0) \in W \Rightarrow g(x, \chi(x)) = (x, f(x, \chi(x))) = (x, 0) \Rightarrow (x, \chi(x)) = h \circ g(x, \chi(x)) = h(x, 0) = (\psi(x), \phi(x)) = (x, \phi(x)) \Rightarrow \chi(x) = \phi(x)$, which proves the uniqueness of ϕ. ϕ is continuously differentiable at $\mathbf{0}$ because h is continuously differentiable at $\mathbf{0}$, and $\phi(0) = (h_{n+1}(0, 0), \ldots, h_{n+m}(0, 0)) = 0$. ●

16. REMARK. The hypothesis that the map $y \mapsto df_\mathbf{0}(0, y)$ be invertible is essential to assure us that the function ϕ exists. For example, if $f(x, y) = x^2 + y^2$, $f(0, 0) = 0$, but for no $x \neq 0$ is there a $y \in \mathbf{R}$ such that $x^2 + y^2 = 0$.

Exercises

1. Let f be the function from $\mathbf{R}^2$ to $\mathbf{R}$ defined by $f(0, 0) = 0$, $f(x, y) = (x^6 + y^6)/(x^6 + y^6 + (y - x^2)^2)$ if $(x, y) \neq (0, 0)$. Show that f has directional derivatives in every direction at every point, that the function $h \mapsto \dfrac{\partial f}{\partial h}(x, y)$ is

linear for every $(x, y) \in \mathbf{R}^2$, but that f is discontinuous at $\mathbf{0}$. *Hint:* f is clearly well-behaved except possible at $(0, 0)$. Show by direct calculation, that $\lim_{t \to 0} (1/t)(f((0, 0) + t(h_1, h_2)) - f(0, 0)) = 0$ for every $(h_1, h_2) \in \mathbf{R}^2$. Show, however, that $f(x, y) = 1$ at all points of the parabola $y = x^2$ excepting $(0, 0)$.

*2. Let $f = (f^1, \ldots, f^m)$ be a function from $\mathbf{R}^n$ to $\mathbf{R}^m$. Show that in order for f to be differentiable, it is necessary and sufficient that $f^1, \ldots, f^m$ all be differentiable, and that in this case $df_v(x) = (df_v^1(x), \ldots, df_v^m(x))$.

*3. Let f be a function from $\mathbf{R}^n$ to $\mathbf{R}$, and suppose that in a neighborhood of v, every one of the partial derivatives $\partial f/\partial x_j$ exists and is continuous. Show that under these conditions f is differentiable at v. *Hint:* Put $f(x) - f(v) = f(x_1, \ldots, x_n) - f(v_1, x_2, \ldots, x_n) + f(v_1, x_2, \ldots, x_n) - f(v_1, v_2, \ldots, x_n) + \cdots - f(v_1, \ldots, v_n)$. Use the Mean Value Theorem to get

$$f(v_1, \ldots, v_{j-1}, x_j, \ldots, x_n) - f(v_1, \ldots, v_j, x_{j+1}, \ldots, x_n)$$
$$= \frac{\partial f}{\partial x_j} (v_1, \ldots, v_{j-1}, \zeta_j, x_{j+1}, \ldots, x_n)(x_j - v_j).$$

Use this expression to show that $f(x) - f(v) - \langle x - v, \nabla f(v) \rangle$ goes to 0 faster than $x - v$.

4. Restate the Inverse Function Theorem for the one-dimensional case so that it involves the condition $f'(x_0) \neq 0$.

5. Do the same for the Implicit Function Theorem for the case $n = m = 1$, so that the theorem involves the condition

$$\frac{\partial f}{\partial y} (0, 0) \neq 0.$$

CHAPTER VII

Compactness

Now that I have given a number of applications of the notions of metric space and normed space, and of our first theorems about them, we are ready to resume our discussion of abstract metric spaces. In this chapter, we shall once more formulate and study in the setting of metric spaces a property that was first discovered in the case of the real number system. One formulation of this property is the Bolzano-Weierstrass Theorem II C 5. Recall from Exercise II C 8 that this theorem could be used to show that every continuous real-valued function attains its maximum on a closed bounded interval, as well as to obtain many other goodies. Its generalization to metric spaces fulfills its promise of being very useful.

A. Definition of compactness, equivalences, first examples

1. DEFINITION. An **open cover** of a metric space $\langle M, d \rangle$ is a family $\mathfrak{F}$ of open subsets of M whose union is M. A **subcover** is a subset of the collection $\mathfrak{F}$ which is also a cover.

2. HEINE-BOREL THEOREM. Let $M = [a, b] = \{x \in \mathbf{R} : a \leq x \leq b\}$ be a closed bounded interval in $\mathbf{R}$, and let d be the usual metric on $\mathbf{R}$. Then every open cover of $\langle M, d \rangle$ has a *finite* subcover.

Proof. Let $\mathfrak{F}$ be an open cover of $[a, b]$. Let S be the set of points $x \in [a, b]$ such that $[a, x]$ is covered by finitely many of the sets in $\mathfrak{F}$. Then S is bounded above by b. S is not empty, because $a \in S$. In fact, the interval $[a, a]$, which contains only a, is covered by any *one* of the sets in $\mathfrak{F}$ that contains a. Because S is nonempty and bounded above, it has a lub, L. I claim that $L = b$ and $L \in S$. In fact, there is some open set $U \in \mathfrak{F}$ that contains L. Because U is open, it contains the interval $M \cap [L - \epsilon, L]$ for some $\epsilon > 0$. By Theorem II C 1 there is an $x \in S$ such that $L - \epsilon < x \leq L$. Then finitely many sets in $\mathcal{F}, V_1, \ldots, V_n$ cover $[a, x]$. The finitely many sets $U, V_1, \ldots, V_n$ thus cover $[a, L]$ so that $L \in S$. If $L < b$, $U, V_1, \ldots, V_n$ would cover an entire neighborhood of L of the form $[L - \delta, L + \delta]$, with $L + \delta < b$. But then $L + \delta \in S$, so L would not be an upper bound for S. Thus $b \in S$. $\bullet$

3. DEFINITION. A metric space $\langle M, d \rangle$ is **compact** if every open cover of M has a finite subcover.

Remark. Let N be a subset of $\langle M, d \rangle$. Then $\langle N, d \rangle$ is a metric subspace of $\langle M, d \rangle$. We say that N is a **compact subset** of M if the metric subspace $\langle N, d \rangle$ is compact. This is equivalent to the following condition: if $\mathcal{F}$ is a family of open sets in M that covers N (its union contains N), then some finite subfamily of $\mathcal{F}$ covers N. This follows at once from the fact that a subset U of N is open in the metric space $\langle N, d \rangle$ if, and only if, there is an open set U^* in M such that $U = U^* \cap N$ (Theorem IV C 17).

Suppose $\langle N, d \rangle$ is compact. Let $\mathcal{F}$ be a family of open sets in $\langle M, d \rangle$ that covers N. Then the family of sets of the form $U \cap N$ for $U \in \mathcal{F}$ is an open cover of N. It has a finite subcover $U_1 \cap N, \ldots, U_k \cap N$. Then $\{U_1, \ldots, U_k\}$ is a finite subfamily of $\mathcal{F}$ that covers N.

Conversely, suppose that whenever $\mathcal{F}^*$ is a family of open subsets of M such that $N \subseteq \bigcup \mathcal{F}^*$, there is a finite subfamily $\mathcal{G}^* \subseteq \mathcal{F}^*$ such that $N \subseteq \bigcup \mathcal{G}^*$. Let $\mathcal{F}$ be an open cover of $\langle N, d \rangle$. For each $U \in \mathcal{F}$, choose an open subset U^* of $\langle M, d \rangle$ such that $U^* \cap N = U$, and let $\mathcal{F}^* = \{U^*: U \in \mathcal{F}\}$. Then $\mathcal{F}^*$ covers N, so we can extract a finite subfamily $\mathcal{G}^* = \{U_1^*, \ldots, U_k^*\}$ from $\mathcal{F}^*$ such that $N \subseteq U_1^* \cup \cdots \cup U_k^* \Rightarrow N = N \cap (U_1^* \cup \cdots \cup U_k^*) = (N \cap U_1^*) \cup \cdots \cup (N \cap U_k^*) = U_1 \cup \cdots \cup U_k$. This shows that $\mathcal{F}$ contains a finite subcover, so $\langle N, d \rangle$ is compact.

4. DEFINITION. A family $\mathcal{F}$ of subsets of a set S has the **Finite Intersection Property** if $U_1 \cap \cdots \cap U_k$ is nonempty for every finite collection $U_1, \ldots, U_k$ of sets in $\mathcal{F}$.

5. PROPOSITION. For the metric space $\langle M, d \rangle$ to be compact, it is necessary and sufficient that whenever $\mathcal{F}$ is a family of closed subsets of M that has the Finite Intersection Property, the intersection of *all* the sets in $\mathcal{F}$ is nonempty.

Proof. Let $\mathcal{F}$ be a family of closed subsets of M, and let $\mathcal{G} = \{M - U: U \in \mathcal{F}\}$. If $B \subseteq \mathcal{F}$, then $\bigcap B = \varnothing \Leftrightarrow M = M - \bigcap B = \bigcup \{M - U: U \in B\}$. Therefore, $\mathcal{F}$ has the Finite Intersection Property if, and only if, no finite subfamily of $\mathcal{G}$ covers M. The intersection of all the sets in $\mathcal{F}$ is nonempty if and only if $\mathcal{G}$ does not cover M. Therefore, to say that whenever $\mathcal{F}$ has the Finite Intersection Property, the intersection of all the sets of $\mathcal{F}$ is nonempty, is the same as to say that if no finite subfamily of $\mathcal{G}$ covers M, then $\mathcal{G}$ does not cover M. But the latter is the definition of compactness.

6.   DEFINITIONS. A metric space $\langle M, d \rangle$ is **totally bounded** if for every $\epsilon > 0$, M can be covered by finitely many open balls of radius ϵ. A point $x \in M$ is a **point of accumulation** of the subset S of M if every neighborhood of x contains *infinitely many* points in S. Let $s \colon \mathbf{N} \to M$ be a sequence of points in M. Then a **subsequence** of s is any sequence obtained as follows. Let $k \mapsto n_k$ be a function from $\mathbf{N}$ to $\mathbf{N}$ that is *strictly increasing*: $k > j \Rightarrow n_k > n_j$. Then the function $k \mapsto s_{n_k}$ from $\mathbf{N}$ into M is a **subsequence** of s.

EXAMPLE. The sequence $k \mapsto 2^{-k}$ is a subsequence of the sequence $n \mapsto 1/n$ obtained by putting $n_k = 2^k$.

7.   PROPOSITION. *If* $\lim\limits_{n \to \infty} s_n = L$ *and* $k \mapsto s_{n_k}$ *is a subsequence of* s, *then* $\lim\limits_{k \to \infty} s_{n_k} = L$.

Proof. Let $\epsilon > 0$ be given, and choose N so large that $n > N \Rightarrow d(s_n, L) < \epsilon$. Because $k \mapsto n_k$ is a *strictly increasing* function, $n_k \geq k$ for every k. ($n_1 \geq 1$ because n_1 is a positive integer. If $n_k \geq k$, then $n_{k+1} > n_k \geq k$, so $n_{k+1} \geq k + 1$.) Thus if $k > N$, $n_k > N$ so that $d(s_{n_k}, L) < \epsilon$.   ●

8.   THEOREM CHARACTERIZING COMPACT METRIC SPACES. The following statements about the metric space $\langle M, d \rangle$ are equivalent.

(a) $\langle M, d \rangle$ is compact.
(b) $\langle M, d \rangle$ is *complete* and *totally bounded*.
(c)   **Generalized Bolzano-Weierstrass Property:** Every infinite subset of M has a point of accumulation.
(d)   Every sequence of points in M has a subsequence that converges.

Proof. (a) $\Rightarrow$ (c). Suppose $\langle M, d \rangle$ is compact, and that S is a subset of M that has no point of accumulation. Then for *every* $x \in M$, there is a ball $B_{r(x)}(x)$ that contains only finitely many points of S (otherwise x would be a point of accumulation). The family of all such balls is a covering of M by open sets. Therefore M can be covered by finitely many of these balls, each of which contains only finitely many points of S; so S is finite.

(c) $\Rightarrow$ (d). Let $n \mapsto s_n$ be a sequence of points in M.
Case 1. For some N, $s_n = s_N$ for infinitely many values of n. Let n_k be the kth value of n at which $s_n = s_N$. Then $k \mapsto s_{n_k}$ is a *constant* subsequence of s, which converges to s_N.

Case 2. $\{s_n \colon n \in \mathbf{N}\}$ is an infinite set of points in M. Then by the Generalized Bolzano-Weierstrass Property, there is a point of accumulation L of this set. Let n_1 be the smallest integer such that $s_{n_1} \in B_1(L)$. Once we have found $n_1, \ldots, n_k$, let n_{k+1} be the smallest integer n such that $n > n_k$ and $s_n \in B_{1/k+1}(L)$. Then $k \mapsto n_k$ is a strictly increasing function, so $k \mapsto s_{n_k}$ is a subsequence of s and $s_{n_k} \in B_{1/k}(L)$ for every k, so that $\lim_{k \to \infty} s_{n_k} = L$.

(d) $\Rightarrow$ (b). Assume (d) is true. Let s be a Cauchy sequence of points in M. Then s has a subsequence $k \mapsto s_{n_k}$ that converges to a point $L \in M$. Given $\epsilon > 0$, choose N so large that $n, m > N \Rightarrow d(s_n, s_m) < \epsilon/2$. Then choose k so large that $n_k > N$ and $d(s_{n_k}, L) < \epsilon/2$. For any $m > N$, $d(s_m, L) \leq d(s_m, s_{n_k}) + d(s_{n_k}, L) < \epsilon/2 + \epsilon/2 = \epsilon$. Thus $\lim_{n \to \infty} s_n = L$. This shows that M is *complete*.

9. COROLLARY. Let s be a Cauchy sequence in the metric space $\langle M, d \rangle$. If s has a subsequence that converges to L, then s converges to L.

Back to the proof that (d) $\Rightarrow$ (b). Suppose M were not totally bounded. Then there would be an $\epsilon > 0$ such that no finite collection of balls of radius ϵ could cover M. Choose $x_1 \in M$. Then $B_\epsilon(x_1)$ does not cover M, so there is an $x_2 \in M - B_\epsilon(x_1)$. Similarly, there is an x_3 that is in neither $B_\epsilon(x_1)$ or $B_\epsilon(x_2)$. Proceeding in this way, we can choose a sequence $n \mapsto x_n$ of points in M such that $x_n \notin B_\epsilon(x_1) \cup \cdots \cup B_\epsilon(x_{n-1})$. Then, for any $n > m$, $x_n \notin B_\epsilon(x_m) \Rightarrow d(x_n, x_m) \geq \epsilon$. Therefore no subsequence of $n \mapsto x_n$ can be a Cauchy sequence, so no subsequence of $n \mapsto x_n$ can converge. Contradiction.

(b) $\Rightarrow$ (a). LEMMA. If $\langle M, d \rangle$ is totally bounded and $S \subseteq M$, then $\langle S, d \rangle$ is totally bounded.

Proof. Given $\epsilon > 0$, we can choose a covering $B_{\epsilon/2}(x_1), \ldots, B_{\epsilon/2}(x_n)$ of $\langle M, d \rangle$ by finitely many balls of radius $\epsilon/2$. We may suppose that the first m of them are those that intersect S, so that $B_{\epsilon/2}(x_1), \ldots, B_{\epsilon/2}(x_m)$ is a covering of S, and for each $j = 1, \ldots, m$, there is a $y_j \in S \cap B_{\epsilon/2}(x_j)$. But $d(y_j, x_j) < \epsilon/2 \Rightarrow B_{\epsilon/2}(x_j) \subseteq B_\epsilon(y_j) \Rightarrow S \subseteq B_{\epsilon/2}(x_1) \cup \cdots \cup B_{\epsilon/2}(x_m) \subseteq B_\epsilon(y_1) \cup \cdots \cup B_\epsilon(y_m) \Rightarrow S = (S \cap B_\epsilon(y_1)) \cup \cdots \cup (S \cap B_\epsilon(y_m))$. For each j, $S \cap B_\epsilon(y_j)$ is a ball of radius ϵ *in* $\langle S, d \rangle$. Thus S can be covered by finitely many balls of radius ϵ in $\langle S, d \rangle$. ● *(End of Lemma)*

Now suppose $\langle M, d \rangle$ were complete and totally bounded, but not compact. Then there would be an open cover $\mathfrak{F}$ of M that has no finite subcover. Since M is totally bounded, it can be covered by finitely many balls $B_1(x(1, 1)), \ldots, B_1(x(1, n(1)))$ of radius 1. If every one of these could be covered by finitely many of the sets in $\mathfrak{F}$, so could M. Thus there must be one of them, say $B_1(x(1, 1))$ that cannot be covered by a finite subfamily of $\mathfrak{F}$. The metric space $\langle B_1(x(1, 1)), d \rangle$ is itself totally bounded—by the Lemma. A repetition of this argument shows that we can find a ball $B_{1/2}(x(2, 1))$ in M about a point $x(2, 1) \in B_1(x(1, 1))$ such that $B_{1/2}(x(2, 1)) \cap B_1(x(1, 1))$ is a ball of radius $\frac{1}{2}$ in $\langle B_1(x(1, 1)), d \rangle$ that cannot be covered by a finite subfamily of $\mathfrak{F}$. Proceeding in this way, we find a sequence $n \mapsto B_{1/n}(x(n, 1))$ of balls of radius $1/n$ in M such that $\bigcap_{k=1}^{n} B_{1/k}(x(k,\,1))$ cannot be covered by finitely many of the sets in $\mathfrak{F}$ and that $x(n,\,1) \in \bigcap_{k=1}^{n-1} B_{1/k}(x(k,\,1))$ for each n. Then $d(x(n,\,1),\,x(m,\,1)) < 1/m$ if $n \geq m$, so that $n \mapsto x(n,\,1)$ is a Cauchy sequence in $\langle M,\,d \rangle$. Since $\langle M,\,d \rangle$ is complete, it has a limit y. But y is in some set $U \in \mathfrak{F}$. Since U is open, there is an $\epsilon > 0$ such that $B_\epsilon(y) \subseteq U$. Now choose n so large that $1/n < \epsilon/2$ and that $d(x(n,\,1),\,y) < \epsilon/2$. $B_{1/n}(x(n,\,1)) \subseteq B_{\epsilon/2}(x(n,\,1)) \subseteq B_\epsilon(y) \subseteq U$. This shows that $B_{1/n}(x(n,\,1))$ can be covered by a single one of the sets in $\mathfrak{F}$, a contradiction. ●

10. COROLLARY. A compact subset of a metric space is closed.

Proof. If S is a compact subset of $\langle M,\,d \rangle$, then $\langle S,\,d \rangle$ is complete. If $n \mapsto s_n$ is a sequence in S that converges to a point L in M, then $n \mapsto s_n$ is a Cauchy sequence. Because $\langle S,\,d \rangle$ is complete, $\lim_{n \to \infty} s_n \in S$.

11. COROLLARY. A closed subset of a compact metric space is compact.

Proof. Let S be a closed subset of the compact space $\langle M,\,d \rangle$. Let $n \mapsto s_n$ be any sequence in S. Then $n \mapsto s_n$ has a subsequence $k \mapsto s_{n_k}$ that converges to a point $L \in M$. Because S is closed, $L \in S$. Thus $\lim_{k \to \infty} s_{n_k} = L$ in $\langle S,\,d \rangle$.

12. COROLLARY. Let $\langle M,\,d \rangle$ be a compact metric space. Then there is a *countable* family $\mathfrak{F}$ of open balls in M with the property that for every open subset U of M and every $x \in U$, there is a ball B in $\mathfrak{F}$ such that $x \in B \subseteq U$.

Proof. For each n, cover M by finitely many balls of radius $1/n$. Let $\mathfrak{F}$ be the family consisting of all these balls. Then $\mathfrak{F}$ is a countable union of finite families of balls, so $\mathfrak{F}$ is a countable family of balls. If U is open and $x \in U$, then there is an $\epsilon > 0$ such that $B_\epsilon(x) \subseteq U$. Let n be so large that $1/n < \epsilon/2$. Choose a ball B of radius $1/n$ in $\mathfrak{F}$ such that $x \in B$. Then $B = B_{1/n}(y)$ for some $y \in M$, so $d(x,\,y) < 1/n < \epsilon/2$. Thus $B = B_{1/n}(y) \subseteq B_\epsilon(x) \subseteq U$.

13. COROLLARY. Every compact metric space $\langle M,\,d \rangle$ contains a countable subset D such that $D^c = M$.

Proof. Take D to be the set of centers of the balls in the family given in Corollary 12. Then D intersects every neighborhood of every point in M.

14. DEFINITION. A metric space M is said to **satisfy the Second Axiom of Countability** if there is a countable family $\mathfrak{F}$ of open balls in M such that for every point $x \in M$ and every open set U containing x, there is a set $B \in \mathfrak{F}$ such that $x \in B \subseteq U$. A subset D of a metric space M is **dense** if $D^c = M$. A metric space is **separable** if it has a countable dense subset. ●

It has just been shown that every compact metric space is separable and satisfies the Second Axiom of Countability.

Exercises

1. Show that every finite metric space is compact.
2. Let M be a metric space, and s a sequence in M converging to L. Show that the subset S of M consisting of L and the points $s_1, s_2, s_3, \ldots$ is compact.
3. Let $\langle M, d \rangle$ be a metric space satisfying the Second Axiom of Countability, and let S be an uncountable subset of M. Show that all but countably many of the points of S are points of accumulation of S. *Hint:* Let $\mathfrak{F}$ be a countable family of open balls in M with the property described in Definition 14. Show that if $x \in S$ but x is not an accumulation point of S, then there is an open subset U_x of M such that $U_x \cap S$ contains only x. Choose a ball B_x in $\mathfrak{F}$ such that $x \in B_x \subseteq U_x$. Show that the function $x \mapsto B_x$ is one-to-one.
*4. Show that any metric space that satisfies the Second Axiom of Countability is separable, and conversely. *Hint:* If D is a countable dense subset of M, look at the family of balls $B_{1/n}(d)$ for $d \in D$.
*5. Show that the metric space $\langle \mathbf{R}, \text{usual metric} \rangle$ is *not* compact, but that it is separable. *Hint:* $\mathbf{Q}$ is dense in $\mathbf{R}$.

B. Continuous functions, product of compact spaces, compact subsets of $\mathbf{R}^n$.

One of the most useful facts about compact sets is the following:

1. THEOREM. Let f be a continuous function from the metric space $\langle M, d \rangle$ to the metric space $\langle N, e \rangle$. If S is a compact subset of M, then $f(S)$ is a compact subset of N.

Proof. Let $\mathfrak{F}$ be a family of open sets in N that covers $f(S)$. Then $\{f^{-1}(U): U \in \mathfrak{F}\}$ is a family of open subsets of M that covers S. Since S is compact, a finite subfamily $f^{-1}(U_1), \ldots, f^{-1}(U_k)$ covers S. That is to say, $U_1, \ldots, U_k$ is a finite subfamily of $\mathfrak{F}$ that covers $f(S)$.

2. THEOREM. For each $i = 1, \ldots, n$, let $\langle M_i, d_i \rangle$ be a compact metric space. Then the product space $\langle M_1 \times \cdots \times M_n, d_1 \times \cdots \times d_n \rangle$ is also compact.

Proof. Put $M = M_1 \times \cdots \times M_n$, $d = d_1 \times \cdots \times d_n$. We shall prove that every sequence in M has a convergent subsequence. Let $s: m \mapsto s_m$ be a sequence of points in M. Then for each j, the jth component $\pi_j(s)$ lies in M_j. $m \mapsto \pi_j(s_m)$ is thus a sequence in M_j. Because $\langle M_1, d_1 \rangle$ is compact, there is a monotonically increasing function $f_1: \mathbf{N} \to \mathbf{N}$ such that $\pi_1 \circ s \circ f_1: \mathbf{N} \to M_1$ is a convergent subsequence of the sequence $\pi_1 \circ s: m \mapsto \pi_1(s_m)$. Likewise, the sequence $\pi_2 \circ s \circ f_1: \mathbf{N} \to M_2$ has a convergent subsequence $\pi_2 \circ s \circ f_1 \circ f_2$, where f_2 is a monotonically increasing function from $\mathbf{N}$ to $\mathbf{N}$. Proceeding in this way, we successively extract subsequences $s \circ f_1$ of s, $s \circ f_1 \circ f_2$ of $s \circ f_1$, $\ldots$, $s \circ f_1 \circ f_2 \circ \cdots \circ f_n$ of $s \circ f_1 \circ f_2 \circ \cdots \circ f_{n-1}$ such that $\pi_i \circ s \circ f_1 \circ \cdots \circ f_i$ converges in M_i for each $i = 1, \ldots, n$. Say $\lim_{m \to \infty} \pi_i \circ s \circ f_1 \circ \cdots \circ f_i(m) = L_i$. Since

$\pi_i \circ s \circ f_1 \circ \cdots \circ f_n$ is evidently a subsequence of $\pi_i \circ s \circ f_1 \circ \cdots \circ f_i$ ($f_{i+1} \circ \cdots \circ f_n$ is a monotonically increasing function from $\mathbf{N}$ to $\mathbf{N}$), $\lim\limits_{m \to \infty} \pi_i \circ s \circ f_1 \circ \cdots \circ f_n$ equals L_i as well (Proposition VII A 7). Thus for each $i = 1, \ldots, n$, $\lim\limits_{m \to \infty}$ $\pi_i \circ s \circ f_1 \circ \cdots \circ f_n = L_i$. It follows that $\lim\limits_{m \to \infty} s \circ f_1 \circ \cdots \circ f_n = (L_1, \ldots, L_n) \in$ $M_1 \times \cdots \times M_n$. This shows that s has a convergent subsequence. ●

This theorem allows us very easily to characterize all the compact subsets of $\mathbf{R}^n$.

3.   THEOREM. For a subset S of the metric space $\langle \mathbf{R}^n, \|\ \|_\infty \rangle$ to be compact, it is necessary and sufficient that S be *closed* and *bounded*.

Proof. Suppose S is a compact subset of $\mathbf{R}^n$. Then S is closed by Corollary A 10. Moreover, S is totally bounded, so it can be covered by finitely many balls of radius 1, say $B_1(x_1), \ldots, B_1(x_k)$. Let B be a bound for the finite set $\{\|x_1\|_\infty, \ldots, \|x_k\|_\infty\}$. Then $B + 1$ is a bound for $\{\|x\|_\infty : x \in S\}$. In fact, $x \in S \Rightarrow x \in B_1(x_j)$ for some $j \Rightarrow \|x\|_\infty \leq \|x_j\|_\infty + \|x - x_j\|_\infty < B + 1$.

Conversely, suppose S is closed and bounded by B. Then for every $x \in S$, $\|x\|_\infty \leq B$. Therefore, x lies in the product $[-B, B] \times \cdots \times [-B, B]$ of the closed intervals $[-B, B]$ in $\mathbf{R}$. ($\|x\|_\infty \leq B \Rightarrow |x_j| \leq B$ for each $j = 1, \ldots, n \Rightarrow$ $x_j \in [-B, B]$ for each $j = 1, \ldots, n$.) According to the Heine-Borel Theorem A 2, the metric space $\langle [-B, B], \text{usual metric} \rangle$ is compact. Therefore, so is the product space. Now the product metric is $d((x_1, \ldots, x_n), (y_1, \ldots, y_n)) = \max\limits_{1 \leq j \leq n} |x_i - y_i| = \|x - y\|_\infty$. That is, the metric on the product space is just the metric that $[-B, B]^n$ gets as a subspace of $\mathbf{R}^n$. By hypothesis, S is a closed subset of $\mathbf{R}^n$ that is contained in $[-B, B]^n$. Therefore, S is actually a closed subset of $[-B, B]^n$. Moreover, the metric that S gets as a subspace of the compact product space $[-B, B]^n$ is just $\|\ \|_\infty$, namely, the metric that S gets as a subset of $\mathbf{R}^n$. But we have shown (Corollary A 11) that a closed subset of a compact space is compact. Since $S \subseteq [-B, B]^n$, S is compact.

4.   LEMMA. A closed bounded nonempty subset of $\mathbf{R}$ contains its lub and its glb.

Proof. If S is a closed bounded nonempty subset of $\mathbf{R}$, and $B = \text{lub } S$, it follows from Theorem II C 1 that every neighborhood of B intersects S. Therefore, $B \in S^c = S$. Similarly for the glb.

5.   THEOREM. A continuous function from a nonempty compact metric space into $\mathbf{R}$ attains its lub and its glb. In other words, if f maps the nonempty compact space $\langle M, d \rangle$ into $\mathbf{R}$, there are points x and y in M such that $f(x) = \text{glb } \{f(z) : z \in M\}$ and $f(y) = \text{lub } \{f(z) : z \in M\}$.

Proof. $f(M)$ is a compact subset of **R** by Theorem 1. According to Theorem 3, $f(M)$ is nonempty, closed, and bounded. The lemma then shows that glb $f(M)$ and lub $f(M)$ are in $f(M)$. ●

As an application, we get the following important fact about finite-dimensional normed spaces. Since every finite-dimensional vector space over **R** is isomorphic to $\mathbf{R}^n$ for some n, we can confine our attention to $\mathbf{R}^n$.

6.   THEOREM. All norms on $\mathbf{R}^n$ are equivalent to the uniform norm in the following sense: if $|\;|$ is a norm on $\mathbf{R}^n$, there are *positive* constants b and B such that $b||x||_\infty \leq |x| \leq B||x||_\infty$ for every $x \in \mathbf{R}^n$. Consequently, a sequence $n \mapsto x_n$ converges with respect to $|\;|$ if, and only if, it converges with respect to $||\;||_\infty$. In particular, the same subsets of $\mathbf{R}^n$ are closed with respect to $|\;|$ and with respect to $||\;||_\infty$.

Proof. Let $e_1, \ldots, e_n$ be the usual basis for $\mathbf{R}^n$ and put $B = |e_1| + \cdots + |e_n|$.

$$(e_i = (0, \ldots, 0, \overset{i\text{th place}}{1}, 0, \ldots, 0)). \text{ Then}$$

$$|x| = |x_1e_1 + \cdots + x_ne_n| \leq |x_1|\,|e_1| + \cdots + |x_n|\,|e_n|$$

$$\leq ||x||_\infty |e_1| + \cdots + ||x||_\infty |e_n| = ||x||_\infty B.$$

In particular, since $|\,|x| - |y|\,| \leq |x - y| \leq B||x - y||_\infty$, $|\;|$ is a continuous function on $\langle \mathbf{R}^n, ||\;||_\infty \rangle$. The set $S_1(0) = \{x \in \mathbf{R}^n : ||x||_\infty = 1\}$ (the sphere of radius 1 about 0 in $\langle \mathbf{R}^n, ||\;||_\infty \rangle$) is closed and bounded in $\langle \mathbf{R}^n, ||\;||_\infty \rangle$, so it is compact. Therefore, the continuous function $|\;|$ attains its minimum b at some point x_0 of $S_1(0)$. Since $||x_0||_\infty = 1$, $x_0 \neq 0$. Therefore $b = |x_0| > 0$, because $|\;|$ is also a norm. The relation $b||x||_\infty \leq |x|$ is surely true when $x = 0$. If $x \neq 0$, then $(1/||x||_\infty)x$ is in $S_1(0)$. Since b is the minimum of $|\;|$ on $S_1(0)$, $b \leq |(1/||x||_\infty)x| = (1/||x||_\infty)|x| \Rightarrow b||x||_\infty \leq |x|$.

Suppose $|x_n - L| \to 0$ as $n \to \infty$. Then $||x_n - L||_\infty \leq (1/b)|x_n - L| \to 0$ as $x \to \infty$. Likewise, if $||x_n - L||_\infty \to 0$ as $n \to \infty$, then $|x_n - L| \leq B||x_n - L||_\infty \to 0$ as $n \to \infty$. Therefore a sequence converges in $\langle \mathbf{R}^n, ||\;||_\infty \rangle$ if, and only if, it converges in $\langle \mathbf{R}^n, |\;| \rangle$. Since closed sets are defined in terms of convergence of sequences, and the same sequences are convergent with respect to the two norms, it follows that the same sets are closed.

7.   COROLLARIES. If $||\;||_A$ and $||\;||_B$ are two norms on $\mathbf{R}^n$, then $||\;||_A$ is a continuous function on $\langle \mathbf{R}^n, ||\;||_B \rangle$, $\langle \mathbf{R}^n, ||\;||_B \rangle$ is complete, a subset of $\langle \mathbf{R}^n, ||\;||_B \rangle$ is compact if and only if it is closed and bounded, $\langle \mathbf{R}^n, ||\;||_A \rangle$ and $\langle \mathbf{R}^n, ||\;||_B \rangle$ have the same open sets, and so on and so forth.

Proof. Obvious. Openness, closure, Cauchyness, and the rest can all be reduced to the corresponding properties in $\langle \mathbf{R}^n, ||\;||_\infty \rangle$ via the preceding theorem.

Exercises

1. Show that for any $x \in \mathbf{R}^n$, $\|x\|_\infty \leq \|x\|_2 \leq \sqrt{n}\|x\|_\infty$.

2. Show by example that it is *not* true that any *metric* on $\mathbf{R}^n$ determines the same convergent sequences as $\|\ \|_\infty$.

*3. Let $\langle M, d \rangle$ be a *compact* metric space, $\langle N, e \rangle$ a metric space, and f a *continuous* function mapping M *one-to-one onto* N. Then there is an inverse function $f^{-1} \colon N \to M$. Show that f^{-1} is continuous. *Hint:* Put $g = f^{-1}$. To show that g is continuous, it must be shown that the inverse image under g of a closed set S in M is closed in N (Theorem IV D 3). But (prove this) a subset S of M is closed if and only if it is compact, and $g^{-1}(S) = f(S)$.

4. Show that the theorem proved in Exercise 3 is false without the condition that M be compact. For example, let M be the half-open interval $[0, 2\pi[= \{x \colon 0 \leq x < 2\pi\}$ with the usual topology, let N be the unit circle in $\langle \mathbf{C}, |\ | \rangle$, and let $f(x) = e^{ix}$. Show that f^{-1} is discontinuous at $e^0 = e^{2\pi i} = 1$, although f is continuous and maps $[0, 2\pi[$ one-to-one onto the unit circle in $\mathbf{C}$.

5. Consider the infinite-dimensional complete normed space of real-valued continuous functions on $[0, 1]$ with the uniform norm $\|\ \|_\infty$. Its closed unit ball $B_1^c(0)$ is closed and bounded. Show that $B_1^c(0)$ is not compact. *Hint:* Let f_n be the function defined by

$$f_n(x) = 0 \quad \text{if} \quad 0 \leq x \leq 2^{-n},$$

$$f_n(x) = 2^{n+1}(x - 2^{-n}) \quad \text{if} \quad 2^{-n} \leq x \leq 2^{-n} + 2^{-(n+1)},$$

$$f_n(x) = 1 - 2^{n+1}(x - 2^{-n} - 2^{-(n-1)})$$

if $2^{-n} + 2^{-(n+1)} \leq x \leq 2^{-n} + 2 \times 2^{-(n+1)} = 2^{-n} + 2^{-n} = 2 \times 2^{-n} = 2^{-(n-1)}$,

$$f_n(x) = 0 \quad \text{if} \quad x \geq 2^{-(n-1)}.$$

Figure 25 illustrates f_2.

    Show that $\|f_n - f_m\|_\infty = 1$ for every $n \neq m$. Conclude that the infinite set $\{f_n\}$ can have no accumulation point, or that the sequence $n \mapsto f_n$ can have no convergent subsequence, although $f_n \in B_1^c(0)$ for every n.

6. Show that there can be no continuous map from a closed disk $B_r^c(x) \subseteq \mathbf{R}^2$ onto the whole plane $\mathbf{R}^2$.

C.   Equicontinuity and uniform continuity

    The $\delta - \epsilon$ definition of continuity of a function says that f is continuous at x if for every $\epsilon > 0$, there is a $\delta > 0$ such that $d(f(y), f(x)) < \epsilon$ if $d(y, x) < \delta$. Ordinarily the choice of δ depends not only on ϵ but also on f and on x. (If δ were independent of ϵ, f would have to be *constant* near x.) For a number of purposes, it is important to be able to choose one δ that works for all x in a certain set (in which case f is said to be *uniformly continuous* on that set), or for all functions f in a certain family of functions $\mathfrak{F}$ (in which case $\mathfrak{F}$ is said to be *equicontinuous* at x), or both (in which case $\mathfrak{F}$ is *uniformly equicontinuous*). Such arguments are important in proving that continuous functions are integrable, and in proving the existence of solutions to differential equations, as we shall see below.

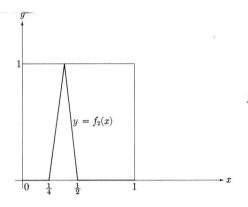

Figure 25

1.   DEFINITION. Let $\langle M, d \rangle$ and $\langle N, e \rangle$ be metric spaces, and let S be a subset of M. A function $f: M \to N$ is **uniformly continuous on S** if for every $\epsilon > 0$, there is a $\delta > 0$ such that $e(f(y), f(x)) < \epsilon$ whenever $d(y, x) < \delta$, *for all $x \in S$*. A family $\mathfrak{F}$ of functions from M to N is **equicontinuous** at the point $x \in M$ if for every $\epsilon > 0$, there is a $\delta > 0$ such that $e(f(y), f(x)) < \epsilon$ whenever $d(y, x) < \delta$, *for all $f \in \mathfrak{F}$*. $\mathfrak{F}$ is **uniformly equicontinuous on S** if for every $\epsilon > 0$, there is a $\delta > 0$ such that $e(f(y), f(x)) < \epsilon$ whenever $d(y, x) < \delta$, *for all $x \in S$ and all $f \in \mathfrak{F}$*.

2.   THEOREM. Let $\langle M, d \rangle$ and $\langle N, e \rangle$ be metric spaces, S a *compact* subset of M, and $\mathfrak{F}$ a family of functions from M to N that is equicontinuous on S. Then $\mathfrak{F}$ is uniformly equicontinuous on S.

Proof. Let $\epsilon > 0$ be given. For each $x \in S$, there is a $\delta(x)$ such that $d(y, x) < \delta(x) \Rightarrow e(f(y), f(x)) < \epsilon/2$ for all $f \in \mathfrak{F}$, by hypothesis. Because S is compact, we can extract a finite subcovering from the covering of S by the open balls $B_{\delta(x)/2}(x)$. Suppose S is covered by the balls $B_{\delta(x(i))/2}(x(i))$ for $i = 1, \ldots, n$. Let $2\delta = \min \{\delta(x(i)): i = 1, \ldots, n\}$. Suppose $x \in S$ and that $d(y, x) < \delta$. Then x is in one of the balls $B_{\delta(x(i))/2}(x(i))$ because they cover S, and therefore y and x are both in the ball $B_{\delta(x(i))}(x(i))$. It follows that $e(f(y), f(x)) \leq e(f(y), f(x(i)) + e(f(x(i)), f(x)) < \epsilon/2 + \epsilon/2 = \epsilon$, for any $f \in \mathfrak{F}$.

3.   COROLLARY. If f is a continuous function on the compact metric space $\langle M, d \rangle$, then f is uniformly continuous.

Proof. Apply the preceding theorem to the set $S = M$ and to the family $\mathfrak{F}$ that consists of the single function f. Because f is continuous and there is no other function in $\mathfrak{F}$, $\mathfrak{F}$ is automatically equicontinuous.

4.   THEOREM. Let $\langle M, d \rangle$ and $\langle N, e \rangle$ be metric spaces, and let M be compact. Let $\mathfrak{F}$ be an equicontinuous family of functions from M into N. If $n \mapsto f_n$ is a

sequence of functions in $\mathfrak{F}$ that converges *pointwise* to the limiting function L, then $n \mapsto f_n$ converges to L uniformly. In particular, L is continuous.

Proof. Let $\epsilon > 0$ be given. Because $\mathfrak{F}$ is equicontinuous, it is uniformly equicontinuous by Theorem 2. There is then a $\delta > 0$ such that $d(x, y) < \delta \Rightarrow e(f(x), f(y)) < \epsilon/3$ for all $f \in \mathfrak{F}$. Since M is compact, it is totally bounded, so it can be covered by finitely many balls of radius δ, $B_\delta(x_1), \ldots, B_\delta(x_n)$. Since $\lim_{k \to \infty} f_k(x_i) = L(x_i)$ for each $i = 1, \ldots, n$, we can choose K so large that $e(f_k(x_i), f_j(x_i)) < \epsilon/3$ for each $i = 1, \ldots, n$ whenever $j, k > K$. Then, if j, $k > K$ and $x \in M$, there is an i such that $x \in B_\delta(x_i)$, because the balls $B_\delta(x_1), \ldots, B_\delta(x_n)$ cover M. Then $e(f_j(x), f_k(x)) \leq e(f_j(x), f_j(x_i)) + e(f_j(x_i), f_k(x_i)) + e(f_k(x_i), f_k(x)) < \epsilon/3 + \epsilon/3 + \epsilon/3 = \epsilon$. This shows that $n \mapsto f_n$ is a uniformly Cauchy sequence. Moreover, for any $x \in M$ and any $k > K$, $e(f_k(x), L(x)) = \lim_{j \to \infty} e(f_k(x), f_j(x)) \leq \epsilon$, so $k \mapsto f_k$ converges uniformly to L.

Being the uniform limit of continuous functions, L is continuous.

5.   PROPOSITION. Let $\langle M, d \rangle$ and $\langle N, e \rangle$ be metric spaces with M compact, and let $\mathfrak{F}$ be an equicontinuous family of functions from M to N. Then the closure $\mathfrak{F}^c$ of $\mathfrak{F}$ in $\mathcal{C}(M, N)$ is also equicontinuous.

Proof. Take any $x \in M$ and any $\epsilon > 0$. Because $\mathfrak{F}$ is equicontinuous, there is a $\delta > 0$ such that $e(f(y), f(x)) < \epsilon/2$ whenever $d(y, x) < \delta$, for every $f \in \mathfrak{F}$. Now if $g = \lim_{n \to \infty} f_n$ is in $\mathfrak{F}^c$, then when $d(y, x) < \delta$, $e(g(y), g(x)) = \lim_{n \to \infty} e(f_n(y), f_n(x)) \leq \epsilon/2 < \epsilon$.   ●

We now have several means by which a subset of a metric space can be recognized to be compact, and by which new compact metric spaces can be constructed from old. We know that closed subsets and products of compact spaces are compact, as is the image of a compact space under a continuous function. We have further characterized the compact subsets of $\langle \mathbf{R}^n, \| \ \| \rangle$ for any norm $\| \ \|$ (since all norms on $\mathbf{R}^n$ are equivalent). The next theorem permits us to recognize a new class of compact sets—the compact subsets of the space of continuous functions from one metric space to another.

6.   DEFINITION. Let $\langle M, d \rangle$ and $\langle N, e \rangle$ be metric spaces, and let M be compact. $\mathcal{C}(M, N)$ is the space of all continuous functions from M to N. Because M is compact, if $f, g \in \mathcal{C}(M, N)$, $e(f(x), g(x))$ is a continuous real-valued function that attains its maximum $e_\infty(f, g)$ on M. e_∞ is a metric on $\mathcal{C}(M, N)$ (see Exercise 7), the **uniform metric.**

7.   ARZELA-ASCOLI THEOREM. Let $\langle M, d \rangle$ and $\langle N, e \rangle$ be metric spaces, and let M be compact. For a family $\mathfrak{F}$ of continuous functions from M to N to be a compact subset of the metric space $\langle \mathcal{C}(M, N), e_\infty \rangle$, it is necessary and sufficient that $\mathfrak{F}$ fulfill the following three conditions.

(a) $\mathfrak{F}$ is closed in $\langle \mathcal{C}(M, N), e_\infty \rangle$.
(b) $\mathfrak{F}$ is equicontinuous on M.
(c) There is a single compact subset S of N in which all the functions in $\mathfrak{F}$ take their values; that is, $\{f(x) : f \in \mathfrak{F}, \ x \in M\}$ must be contained in a compact subset S of N.

Proof. Suppose $\mathfrak{F}$ is a compact subset of $\mathcal{C}(M, N)$. Then $\mathfrak{F}$ is closed according to Corollary A 10. Let $\epsilon > 0$ be given. Then $\mathfrak{F}$ can be covered by finitely many balls of radius $\epsilon/3$ in $\mathcal{C}(M, N)$, say $B_{\epsilon/3}(f_1), \ldots, B_{\epsilon/3}(f_n)$. Let $x \in M$. Then we can find a $\delta > 0$ such that $d(y, x) < \delta \Rightarrow (f_i(y), f_i(x)) < \epsilon/3$ for each of the *finitely many* functions $f_1, \ldots, f_n$. Now take any $f \in \mathfrak{F}$, and any $y \in M$ such that $d(y, x) < \delta$. Then there is an $i \in \{1, \ldots, n\}$ such that $e_\infty(f, f_i) < \epsilon/3$. $e(f(y), f(x)) \leq e(f(y), f_i(y)) + e(f_i(y), f_i(x)) + e(f_i(x), f(x)) \leq e_\infty(f, f_i) + e(f_i(y), f_i(x)) + e_\infty(f, f_i) < \epsilon/3 + \epsilon/3 + \epsilon/3 = \epsilon$. This proves (b). Finally, consider the product metric space $M \times \mathfrak{F}$. There is a natural **evaluation function** Eval: $M \times \mathfrak{F} \to N$ defined by Eval $(x, f) = f(x)$ for each pair $\langle x, f \rangle$ with $x \in M$ and $f \in \mathfrak{F}$. I claim that Eval is continuous. In fact, if $x_0 \in M$, $f_0 \in \mathfrak{F}$, and $\epsilon > 0$ are given, we can choose $\eta > 0$ so that $d(y, x_0) < \eta \Rightarrow e(f_0(y), f_0(x_0)) < \epsilon/2$. Let $\delta = \min(\eta, \epsilon/2)$. Then if $\max(d(y, x_0), e_\infty(f, f_0)) < \delta$, we have $e(\text{Eval }(y, f), \text{Eval }(x_0, f_0)) = e(f(y), f_0(x_0)) \leq e(f(y), f_0(y)) + e(f_0(y), f_0(x_0)) \leq e_\infty(f, f_0) + e(f_0(y), f_0(x_0)) < \epsilon/2 + \epsilon/2 = \epsilon$. (Note that $\max(d(y, x_0), e_\infty(f, f_0))$ is the distance between the pairs (y, f) and (x_0, f_0) with respect to the product metric on $M \times \mathfrak{F}$.) Now $M \times \mathfrak{F}$ is the product of two compact metric spaces, so it is compact. Therefore, its image under the continuous function Eval is also compact. That is, $S = \{\text{Eval }(x, f) : (x, f) \in M \times \mathfrak{F}\} = \{f(x) ; x \in M, f \in \mathfrak{F}\}$ is compact. This proves (c), and more—namely, that if $\mathfrak{F}$ is compact, $\{f(x) : x \in M, f \in \mathfrak{F}\}$ not only is contained in a compact set, but that it is itself compact.

To prove the converse, let us suppose that $\mathfrak{F}$ fulfills the conditions (a), (b), and (c). We shall show that $\mathfrak{F}$ is complete and totally bounded. First of all, because all the functions in $\mathfrak{F}$ take their values in S, $\mathfrak{F}$ is actually a closed subset of the metric space $\langle \mathcal{C}(M, S), e_\infty \rangle$, which is a subspace of $\langle \mathcal{C}(M, N), e_\infty \rangle$. Because S is compact, S is complete. Therefore $\langle \mathcal{C}(M, S), e_\infty \rangle$ is complete as well (for a proof, see section VI A, where a proof is given under the additional, and unnecessary, assumption that S is a vector space and e is a norm on S). Since $\mathfrak{F}$ is a closed subset of a complete space, it also is complete (Theorem IV C 10). It remains only to show that $\mathfrak{F}$ is totally bounded. Let $\epsilon > 0$ be given. Because $\mathfrak{F}$ is equicontinuous on the compact metric space M, it is uniformly equicontinuous. Therefore there is a $\delta > 0$ such that $e(f(x), f(y)) < \epsilon/3$ whenever $d(x, y) < \delta$, for all $f \in \mathfrak{F}$. Because M is totally bounded, it can be covered by finitely many open balls of radius δ, say $B_\delta(x_1), \ldots, B_\delta(x_n)$. Now consider the product space

$$\langle S, e \rangle^n = \underbrace{\langle S, e \rangle \times \cdots \times \langle S, e \rangle}_{n \text{ factors}}.$$

Because S is compact, $\langle S, e \rangle^n$ is also compact, and in particular is totally bounded. Now consider the subset Φ of $\langle S, e \rangle^n$ defined by $\Phi = \{(f(x_1), \ldots, f(x_n)): f \in \mathfrak{F}\}$. Being a subset of a totally bounded space, Φ also is totally bounded (see the Lemma in the proof of Theorem A 8). Therefore there exist finitely many balls of radius $\epsilon/3$ in $\langle S, e \rangle^n$ that cover Φ, say $B_{\epsilon/3}((f_1(x_1), \ldots, f_1(x_n))), \ldots, B_{\epsilon/3}((f_k(x_1), \ldots, f_k(x_n))$, where for $j = 1, \ldots, k$, $f_j \in \mathfrak{F}$. What is such a ball in Φ? It is just the collection of points $(f(x_1), \ldots, f(x_n))$ in Φ such that $\max \{e(f(x_i), f_j(x_i)): i = 1, \ldots, n\} < \epsilon/3$. Now I claim that the balls of radius ϵ in $\mathcal{C}(M, S)$ about the points $f_1, \ldots, f_k$ cover $\mathfrak{F}$. In fact, if $f \in \mathfrak{F}$, then there is a $j \in \{1, \ldots, k\}$ such that $(f(x_1), \ldots, f(x_n))$ lies in the ball of radius $\epsilon/3$ about $(f_j(x_1), \ldots, f_j(x_n))$ in Φ. To show that $e_\infty(f, f_j) < \epsilon$, take any $x \in M$. Then x lies in one of $B_\delta(x_1), \ldots, B_\delta(x_n)$, say $x \in B_\delta(x_i)$. Then $e(f(x), f_j(x)) \le e(f(x), f(x_i)) + e(f(x_i), f_j(x_i)) + e(f_j(x_i), f_j(x))$. The first and third of these terms are less than $\epsilon/3$ by the choice of δ. The middle one is $e(f(x_i), f_j(x_i)) \le \max \{e(f(x_q), f_j(x_q)): q = 1, \ldots, n\} < \epsilon/3$ because $(f(x_1), \ldots, f(x_n))$ lies in the ball of radius $\epsilon/3$ about $(f_j(x_1), \ldots, f_j(x_n))$ in Φ.

8.   COROLLARY. Let $\langle M, d \rangle$ be a compact metric space. For a family $\mathfrak{F}$ of functions from M into $\langle \mathbf{R}^n, \|\ \|_\infty \rangle$ to be compact in the Banach space $\langle \mathcal{C}(M, \mathbf{R}^n), |\ |_\infty \rangle$ it is necessary and sufficient that $\mathfrak{F}$ fulfill the following three conditions.

(a) $\mathfrak{F}$ is closed in $\langle \mathcal{C}(M, \mathbf{R}^n), |\ |_\infty \rangle$.
(b) $\mathfrak{F}$ is equicontinuous on M.
(c)   There is a uniform bound B for the functions in $\mathfrak{F}$: $\|f(x)\|_\infty \le B$ for all $x \in M$ and $f \in \mathfrak{F}$.

Proof. The only thing at stake is to show that condition (c) of the Corollary is equivalent in this case to condition (c) of Theorem 7. But Theorem B 3 shows that a subset of $\langle \mathbf{R}^n, \|\ \|_\infty \rangle$ is contained in a compact subset of $\mathbf{R}^n$ if, and only if, it is bounded. (Clearly, the closure of a bounded set is bounded.) ●

As an application of the Arzela-Ascoli Theorem, I shall prove an existence theorem for ordinary differential equations under weaker hypotheses than those of Picard's Theorem. In particular, I shall prove that the initial value problem $f'(x) = \Phi(x, f(x))$, $f(x_0) = y_0$ has a solution if Φ is merely required to be continuous. Mere continuity is *not*, of course, sufficient to guarantee uniqueness of the solution, as is shown by the old example $\Phi(x, y) = 3y^{2/3}$. The initial value problem $f'(x) = [f(x)]^{2/3}$, $f(0) = 0$ has the two solutions $f(x) = x^3$ and $f(x) = 0$. That is, we get existence without uniqueness.

The idea behind the proof of the forthcoming existence theorem is this: take a simple numerical method for the approximate computation of solutions to a differential equation whose error depends on a *step length* h, and show that as $h \to 0$ the approximate solutions actually converge to a solution.

The numerical method to be used is motivated by the physical interpre-

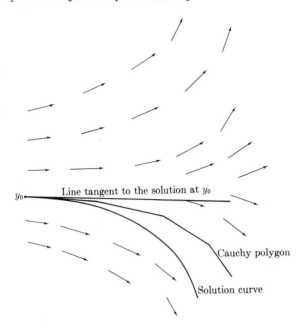

Figure 26

tation of a differential equation offered in Figure 21 in Section VI E. A solution to the differential equation $f'(x) = \Phi(f(x))$ is a curve that is tangent to the velocity vector attached to each point through which the solution passes (Figure 26). We obtain an approximate solution by approximating the solution passing through y_0 by its tangent line, the line through y_0 in the direction of the velocity vector at y_0. Of course, as the solution moves away from y_0, the direction of the velocity vector changes. The farther we progress along the tangent line, the more error there is in the approximation, and presumably, the more the direction of the tangent line departs from the direction of the true solution curve.

We can improve the accuracy of the approximation by stopping from time to time to observe which way the velocity vectors are pointing. In this way, we get a **Cauchy polygon approximation** *with mesh size* (or *step length*) h, defined as follows. To get an approximate solution $a(x)$ to the initial value problem $f'(x) = \Phi(x, f(x))$, $f(x_0) = y_0$, we put $a(x) = y_0 + (x - x_0)\Phi(x_0, y_0)$ for $x_0 \leq x \leq x_0 + h = x_1$. Thus $a(x)$ describes a segment of the tangent line to the true solution at (x_0, y_0). Next, we use the tangent line to the true solution at (x_1, y_1), where $y_1 = a(x_1)$, namely $a(x) = a(x_1) + (x - x_1)\Phi(x_1, y_1)$ for $x_1 \leq x \leq x_2 = x_1 + h$. And so on. What we might hope is that as $h \to 0$ the Cauchy polygons would converge to a true solution. This is actually the case if Φ satisfies a Lipschitz condition. Even without a Lipschitz condition, however, it is not hard to show that from every sequence of Cauchy polygons with mesh sizes decreasing to 0, a subsequence can be extracted that converges to a true solution. We now proceed to do so.

Suppose $\Phi\colon [x_0,\, \bar{x}] \times B_r^c(y_0) \to \mathbf{R}^n$. Let us define a finite sequence of points $k \mapsto x_k(h)$ in $[x_0,\, \bar{x}]$ by putting $x_k(h) = x_0 + kh$ for each integer k such that $x_0 + kh \in [x_0,\, \bar{x}]$. Let us also define a finite sequence $k \mapsto y_k(h)$ of points in $B_r^c(y_0)$ as well as an approximate solution $x \mapsto A(h,\, x)$ to the DE $f'(x) = \Phi(x,\, f(x))$ by putting $y_0(h) = y_0$, and, for every k such that $(x_k(h),\, y_k(h))$ is in the set $[x_0,\, \bar{x}] \times B_r^c(y_0)$, putting $A(h,\, x) = y_k(h) + (x - x_k(h))\Phi(x_k(h),\, y_k(h))$ when $x_k(h) \leq x \leq x_k(h) + h$ and $y_{k+1}(h) = A(h,\, x_k(h) + h)$. We should remark now that $A(h,\, x)$ can be expressed as an integral, using the *discontinuous step function* $S(h,\, x)$, defined as follows: if $x_k(h) \leq x < x_k(h) + h$ and if $y_k(h)$ is defined, we put $S(h,\, x) = y_k(h)$. Observe that on the interval $x_k(h) < x < x_{k+1}(h)$, $A(h,\, x)$ has a derivative with respect to x constantly equal to $\Phi(x_k(h),\, y_k(h)) = \Phi(x_k(h),\, S(h,\, x)) = \Phi(x^*(h,\, x),\, S(h,\, x))$, where $x^*(h,\, x)$ is defined to be $x_k(h)$ if $x_k(h) \leq x < x_{k+1}(h)$. It is then clear that $A(h,\, x) = y_0 + \int_{x_0}^{x} (x^*(h,\, t),\, S(h,\, t))\, dt$.

9.   PEANO'S EXISTENCE THEOREM FOR ODE's. Let Φ be a continuous function from $[x_0,\, \bar{x}] \times B_r^c(y_0)$ into the ball of radius M about 0 in $\langle \mathbf{R}^n,\, ||\ ||_\infty \rangle$, $B_M^c(0)$. Let ϵ be the smaller of the numbers $\bar{x} - x_0$ and r/M. Then there is a differentiable function f from $[x_0,\, x_0 + \epsilon]$ into $B_r^c(y_0)$ such that $f'(x) = \Phi(x,\, f(x))$ and $f(x_0) = y_0$.

Remark. Since $[x_0,\, \bar{x}] \times B_r^c(y_0)$ is compact, $||\Phi||_\infty$ is automatically bounded by some constant, M (Theorem B 5). M is mentioned in the statement of the theorem only to facilitate the description of ϵ.

Proof. We consider the approximate solutions $x \mapsto A(h,\, x)$ defined above. The first step is to show that so long as $x_0 \leq x \leq x_0 + \epsilon$, $A(h,\, x)$ stays in the ball $B_r^c(y_0)$, so that $\Phi(x,\, A(h,\, x))$ is defined. Since $A(h,\, x)$ is defined inductively, we must make a proof by induction. It is clear that $y_0 \in B_r^c(y_0)$. Suppose that $x_0 \leq x \leq x_0 + \epsilon$ and, for the induction hypothesis, that $x_k(h) \leq x \leq x_k(h) + h$ and that $(x_k(h),\, y_k(h)) \in [x_0,\, x_0 + \epsilon] \times B_r^c(y_0)$. We shall need to use the following fact: if $A(h,\, x)$ and $A(h,\, x')$ are defined, then $||A(h,\, x) - A(h,\, x')||_\infty \leq M|x - x'|$. That is, $x \mapsto A(h,\, x)$ satisfies a Lipschitz condition with constant M. This follows from the representation of $A(h,\, \cdot)$ as an integral:

$$||A(h,\, x) - A(h,\, x')||_\infty = \left\|\int_{x'}^{x} \Phi(x^*(h,\, t),\, S(h,\, t))\, dt\right\|_\infty$$

$$\leq \int_{x'}^{x} ||\Phi(x^*(h,\, t),\, S(h,\, t))||_\infty\, dt \leq \int_{x'}^{x} M\, dt \leq M(x - x')$$

(supposing that $x \geq x'$). Therefore, using the induction hypothesis we can conclude that $||A(h,\, x) - y_0||_\infty = ||A(h,\, x) - A(h,\, x_0)||_\infty \leq M(x - x_0) \leq M\epsilon \leq Mr/M = r$.

The information that for each h, $x \mapsto A(h,\, x)$ satisfies a Lipschitz condition with constant M is also important for the following reason. Let $\mathfrak{F}$ be the family of all the functions $x \mapsto A(h, x)$ from $[x_0, x_0 + \epsilon]$ into $B_r^c(y_0)$. The fact that these functions all satisfy the same Lipschitz condition shows that $\mathfrak{F}$ is an *equicontinuous* family. Let $\mathfrak{F}^c$ be the closure of $\mathfrak{F}$ in $\mathcal{C}([x_0, x_0 + \epsilon], B_r^c(y_0))$. According to Proposition 5, $\mathfrak{F}^c$ is also an equicontinuous family, which is closed in $\mathcal{C}([x_0, x_0 + \epsilon], B_r^c(y_0))$. Because $\mathfrak{F}^c$ is a closed family of functions from the compact metric space $[x_0, x_0 + \epsilon]$ into the compact metric space $B_r^c(y_0)$, the Arzela-Ascoli Theorem shows that $\mathfrak{F}^c$ is *compact* in $\mathcal{C}([x_0, x_0 + \epsilon], B_r^c(y_0))$.

Let $n \mapsto h_n$ be any sequence of positive real numbers converging to 0. Then we can extract from the sequence $n \mapsto (x \mapsto A(h_n, x))$ of functions in $\mathfrak{F}^c$ a subsequence $j \mapsto g_j$ (where $g_j(x) = A(h_{n_j}, x)$), that converges to a function f in $\mathfrak{F}^c$ (Theorem A 8). I claim that f is a solution to our differential equation. The metric on $\mathfrak{F}^c$ is of course obtained from the uniform norm on $\mathcal{C}([x_0, x_0 + \epsilon], B_r^c(y_0))$. Thus the sequence of functions $j \mapsto g_j$ converges to f uniformly on $[x_0, x_0 + \epsilon]$ as $j \to \infty$. To show that $f'(x) = \Phi(x, f(x))$ and that $f(x_0) = y_0$, it will be enough to show that f satisfies the integral equation $f(x) = y_0 + \int_{x_0}^x \Phi(t, f(t)) \, dt$. The proof uses a lemma about *piecewise constant* functions. A **piecewise constant** function p from $[x_0, x_0 + \epsilon]$ into a set T is a function $p \colon [x_0, x_0 + \epsilon] \to T$ with the following property. There are finitely many *points of discontinuity* $x_0 = d_0 < d_1 < d_2 < \cdots < d_n = x_0 + \epsilon$ such that p is constant on each of the intervals $[d_j, d_{j+1}[= \{x \colon d_j \leq x < d_{j+1}\}$ for $j = 0, \ldots, n - 1$.

LEMMA. Let $n \mapsto p_n$ be a sequence of piecewise constant functions from $[x_0, x_0 + \epsilon]$ to $\mathbf{R}^n$ converging uniformly to a continuous limiting function $L \colon [x_0, x_0 + \epsilon] \to \mathbf{R}^n$. Let $n \mapsto v_n$ be a sequence of piecewise constant functions from $[x_0, x_0 + \epsilon]$ to $\mathbf{R}$ converging uniformly to the identity function $t \mapsto t$. That is, $\lim_{n \to \infty} v_n(t) = t$. Then $\lim_{n \to \infty} \int_{x_0}^x \Phi(v_n(t), p_n(t)) \, dt = \int_{x_0}^x \Phi(t, L(t)) \, dt$ for every $x \in [x_0, x_0 + \epsilon]$.

Proof. Let $\eta > 0$ be given. Because Φ is continuous on the *compact* product space $[x_0, x_0 + \epsilon] \times B_r^c(y_0)$, Φ is uniformly continuous. Therefore, there is a $\delta > 0$ such that $\|\Phi(x, y) - \Phi(x', y')\|_\infty < \eta$ whenever $\max(|x - x'|, \|y - y'\|_\infty) < \delta$. Because the sequences $n \mapsto v_n$ and $n \mapsto p_n$ converge *uniformly*, there is an $N \in \mathbf{N}$ so large that $|v_n(t) - t| < \delta$ and $\|p_n(t) - L(t)\|_\infty < \delta$ for *all* $t \in [x_0, x_0 + \epsilon]$ whenever $n > N$. Then if $n > N$, $\|\Phi(v_n(t), p_n(t)) - \Phi(t, L(t))\|_\infty < \eta$ for all $t \in [x_0, x_0 + \epsilon]$. For each n, the function $t \mapsto \Phi(v_n(t), p_n(t))$ is piecewise constant on $[x_0, x_0 + \epsilon]$, so it can be integrated. If $x \in [x_0, x_0 + \epsilon]$,

$$\left\| \int_{x_0}^x \Phi(v_n(t), p_n(t)) \, dt - \int_{x_0}^x \Phi(t, L(t)) \, dt \right\|_\infty$$

$$\leq \int_{x_0}^x \|\Phi(v_n(t), p_n(t)) \, dt - \Phi(t, L(t)) \, dt\|_\infty$$

$$\leq \int_{x_0}^x \eta \, dt = \eta(x - x_0) \leq \eta\epsilon. \quad \bullet$$

Finally, observe that as $j \to \infty$, the piecewise constant functions $t \mapsto x^*(h_{n_j}, t)$ converge uniformly to t on $[x_0, x_0 + \epsilon]$, since if $x_k(h_{n_j}) \leq t < x_{k+1}(h_{n_j})$, $t - x^*(h_{n_j}, t) = t - x_k(h_{n_j}) < x_{k+1}(h_{n_j}) - x_k(h_{n_j}) = h_{n_j} \to 0$. Likewise, the piecewise constant functions $S(h_{n_j}, t)$ converge uniformly to $f(t)$, since $||f(t) - S(h_{n_j}, t)||_\infty \leq ||f(t) - A(h_{n_j}, t)||_\infty + ||A(h_{n_j}, t) - S(h_{n_j}, t)||_\infty$. The first term on the right converges uniformly to 0 by the choice of f and the sequence $j \mapsto g_j$. As to the second, if $x_k(h_{n_j}) \leq t < x_{k+1}(h_{n_j})$, $||A(h_{n_j}, t) - S(h_{n_j}, t)||_\infty = ||A(h_{n_j}, t) - y_k(h_{n_j})||_\infty = ||A(h_{n_j}, t) - A(h_{n_j}, x_k(h_{n_j}))||_\infty \leq M(t - x_k(h_{n_j})) < Mh_{n_j} \to 0$. By the Lemma, if $x_0 \leq x \leq x_0 + \epsilon$,

$$f(x) = \lim_{j \to \infty} g_j(x) = \lim_{j \to \infty} \left[y_0 + \int_{x_0}^x \Phi(x^*(h_{n_j}, t), S(h_{n_j}, t)) \, dt \right] = y_0$$

$$+ \int_{x_0}^x \Phi(\lim_{j \to \infty} x^*(h_{n_j}, t), \lim_{j \to \infty} S(h_{n_j}, t)) \, dt = y_0 + \int_{x_0}^x \Phi(t, f(t)) \, dt. \quad \bullet$$

Exercises

1. Show that the function $x \mapsto x^2$ is uniformly continuous on every finite interval in **R**, but not on all of **R**. (**R**, of course, is not compact.)
2. Show that the open interval $]0, 1[= \{x \in \mathbf{R} : 0 < x < 1\}$ has an open covering from which no finite subcover can be extracted.
3. Let $\mathcal{F}$ be the family of continuous functions from $\langle [0, 1]$, usual metric$\rangle$ to $\langle [0, 1]$, usual metric$\rangle \subseteq \langle \mathbf{R}$, usual metric$\rangle$. Show that $\mathcal{F}$ is closed in $\mathcal{C}([0, 1], \mathbf{R})$ and that all the functions in $\mathcal{F}$ take their values in a compact subset of **R** (namely $[0, 1]$), but that $\mathcal{F}$ is not compact. *Hint:* Use Exercise B 5.
4. Let $\mathcal{F}$ be the family of *constant* functions from the compact metric space $\langle M, d \rangle$ into the metric space $\langle N, e \rangle$. Take any $x \in M$. Show that $\mathrm{Eval}_x : f \mapsto f(x)$ is an **isometry** of $\mathcal{F}$ onto N in the sense, that if $f, g \in \mathcal{F}$, $e_\infty(f, g) = e(\mathrm{Eval}_x(f), \mathrm{Eval}_x(g))$. Show that a subfamily $\mathcal{G}$ of $\mathcal{F}$ is compact in $\mathcal{C}(M, N)$ if, and only if, $\mathrm{Eval}_x(\mathcal{G}) = \{g(x) : g \in \mathcal{G}\}$ is compact in N.
5. Let $n \mapsto f_n$ be a sequence of continuous functions from a compact metric space $\langle M, d \rangle$ to a metric space $\langle N, e \rangle$, which converges uniformly to a continuous limit function L. Make two proofs that $\{L, f_1, f_2, f_3, \ldots\}$ is an equicontinuous family of functions, one using the Arzela-Ascoli Theorem, the other directly from the definitions.
6. Let Φ be the *discontinuous* function from **R** to **R** defined by $\Phi : y \mapsto 1$ if $y \leq 0$, $\Phi : y \mapsto 0$ if $y > 0$. Show that there can be no differentiable real-valued function f on a neighborhood $U \subseteq \mathbf{R}$ of 0 such that $f(0) = 0$ and $f'(x) = \Phi(f(x))$ for all x in U. *Hint:* Using the equation $f'(0) = \Phi(f(0)) = \Phi(0) = 1$, show that there is an $\epsilon > 0$ such that $0 < x < \epsilon \Rightarrow 0 < f(x)$. Then use the Mean Value Theorem to show that if $0 < x < \epsilon$, $f(x) - f(0) = 0$.
7. Prove that e_∞ is a metric on $\mathcal{C}(M, N)$.

CHAPTER VIII

Connectedness

Intuitively, it appears that a metric space such as M_1 in Figure 27, the union of $B_{1/2}(-1)$ and $B_{1/2}(1)$, is *disconnected*, since it can be taken apart into two pieces without tearing it anywhere, but M_2, a single disk, is *connected*. It should not be hard to imagine that this notion of connectedness is a useful idea, and in fact it is. As we shall see, it is what lies behind the Intermediate Value Theorem of calculus, and it also can be used conveniently to show, for example, that **R** and **R**² are not homeomorphic ($\langle M, d \rangle$ is **homeomorphic** to $\langle N, e \rangle$ if there is a homeomorphism $f: M \to N$, namely a continuous function f mapping M one-to-one onto N, such that f^{-1} is also continuous). As the idea presently stands, it is too vague to do any good. The following definition is quite workable, and leads to a good theory.

A. Definition and general properties

1.   DEFINITION. A metric space $\langle M, d \rangle$ is **connected** if M is *not* the union of any two *disjoint* nonempty open sets. Otherwise, M is **disconnected.**

2.   REMARKS. (a) Suppose M is the union of U and V, where U and V are both open, and $U \cap V$ is empty. Then since $V = M - U$ is open, U is also closed. If U and V are both nonempty, then U is a subset of M that is both open and closed, and is distinct from the two subsets of M that trivially have this property, namely M and the empty subset. That is, we can rephrase Definition 1 as follows: $\langle M, d \rangle$ *is connected if, and only if, the only subsets of M that are both open and closed are M itself and the empty set.*

(b)   As usual, we say that a subset S of a metric space $\langle M, d \rangle$ is connected if the metric space $\langle S, d \rangle$ is connected. As in the case of compactness, it should

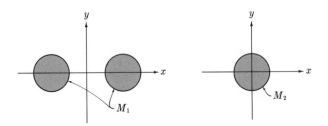

Figure 27

be possible to formulate this condition in terms of open coverings of S by open sets in M, using the correspondence between open sets in M and those in S. However, in this connection, you should observe the following warning.

WARNING: There is a notion of *topological space* more general than that of metric space, which you may later encounter. Connectedness can be formulated in topological spaces just as we have done for metric spaces, but *the forthcoming characterization of connected subsets of a metric space is valid only in metric spaces (and their relatives), and not in topological spaces in general.*

(Topological spaces for which the following characterization of connected subsets is valid are called *completely normal*.)

Here then is the alternative characterization of the connected subsets of a metric space: *a subset S of a metric space $\langle M, d \rangle$ is disconnected if, and only if, S can be covered by two disjoint open subsets of M whose intersections with S are nonempty.*

Proof. If U and V are disjoint open subsets of M whose intersections with S are nonempty and, if $S \subseteq U \cup V$, then $U \cap S$ and $V \cap S$ are two disjoint open subsets of $\langle S, d \rangle$ that are nonempty, and whose union is S. Therefore S is disconnected. (*This much is true in topological spaces in general. What follows is not.*) Conversely, suppose S is disconnected. Let U and V be disjoint nonempty open subsets of $\langle S, d \rangle$ whose union is S. We shall define two subsets U^* and V^* of M as follows. The distance from a point $x \in M$ to a subset T of M is $d(x, T) = \text{glb } \{d(x, y) : y \in T\}$. Put $U^* = \{y \in M : d(y, U) < d(y, V)\}$ and $V^* = \{y \in M : d(y, V) < d(y, U)\}$. Then it is clear from their definitions that U^* and V^* are disjoint.

LEMMA. U^* and V^* are open.

Proof. Choose $y \in U^*$, for example. Let $\epsilon = d(y, V) - d(y, U) > 0$. Let us show that $B_{\epsilon/2}(y) \subseteq U^*$. If $x \in B_{\epsilon/2}(y)$, then $d(x, y) < \epsilon/2$, so that for every

$z \in V$, $d(x, z) \geq d(y, z) - d(y, x) > d(y, z) - \epsilon/2 \geq d(y, V) - \epsilon/2$. If we take the glb with respect to all such $z \in V$, we get $d(x, V) \geq d(y, V) - \epsilon/2$. On the other hand, if $\delta = \epsilon/2 - d(x, y) > 0$, we can choose $u \in U$ such that $d(y, u) < d(y, U) + \delta$ by Theorem II C 1. Then $d(x, U) \leq d(x, u) \leq d(x, y) + d(y, u) < d(x, y) + (d(y, U) + \delta) = d(x, y) + d(y, U) + (\epsilon/2 - d(x, y)) = d(y, U) + \epsilon/2$. Taking into account the definition of ϵ, we see that $d(x, U) < d(y, U) + \epsilon/2 = d(y, V) - \epsilon/2$, and we have shown above that $d(y, V) - \epsilon/2 \leq d(x, V)$. Thus $d(x, U) < d(x, V)$, so $x \in U^*$.

LEMMA. $U \subseteq U^*$ and $V \subseteq V^*$.

Proof. Suppose $u \in U$, for example. Then $d(u, U) \leq d(u, u) = 0$. On the other hand, U is open in $\langle S, d \rangle$, so there is an $\epsilon > 0$ such that the ball of radius ϵ about u *in* S is contained in U. Then if $x \in V$, $x \notin B_\epsilon(u) \Rightarrow d(x, u) \geq \epsilon \Rightarrow d(u, V) \geq \epsilon > 0 = d(u, U)$. Thus $u \in U^*$.

In particular, $U \subseteq U^* \cap S$ and $V \subseteq V^* \cap S$, so that $U^* \cap S$ and $V^* \cap S$ are nonempty and $U^* \cup V^*$ covers S. This completes the proof of the alternate characterization of connected subsets of a metric space.

(c) *Illustration.* Figure 28 illustrates the proof of assertion (b). S_{-1} and S_1 are respectively the closed balls of radius 1 about the points -1 and 1 in the complex plane **C**, *with the origin 0 omitted from each of them:* $S_1 = \{z \in \mathbf{C}: |z - 1| \leq 1, z \neq 0\}$, $S_{-1} = \{z \in \mathbf{C}: |z + 1| \leq 1, z \neq 0\}$. Then $S = S_{-1} \cup S_1$ is a *disconnected* subset of the plane, since it can be covered by the

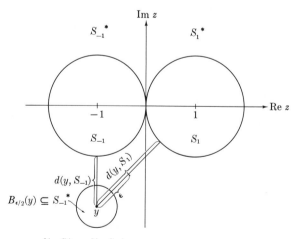

$\epsilon = d(y, S_1) - d(y, S_{-1})$
Neither S_1 nor S_{-1} contains 0
Points on the axis Re $z = 0$ are equidistant from S_1 and S_{-1}

Figure 28

two disjoint open sets $S_1^* = \{z \in \mathbf{C} : \operatorname{Re} z > 0\}$ and $S_{-1}^* = \{z \in \mathbf{C} : \operatorname{Re} z < 0\}$. Note that $S \cup \{0\}$ is connected.

3. DEFINITION. A **(possibly degenerate) interval** in $\mathbf{R}$ is any subset $I \subseteq \mathbf{R}$ with the following property: if x, $y \in I$ and $x \leq z \leq y$, then $z \in I$. Among the intervals are the sets $\varnothing$, $\mathbf{R}$, and the following (in what follows, a, $b \in \mathbf{R}$ and $a \leq b$):

$$]-\infty, b[\; = \{x \in \mathbf{R} : x < b\},$$
$$]-\infty, b] \; = \{x \in \mathbf{R} : x \leq b\},$$
$$[a, b[\; = \{x \in \mathbf{R} : a \leq x < b\},$$
$$[a, b] \; = \{x \in \mathbf{R} : a \leq x \leq b\},$$
$$]a, b[\; = \{x \in \mathbf{R} : a < x < b\},$$
$$]a, b] \; = \{x \in \mathbf{R} : a < x \leq b\},$$
$$[a, \infty[\; = \{x \in \mathbf{R} : a \leq x\},$$
$$]a, \infty[\; = \{x \in \mathbf{R} : a < x\}. \quad \bullet$$

It is not hard to see that this is a complete list of intervals. In fact, if I is a nonempty interval that is bounded above and below, then I must be one of the intervals $]\operatorname{glb} I, \operatorname{lub} I]$, $[\operatorname{glb} I, \operatorname{lub} I]$, $]\operatorname{glb} I, \operatorname{lub} I[$, or $[\operatorname{glb} I, \operatorname{lub} I[$, depending on whether or not $\operatorname{glb} I$ and $\operatorname{lub} I$ are in I. If I is bounded above but not below, I must be either $]-\infty, \operatorname{lub} I[$ or $]-\infty, \operatorname{lub} I]$. And so on. The proofs are all routine. Here is a representative one.

LEMMA. If I is a nonempty bounded interval, $\operatorname{glb} I \notin I$, and $\operatorname{lub} I \in I$, then $I = \,]\operatorname{glb} I, \operatorname{lub} I]$.

Proof. It is evident from the hypotheses that $I \subseteq \,]\operatorname{glb} I, \operatorname{lub} I]$. Thus only the reverse inclusion has to be proved. If $x \in \,]\operatorname{glb} I, \operatorname{lub} I]$, then by Theorem II C 1 there is a $y \in I$ such that $y < \operatorname{glb} I + (x - \operatorname{glb} I) = x$, since $x > \operatorname{glb} I$. Because I is an interval containing y and $\operatorname{lub} I$, and $y < x \leq \operatorname{lub} I$, it follows by definition that $x \in I$. Thus $]\operatorname{glb} I, \operatorname{lub} I] \subseteq I$.

4. THEOREM. A subset S of $\mathbf{R}$ is connected if, and only if, it is a (possibly degenerate) interval.

Proof. If S is not an interval, there must be an $a \in S$ and a $b \in S$ such that $a \leq b$ and $[a, b]$ is not included in S. Then there is an $x \notin S$ such that $a < x < b$. But then $]-\infty, x[$ and $]x, \infty[$ are two open subsets of $\mathbf{R}$ that cover S. $a \in S \cap \,]-\infty, x[$ and $b \in S \cap \,]x, \infty[$. Thus S is disconnected.

Conversely, let us begin our proof by supposing that S is disconnected. There must be two open subsets U and V of $\mathbf{R}$ that are disjoint, cover S, and intersect S in nonempty sets. We may suppose that there exist $a \in U$ and

$b \in V$ such that $a < b$. Let $x = \text{lub } (U \cap [a, b])$. Because U contains a neighborhood of a, $a < x$. Likewise, $x < b$. x cannot be in U, for if it were U would contain an entire neighborhood of x of the form $]x - \epsilon, x + \epsilon[$ (where $\epsilon > 0$). But then x could not be an upper bound for U. Likewise, x cannot be in V, for if it were, V would contain an entire neighborhood of x of the form $]x - \epsilon, x + \epsilon[$ (where $\epsilon > 0$). But then $x - \epsilon$ would be an upper bound for $U \cap [a, b]$ smaller than $\text{lub } (U \cap [a, b]) = x$. This shows that $x \notin S \subseteq U \cup V$, so that S is not an interval.

5.   THEOREM. The continuous image of a connected space is *connected*. In greater detail, if $\langle M, d \rangle$ is a connected metric space, and f is a continuous map from M onto the metric space $\langle N, e \rangle$, then N is connected.

Proof. If U and V were disjoint nonempty open subsets of N whose union is N, then $f^{-1}(U)$ and $f^{-1}(V)$ would be disjoint nonempty open subsets of M whose union is M.

6.   COROLLARY (INTERMEDIATE VALUE THEOREM). Let f be a continuous real-valued function on the connected metric space $\langle M, d \rangle$. If there are points x and y in M at which f takes the values a and b, with $a \leq c \leq b$, then there is a point $z \in M$ such that $f(z) = c$.

Proof. $f(M)$ is a connected subset of $\mathbf{R}$ containing a and b. Therefore, $f(M)$ is an interval containing a and b, so it must contain every real number c between them.   ●

I shall next prove a series of theorems that will enable us to recognize a large class of metric spaces to be connected. Of course, we already can do so using the two preceding theorems. Thus any *curve* in $\mathbf{R}^n$ is connected, since by definition a *curve* is the image of the interval $[0, 1]$ by a continuous function.

7.   THEOREM. If S is a connected subset of $\langle M, d \rangle$, so is its closure S^c.

Proof. If U and V are disjoint open subsets of M that cover S^c, then one of them must have an empty intersection with S^c. In fact, if $U \cap S^c$ is nonempty, then by Theorem IV C 8 $U \cap S$ cannot be empty. We know that one of $U \cap S$ and $V \cap S$ must be empty because S is connected.

8.   THEOREM. Let $\mathfrak{F}$ be a family of connected subsets of $\langle M, d \rangle$, all of which contain the point x. Then their union, $\mathbf{U} \, \mathfrak{F}$, is also connected.

Proof. Let U and V be disjoint open subsets of M that cover $\mathbf{U} \, \mathfrak{F}$. Suppose $x \in U$. Since for each $S \in \mathfrak{F}$, one of the $U \cap S$ and $V \cap S$ must be empty, and $x \in U \cap S$, we must have $S \subseteq U$ for each $S \in \mathfrak{F}$. That is, the intersection of V with $\mathbf{U} \, \mathfrak{F}$ must be empty. This shows that $\mathbf{U} \, \mathfrak{F}$ cannot be covered by two disjoint open subsets of M, both of which have nonempty intersection with $\mathbf{U} \, \mathfrak{F}$. That is, $\mathbf{U} \, \mathfrak{F}$ is connected.

9.   COROLLARY. Let $\langle M, d \rangle$ be a metric space, and suppose that every two points of M belong to some connected subset of M. Then M is connected.

Proof. Pick any $x \in M$. By hypothesis, for each $y \in M$, there is a connected subset S_y of M such that $x, y \in S_y$. Then M is the union of the connected sets S_y, all of which contain x. According to Theorem 8, $M = \bigcup_{y \in M} S_y$ is connected.

10.   COROLLARY. If $\langle M_1, d_1 \rangle, \cdots, \langle M_n, d_n \rangle$ are connected metric spaces, so is their product.

Proof. Let $(x_1, \ldots, x_n)$ and $(y_1, \ldots, y_n)$ be points in the product space $\langle M, d \rangle$. For each $j = 1, \ldots, n$, the set

$$\{(x_1, \ldots, x_{j-1}, z, y_{j+1}, \ldots, y_n) \in M : z \in M_j\}$$

is connected, since it is the image of the connected space M_j under the continuous map

$$z \mapsto (x_1, \ldots, x_{j-1}, z, y_{j+1}, \ldots, y_n).$$

Let us call this set S_j. Then for each $j = 1, \ldots, n-1$, $S_j \cap S_{j+1}$ contains the point $(x_1, \ldots, x_j, y_{j+1}, \ldots, y_n)$. Therefore, by Theorem 8, $S_1 \cup S_2$ is connected, and then $S_1 \cup S_2 \cup S_3$ is connected, and so on, by induction. Finally $S_1 \cup \cdots \cup S_n$ is connected and contains both $(x_1, \ldots, x_n)$ and $(y_1, \ldots, y_n)$. It follows then from Corollary 9 that $\langle M, d \rangle$ is connected.

11.   COROLLARY. $\mathbf{R}^n$ is connected for every n.

12.   DEFINITION. Let (M, d) be a metric space. An *arc* in M is any subset S of M that is the image of $[0, 1]$ by a continuous function $f: [0, 1] \to M$. $\langle M, d \rangle$ is *arcwise connected* if every two points of M can be joined by an arc.

13.   PROPOSITION. Every arcwise connected space is connected.

Proof. An arc, being the continuous image of a connected interval, is connected. Thus Corollary 9 can be applied.

14.   THEOREM. Let S be an *open* subset of the *normed space* $\langle V, \| \ \| \rangle$. Then S is connected if, and only if, it is arcwise connected.

Proof. Arcwise connectedness implies connectedness by Proposition 13. To prove the converse, suppose S is connected, and let $x \in S$. Let U be the set of points in S that can be joined to x by an arc, and let W be the set of points in S that cannot be joined to x by an arc. We must show that W is empty. Since U and W are obviously disjoint and S is connected, it will be enough to show that U and W are open. U is, of course, nonempty because $x \in U$. Let $y \in U$. By hypothesis, there is an arc A joining x to y. Because S is open in V, there is a ball $B_\epsilon(y)$ contained in S. But any point $z \in B_\epsilon(y)$ can be joined to the center y by a straight line segment, B. I claim that $A \cup B$

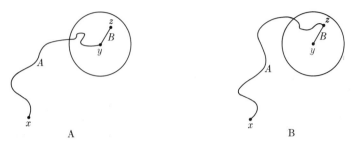

Figure 29

is an arc. In fact, if $A = f([0, 1])$, we can define a continuous function g from $[0, 1]$ to S whose range is $A \cup B$ by putting $g(t) = f(2t)$ if $0 \le x \le \frac{1}{2}$, $g(t) = y + (2t - 1)(z - y)$ if $\frac{1}{2} \le t \le 1$. This shows that U is open (Figure 29, A).

Likewise, if $y \in W$, there is a ball $B_\epsilon(y) \subseteq S$. If any point $z \in B_\epsilon(y)$ were in U, there would be an arc A joining z to x, and a straight line segment $B \subseteq B_\epsilon(y)$ joining z to y, and then $A \cup B$ would be an arc joining x to y. Since $y \in W$, this is impossible. Thus $B_\epsilon(y) \subseteq W$. This shows that W, too, is open (Figure 29, B). ●

Exercises

1. Show that if $n > 1$, $\mathbf{R}^n - \{0\}$ is connected, but that $\mathbf{R} - \{0\}$ is disconnected. Conclude that $\mathbf{R}^n$ and $\mathbf{R}$ are not homeomorphic.
*2. Let $S = \{(0, y) \in \mathbf{R}^2 : -1 \le y \le 1\} \cup \{(x, \sin(1/x)) \in \mathbf{R}^2 : 0 < x \le 1\}$. (Figure 30.) Show that S is compact and connected, but that S is not arcwise connected.

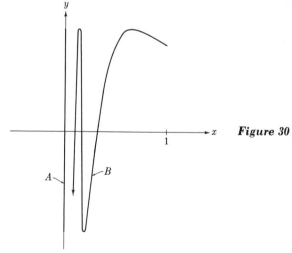

Figure 30

$$S = A \cup B = \{(0, y) : -1 \le y \le 1\} \cup \{(x, \sin(1/x)) : 0 < x \le 1\}$$

Hint: Let $A = \{(0, y) : -1 \leq y \leq 1\}$ and $B = \{(x, \sin (1/x)) : 0 < x \leq 1\}$, so that $S = A \cup B$. For each $n \in \mathbf{N}$, let $B_n = \{(x, \sin (1/x)) : 1/n \leq x \leq 1\}$. To show that S is compact, show that if $k \mapsto s_k$ is a sequence in S and, for every $n \in \mathbf{N}$, there is a $k \in \mathbf{N}$ such that $s_k \notin A \cup B_n$, then S has a subsequence that converges to a point of A. To show that S is connected, let U and V be open subsets of $\mathbf{R}^2$ that cover S, and suppose $(1, \sin 1) \in U$. Show that each B_n is connected, so that $B_n \subseteq U$ for every $n \in \mathbf{N}$. Conclude that $(\bigcup B_n)^c = S \subseteq S \cap U^c = S \cap U$. Finally, to show that S is not arcwise connected, show that any arc contained in S that contains $(1, \sin 1)$ is entirely contained in B. Indeed, if $f : [0, 1] \to S$, $f(1) = (1, \sin 1)$, and $f^{-1}(A) \neq \phi$, then f must be discontinuous at the point $x = \text{lub} f^{-1}(A) \in [0, 1]$.

3. Show that there can be no one-to-one continuous map from a circle onto an interval. *Hint:* See Exercise 1.

4. Use the Baire Category Theorem to show that a countable nonempty connected complete metric space consists of just one point. *Hint:* Show that some point of a countable complete metric space must be both open and closed. *Corollary:* $\mathbf{R}$ is uncountable.

5. Show that any connected metric space $\langle M, d \rangle$ that contains at least two points contains uncountably many points. *Hint:* Choose $x \in M$. Show that the continuous function $y \mapsto d(x, y)$ maps M onto an uncountable subset of $\mathbf{R}$.

*6. A **connected component** of a metric space $\langle M, d \rangle$ is a connected subset $C \subseteq M$ with the property that if $C \subseteq S \subseteq M$ and S is connected, then $S = C$.

   (a) Show that every point in M is contained in a connected component of M. *Hint:* Look at the union of all the connected subsets of M that contain x.

   (b) Show that if C and D are connected components of M and $C \cap D \neq \phi$, then $C = D$.

   (c) Show that if $\langle M, d \rangle$ is a subspace of a normed space $\langle V, \| \ \| \rangle$, and if M is *open* in V, then every connected component of M is open in V as well. *Hint:* Every ball in V is arcwise connected.

   (d) Show that every open subset U of $\mathbf{R}$ is the union of countably many disjoint open intervals. *Hint:* Every connected component of U must be an open interval. (b) shows that distinct components of U are disjoint. There is a map from $\mathbf{Q} \cap U$ onto the set of components of U, which sends each rational number $x \in \mathbf{Q} \cap U$ to the connected component of U containing x.

CHAPTER IX

Further applications

A. The integral as a linear map—construction of the elementary integral

In Greek geometry, areas were computed as shown in Figure 31. Initially, the area of a rectangle is given as the product of the lengths of two of its adjacent sides (Figure 31, A). Areas of polygons can be computed by cutting and piecing (Figure 31, B–G). Areas of curvilinear figures can be approximated by the **Method of Exhaustion** (Figure 31, H–J): a curvilinear figure can be nearly filled up by a polygon, whose area then approximates the area of the curvilinear figure. It was *assumed* that each figure has an area, and that the area function obeys certain laws such as: *the area of the whole is equal to the sum of the areas of its parts*. Modern criticism has discovered faults in the Greeks' reasoning. The modern approach is to *define* the area for a suitable class of geometric figures, and *prove* that the area function has certain properties—like the additive law crudely set forth above.

The modern approach is based on the systematic use of the Method of Exhaustion (approximation of figures by unions of rectangles) without the use of the scissors-and-paste technique (Figure 31, A–G), which works for polygons. The validity of the scissors-and-paste technique can then be *proved*.

In this section, we shall address ourselves primarily to the question posed by Figure 31, J: How do we know that if we express a figure such as I in two ways as a union of rectangles, we get the same value for its area in each case? In Greek geometry, the answer was obvious: I already has an area, and that area is equal to the sum of the areas of its parts by an axiom of geometry. In our case, we want to *define* the area of I to be Area $(R_1) + \cdots +$ Area (R_5) and also to be Area $(R_6) + \cdots +$ Area (R_8), so we have to *prove* that these two sums are equal.

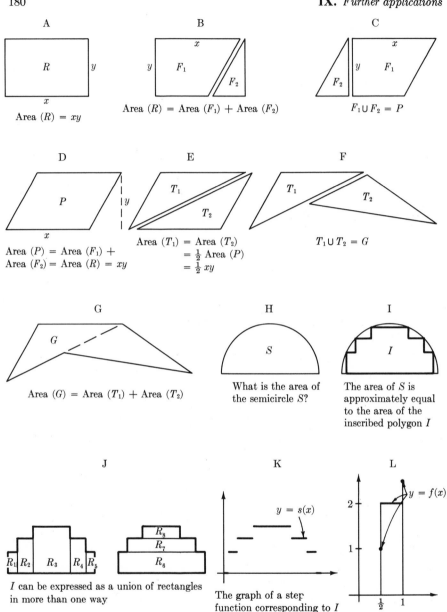

Figure 31

Instead of working with areas, we shall actually work with *integrals*. The connection between the integral of a function and the area of the figure bounded by its graph is well known. The simple functions whose graphs correspond to figures like *I* in Figure 31, I are called *step* functions (Figure

31, K). We now make formal definitions, and increase the generality of the discussion. We have already seen when we studied differential equations that it is useful to be able to integrate vector-valued functions. In the approach to integration taken below, vector-valued functions can be treated in the *same way* as real-valued functions, at the cost of adding one more paragraph to the proof of Theorem B 10.

1.   DEFINITIONS. Let S be a set, and let V be a vector space over **R**. Then $\mathcal{F}(S, V)$ is the vector space of all functions from S to V, with the usual operations of addition and scalar multiplication for functions.

If $a_i < b_i$ for $i = 1, \ldots, n$, then the set $R(a, b) = \{x \in \mathbf{R}^n : a_i < x_i < b_i$ for $i = 1, \ldots, n\}$ is an **open rectangle** in $\mathbf{R}^n$. The **n-dimensional unoriented volume** of $R(a, b)$ is

$$\lambda(R(a, b)) = (b_1 - a_1)(b_2 - a_2) \ldots (b_n - a_n).$$

For each $k = 1, \ldots, n$, $A_k(R(a, b)) = a_k$ and $B_k(R(a, b)) = b_k$.

A **grid** $\mathcal{G}$ on $\mathbf{R}^n$ is an n-tuple of finite subsets of $\mathbf{R}^n$, $(E_1(\mathcal{G}), \ldots, E_n(\mathcal{G}))$, the sets of **end points** of the grid (see Figure 32). Each $E_k(\mathcal{G})$ is required to contain at least two distinct real numbers. $R(\mathcal{G})$ is the rectangle $R(a, b)$, where for each $k = 1, \ldots, n$, $a_k = \mathrm{glb}\, E_k(\mathcal{G})$ and $b_k = \mathrm{lub}\, E_k(\mathcal{G})$. x and y are **adjacent** end points in $E_k(\mathcal{G})$ if $x < y$ and there is no element z of $E_k(\mathcal{G})$ such that $x < z < y$. A **cell** of the grid $\mathcal{G}$ is a rectangle $R(a, b)$, where for each $k = 1, \ldots, n$, a_k and b_k are adjacent end points of $E_k(\mathcal{G})$. $C(\mathcal{G})$ is the set of all cells of $\mathcal{G}$. A grid $\mathcal{G}'$ is a **refinement** of the grid $\mathcal{G}$ if $E_k(\mathcal{G}) \subseteq E_k(\mathcal{G}')$ for each $k = 1, \ldots, n$. Thus if $\mathcal{G}_1, \ldots, \mathcal{G}_N$ are grids, the formula $E_k(\mathcal{G}) = E_k(\mathcal{G}_1) \cup \cdots \cup E_k(\mathcal{G}_N)$ defines a grid $\mathcal{G}$ that is a common refinement of $\mathcal{G}_1, \ldots, \mathcal{G}_N$ (see Figure 33).

A function $f \in \mathcal{F}(\mathbf{R}^n, V)$ is a **step function** if there exists a grid $\mathcal{G}$ with the following properties.

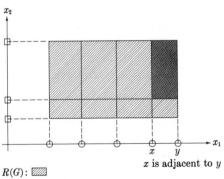

$R(G)$: ▨
A cell of G: ▤
Points in $E_1(G)$: ○
Points in $E_2(G)$: □

Figure 32
A grid on $\mathbf{R}^2$.

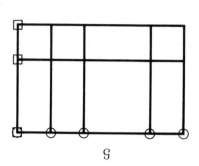
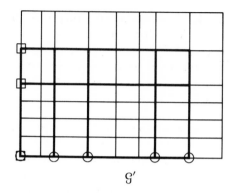

Figure 33. A grid $\mathcal{G}$ and a refinement $\mathcal{G}'$ of $\mathcal{G}$.

(a) If R is a cell of $\mathcal{G}$, then f is a *constant* function on R: f: $x \mapsto f(R)$ for every $x \in R$.

(b) If $x \in \mathbf{R}^n - (R(\mathcal{G}))^c$ then $f(x) = \mathbf{0}$.

$\mathcal{G}$ is a **presentation** of f. Step (n, V) is the set of all step functions from $\mathbf{R}^n$ to V.

2. LEMMA. Let $\mathcal{G}$ be a grid on $\mathbf{R}^n$ and let $\mathcal{G}'$ be a refinement of $\mathcal{G}$.

(a) If R' is a cell of $\mathcal{G}'$ such that $R' \cap (R(\mathcal{G}))^c \neq \varnothing$, then R' is contained in a cell of $\mathcal{G}$.

(b) If R is a cell of $\mathcal{G}$, then $\lambda(R) = \sum \{\lambda(R'): R' \in C(\mathcal{G}') \text{ and } R' \subseteq R\}$.

Proof. (a) $(R(a, b))^c = \{x \in \mathbf{R}^n: a_k \leq x_k \leq b_k \text{ for } k = 1, \ldots, n\}$. Thus if R' is a cell of $\mathcal{G}'$ that intersects $(R(\mathcal{G}))^c$, there must be for each $k = 1, \ldots, n$, an x_k such that $\mathrm{glb}\, E_k(\mathcal{G}) \leq x_k \leq \mathrm{lub}\, E_k(\mathcal{G})$ and $A_k(R') < x_k < B_k(R')$. If $A_k(R')$ were less than $\mathrm{glb}\, E_k(\mathcal{G})$, then since $B_k(R')$ is adjacent to $A_k(R')$ in $E_k(\mathcal{G}')$ and $\mathrm{glb}\, E_k(\mathcal{G}) \in E_k(\mathcal{G}')$, it must be that $B_k(R') \leq \mathrm{glb}\, E_k(\mathcal{G})$. Thus $x_k < B_k(R') \leq \mathrm{glb}\, E_k(\mathcal{G}) \leq x_k$, a contradiction. Similarly, we see that $B_k(R')$ cannot be larger than $\mathrm{lub}\, E_k(\mathcal{G})$. Let a_k be the largest element of $E_k(\mathcal{G})$ that is $\leq A_k(R')$ and let b_k be the smallest element of $E_k(\mathcal{G})$ that is $\geq B_k(R')$. Since $A_k(R')$ and $B_k(R')$ are adjacent in $E_k(\mathcal{G}')$, a_k and b_k are adjacent in $E_k(\mathcal{G})$. If we put $a = (a_1, \ldots, a_k)$ and $b = (b_1, \ldots, b_n)$, then $R(a, b)$ is a cell of $\mathcal{G}$ that contains R'.

(b) Let R be a cell of $\mathcal{G}$ and for each $k = 1, \ldots, n$, let $E_k(\mathcal{G}'') = \{x \in E_k(\mathcal{G}'): A_k(R) \leq x \leq B_k(R)\}$. Then $\mathcal{G}''$ is a grid on $\mathbf{R}^n$, $R = R(\mathcal{G}'')$, and the cells of $\mathcal{G}'$ that are contained in R are exactly the cells of $\mathcal{G}''$. Thus what we have to prove is that $\lambda(R(\mathcal{G}'')) = \sum \{\lambda(R'): R' \in C(\mathcal{G}'')\}$. If $E_k(\mathcal{G}'')$ contains just two points for each $k = 1, \ldots, n$, then $\mathcal{G}''$ has a single cell, $R(\mathcal{G}'')$, and the assertion is clearly true. Otherwise, let m be the largest integer such that $E_m(\mathcal{G}'')$ contains more than two points, and let us prove (b) by induction on m. We may put $m = 0$ when $E_k(\mathcal{G}'')$ contains just two points for every $k = 1, \ldots, n$. By assumption, if $n \geq k > m$, $E_k(\mathcal{G}'') = \{A_k(R(\mathcal{G}'')), B_k(R(\mathcal{G}''))\}$, and $E_m(\mathcal{G}'') = \{x_0, x_1, \ldots, x_p\}$, where $x_0 < x_1 < \cdots < x_p$, $x_0 = A_m(R(\mathcal{G}''))$, $x_p = B_m(R(\mathcal{G}''))$. Let $\mathcal{G}_i''$ be the grid defined by $E_k(\mathcal{G}_i'') = E_k(\mathcal{G}'')$ if $k \neq m$, $E_m(\mathcal{G}_i'') = \{x_i, x_{i+1}\}$ for $i = 0, \ldots, p - 1$. By the induction hypothesis, $\lambda(R(\mathcal{G}_i'')) = \sum \{\lambda(R') : R' \in C(\mathcal{G}_i'')\}$. Put $A_k = A_k(R(\mathcal{G}''))$, $B_k = B_k(R(\mathcal{G}''))$. Then

$$\lambda(R(\mathcal{G}'')) = (B_1 - A_1) \ldots (B_m - A_m) \ldots (B_n - A_n)$$
$$= (B_1 - A_1) \ldots ((x_p - x_{p-1}) + (x_{p-1} - x_{p-2})$$
$$+ \cdots + (x_1 - x_0)) \ldots (B_n - A_n)$$
$$= \sum_{i=0}^{p-1} (B_1 - A_1) \ldots (x_{i+1} - x_i) \ldots (B_n - A_n)$$
$$= \sum_{i=0}^{p-1} \lambda(R(\mathcal{G}_i''))$$
$$= \sum_{i=0}^{p-1} [\sum \{\lambda(R') : R' \in C(\mathcal{G}_i'')\}]$$
$$= \sum \{\lambda(R') : R' \in C(\mathcal{G}'')\}.$$

This completes the induction step.

3. COROLLARY. If $f \in$ Step (n, V), $\mathcal{G}$ is a presentation of f and $\mathcal{G}'$ is a refinement of $\mathcal{G}$, then $\mathcal{G}'$ is also a presentation of f.

Proof. This follows from the definition of "step function" and Lemma 2(a).

4. COROLLARY. If V and W are vector spaces over $\mathbf{R}$, T is any function from V^N into W such that $T(0, \ldots, 0) = 0$, and $f = (f_1, \ldots, f_N)$ is a function from $\mathbf{R}^n$ to V^N, each of whose components is a step function, then $T \circ f$ is a step function.

Proof. Let $\mathcal{G}_i$ be a presentation of f_i for $i = 1, \ldots, N$. Then there is a grid $\mathcal{G}$ that is a common refinement of $\mathcal{G}_1, \ldots, \mathcal{G}_N$, and according to Corollary 3, $\mathcal{G}$ is a presentation of each of $f_1, \ldots, f_N$. It is clear, since $T(0, \ldots, 0) = 0$, that $\mathcal{G}$ is also a presentation of $T \circ f$.

5. COROLLARY. Step (n, V) is a linear subspace of $\mathcal{F}(\mathbf{R}^n, V)$. If $\| \ \|$ is a norm on V, and $f \in$ Step (n, V), then the function $\|f\| : x \mapsto \|f(x)\|$ is a step function from $\mathbf{R}^n$ to $\mathbf{R}$. If $f, g \in$ Step $(n, \mathbf{R})$, so are the functions $f \vee g : x \mapsto \max (f(x), g(x))$ and $f \wedge g : x \mapsto \min (f(x), g(x))$.

Proof. All these facts follow at once from Corollary 3. For example, to show that $f + g \in$ Step (n, V) when $f, g \in$ Step (n, V), use the function $T : (v, w) \mapsto v + w$. $f + g = T \circ (f, g)$.

6. COROLLARY. If $\mathcal{G}$ and $\mathcal{G}'$ are presentations of the function $f \in$ Step (n, V), then $\sum \{\lambda(R)f(R) : R \in C(\mathcal{G})\} = \sum \{\lambda(R')f(R') : R' \in C(\mathcal{G}')\}$.

Proof. Let $\mathcal{G}''$ be a common refinement of $\mathcal{G}$ and $\mathcal{G}'$. It will be sufficient to show that both of these sums are equal to $\sum \{\lambda(R'')f(R'') : R'' \in C(\mathcal{G}'')\}$. Obviously, if $f(R'') = \mathbf{0}$ for each $R'' \in C(\mathcal{G}'')$, then all three sums are $\mathbf{0}$. If R'' is a cell of $\mathcal{G}''$ such that $f(R'') \neq \mathbf{0}$, then by Definition 1(b), $R'' \cap (R(\mathcal{G}))^c \neq \varnothing$. According to Lemma 2(a), R'' is contained in a cell of $\mathcal{G}$. Since f is constant on each cell of $\mathcal{G}$, if there is an $R'' \in C(\mathcal{G}'')$ such that $f(R'') \neq \mathbf{0}$, then

$$\sum \{\lambda(R'')f(R'') : R'' \in C(\mathcal{G}'')\} = \sum \{\lambda(R'')f(R'') : R'' \in C(\mathcal{G}'') \text{ and } f(R'') \neq \mathbf{0}\}$$

$$= \sum_{R \in C(\mathcal{G})} [\sum \{\lambda(R'')f(R'') : R'' \in C(\mathcal{G}'') \text{ and } R'' \subseteq R\}]$$

$$= \sum_{R \in C(\mathcal{G})} [\sum \{\lambda(R'')f(R) : R'' \in C(\mathcal{G}'') \text{ and } R'' \subseteq R\}]$$

$$= \sum_{R \in C(\mathcal{G})} [\sum \{\lambda(R'') : R \in C(\mathcal{G}'') \text{ and } R'' \subseteq R\}]f(R).$$

Lemma 2(b) shows that

$$\sum \{\lambda(R'') : R'' \in C(\mathcal{G}'') \text{ and } R'' \subseteq R\} = \lambda(R).$$

Substitution in the last preceding equation yields

$$\sum \{\lambda(R'')f(R'') : R'' \in C(\mathcal{G}'')\} = \sum_{R \in C(\mathcal{G})} \lambda(R)f(R) = \sum \{\lambda(R)f(R) : R \in C(\mathcal{G})\}.$$

7. DEFINITION. If $f \in \text{Step } (n, V)$, then $\int f$ is the common value of the expressions $\sum \{\lambda(R)f(R) : R \in C(\mathcal{G})\}$ for all presentations $\mathcal{G}$ of f. In other words, the relation

$$\int = \{\langle f, v \rangle \in \text{Step } (n, V) \times V : \text{there is a presentation } \mathcal{G} \text{ of } f \text{ such that}$$

$$v = \sum \{\lambda(R)f(R) : R \in C(\mathcal{G})\}\}$$

is a *function* (Corollary 6).

8. REMARKS. If R is a single open rectangle and $f \in \text{Step } (n, \mathbf{R})$ is $\mathbf{0}$ outside R and nonnegative, then it is obvious what $\int f$ ought to be. As shown in Figure 31, L, for the case $n = 1$, $R =]\frac{1}{2}, 1[$, and $f(R) = 2$, the graph of f bounds a rectangle in $\mathbf{R}^{n+1}$ whose volume is $\lambda(R)f(R)$. This is not quite accurate, since the definition of a step function does not restrict the values of f on the boundary ∂R of R. However, ∂R has volume 0, and therefore so does that part of the figure bounded by the graph of f that lies above ∂R. In Figure 31, L, $\partial R = \{\frac{1}{2}, 1\}$ and the part of the figure bounded by the graph of f that lies above ∂R consists of two line segments. Their area is 0.

It is natural to define the integral of a step function f with presentation $\mathcal{G}$ to be $\sum \{\lambda(R)f(R) : R \in C(\mathcal{G})\}$, but we have to be sure that this sum depends only on f, not on $\mathcal{G}$. This fact is precisely the content of Corollary 6. Moreover, the expression $\sum \{\lambda(R)f(R) : R \in C(\mathcal{G})\}$ makes just as good sense when f takes its values in a vector space as it does when f maps $\mathbf{R}^n$ into $\mathbf{R}$. When we discussed differential equations, we simply integrated vector-valued functions one component at a time. In case V is finite-dimensional, that approach yields the same result as the present one. However, if V does not have a finite basis, the earlier definition is not available.

9.　THEOREM. (a) $\int$ is a linear map from Step (n, V) into V.

(b)　If W is another vector space over **R** and $L: V \to W$ is a linear map, then for each $f \in$ Step (n, V), $L \circ f \in$ Step (n, W) and $L \int f = \int (L \circ f)$.

(c)　If $\| \; \|$ is a norm on V and $f \in$ Step (n, V), then $\|f\| \in$ Step $(n, \mathbf{R})$ and $\|\int f\| \leq \int \|f\|$.

(d)　If $f, g \in$ Step $(n, \mathbf{R})$ and $f(x) \leq g(x)$ for all $x \in \mathbf{R}^n$, then $\int f \leq \int g$.

Proof. (a) If $f, g \in$ Step (n, V) and $r, s \in \mathbf{R}$, we can choose a common presentation $\mathcal{G}$ for f and g. Then $\mathcal{G}$ also presents $rf + sg$, and

$$\int (rf + sg) = \sum \{\lambda(R)[rf(R) + sg(R)]: R \in C(\mathcal{G})\}$$

$$= r \sum \{\lambda(R)f(R): R \in C(\mathcal{G})\} + s \sum \{\lambda(R)g(R): R \in C(\mathcal{G})\}$$

$$= r \int f + s \int g.$$

(b)　If $\mathcal{G}$ is a presentation of f, then $\mathcal{G}$ also presents $L \circ f$, and because L is linear, $L \int f = L(\sum \{\lambda(R)f(R): R \in C(\mathcal{G})\}) = \sum \{\lambda(R)L \circ f(R): R \in C(\mathcal{G})\} = \int (L \circ f)$.

(c)　If $\mathcal{G}$ is a presentation of f, then $\mathcal{G}$ also presents $\|f\|$, and by the triangle inequality, $\|\int f\| = \|\sum \{\lambda(R)f(R): R \in C(\mathcal{G})\}\| \leq \sum \{\|\lambda(R)f(R)\|: R \in C(\mathcal{G})\} = \sum \{\lambda(R)\|f(R)\|: R \in C(\mathcal{G})\} = \int \|f\|$.

(d)　If $\mathcal{G}$ is a common presentation of f and g, $\int f = \sum \{\lambda(R)f(R): R \in C(\mathcal{G})\} \leq \sum \{\lambda(R)g(R): R \in C(\mathcal{G})\} = \int g$.

10.　LEMMA. If $a_1, \ldots, a_N, b_1, \ldots, b_N \in \mathbf{R}$ $(N \geq 2)$, then

$$|\min \{a_1, \ldots, a_N\} - \{\min b_1, \ldots, b_N\}| \leq |a_1 - b_1| + \cdots + |a_N - b_N|.$$

Proof, by induction on N. $\min (a, b) = \frac{1}{2}(a + b - |a - b|)$. Thus

$$|\min \{a_1, a_2\} - \min \{b_1, b_2\}| = \frac{1}{2}|(a_1 + a_2 - |a_1 - a_2|) - (b_1 + b_2 - |b_1 - b_2|)|$$

$$\leq \frac{1}{2}(|a_1 - b_1| + |a_2 - b_2| + ||a_1 - a_2| - |b_1 - b_2||)$$

$$\leq \frac{1}{2}(|a_1 - b_1| + |a_2 - b_2| + |(a_1 - a_2) - (b_1 - b_2)|)$$

$$\leq \frac{1}{2}(|a_1 - b_1| + |a_2 - b_2| + |a_1 - b_1| + |a_2 - b_2|) = |a_1 - b_1| + |a_2 - b_2|.$$

Now suppose the lemma is true when $N = k$. Let us prove it when $N = k + 1$:

$$|\min \{a_1, \ldots, a_{k+1}\} - \min \{b_1, \ldots, b_{k+1}\}| = |\min \{\min \{a_1, \ldots, a_k\}, a_{k+1}\}$$

$$- \min \{\min \{b_1, \ldots, b_k\}, b_{k+1}\}| \leq |\min \{a_1, \ldots, a_k\} - \min \{b_1, \ldots, b_k\}|$$

$$+ |a_{k+1} - b_{k+1}| \leq (|a_1 - b_1| + \cdots + |a_k - b_k|) + |a_{k+1} - b_{k+1}|$$

by the induction hypothesis and the case $N = 2$.

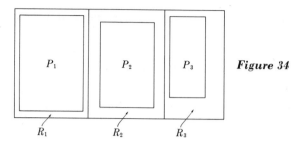

Figure 34

R_1 R_2 R_3

11. LEMMA. Let f be a nonnegative function in Step $(n, \mathbf{R})$. Then for each $\epsilon > 0$, there is a $g \in \text{Step } (n, \mathbf{R})$ such that

(a) $0 \le g(x) \le f(x)$ for all $x \in \mathbf{R}^n$,

(b) $\int f < \epsilon + \int g$,

(c)   for each $r \in \mathbf{R}^+$, $g^{-1}([r, \infty[)$ is a closed set in $\mathbf{R}^n$.

Proof. Let $\mathcal{G}$ be a presentation of f and let $C(\mathcal{G}) = \{R_1, \ldots, R_N\}$. For each $i = 1, \ldots, N$, define a step function g_i as follows. Let P_i be an open rectangle such that $(P_i)^c \subseteq R_i$ and $f(R_i)(\lambda(R_i) - \lambda(P_i)) < 2^{-i}\epsilon$ (see Figure 34). Put $g_i(x) = f(R_i)$ if $x \in (P_i)^c$ and $g_i(x) = 0$ if $x \in \mathbf{R}^n - (P_i)^c$. Then put $g = \sum_{i=1}^{N} g_i$. Since for each $x \in \mathbf{R}^n$, at most one of $g_1(x), \ldots, g_N(x)$ is non-zero, $0 \le g(x) = \sum_{i=1}^{N} g_i(x) \le f(x)$. $\int f = \sum \{\lambda(R)f(R) : R \in C(\mathcal{G})\} = \sum_{i=1}^{N} \lambda(R_i)f(R_i) < \sum_{i=1}^{N} (2^{-i}\epsilon + \lambda(P_i)f(R_i)) = (1 - 2^{-N})\epsilon + \sum_{i=1}^{N} \int g_i = (1 - 2^{-N})\epsilon + \int g < \epsilon + \int g$. Finally, for each $r \in \mathbf{R}^+$, $g^{-1}([r, \infty[) = \bigcup \{(P_i)^c : f(R_i) \ge r\}$, which is a finite union of closed sets.

12. THEOREM. Let $k \mapsto f_k$ be a sequence in Step $(n, \mathbf{R})$ such that for each $x \in \mathbf{R}^n$,

(a) $0 \le f_{k+1}(x) \le f_k(x)$ for every $k \in \mathbf{N}$, and

(b) $\lim_{k \to \infty} f_k(x) = 0$.

Then $\lim_{k \to \infty} \int f_k = 0$.

Proof. Let $\epsilon > 0$ be given. For each $k \in \mathbf{N}$, choose $g_k \in \text{Step } (n, \mathbf{R})$ such that $0 \le g_k \le f_k$, $\int f_k < 2^{-k}\epsilon + \int g_k$, and that for each $r \in \mathbf{R}^+$, $g_k^{-1}([r, \infty[)$ is closed in $\mathbf{R}^n$, as in Lemma 11. Let $h_k = g_1 \wedge \cdots \wedge g_k$. Then for each $k \in \mathbf{N}$ and each $x \in \mathbf{R}^n$, $0 \le h_k(x) \le g_k(x) \le f_k(x)$, so that $\lim_{k \to \infty} h_k(x) = 0$, and $h_{k+1}(x) \le h_k(x)$. Note that $f_k = f_k \wedge f_{k-1} \wedge \cdots \wedge f_1$. Therefore, by Lemma 10, for each $x \in \mathbf{R}^n$,

$$|f_k(x) - h_k(x)| = |\min \{f_1(x), \ldots, f_k(x)\} - \min \{g_1(x), \ldots, g_k(x)\}|$$
$$\le |f_1(x) - g_1(x)| + \cdots + |f_k(x) - g_k(x)|.$$

It follows that

$$0 \le \int f_k - \int h_k \le \int (f_1 - g_1) + \cdots + \int (f_k - g_k)$$
$$< 2^{-1}\epsilon + 2^{-2}\epsilon + \cdots + 2^{-k}\epsilon = (1 - 2^{-k})\epsilon < \epsilon.$$

Thus it will suffice to show that $\lim_{k \to \infty} \int h_k = 0$.

For each $k \in \mathbf{N}$,

$$h_k^{-1}([\epsilon, \infty[) = \{x \in \mathbf{R}^n : h_k(x) \geq \epsilon\} = \{x \in \mathbf{R}^n : g_i(x) \geq \epsilon \text{ for } i = 1, \ldots, k\}$$

$$= \bigcap \{g_i^{-1}([\epsilon, \infty[) : i = 1, \ldots, k\},$$

which is an intersection of closed sets. Let $\mathcal{G}$ be a presentation of h_1. Then $(R(\mathcal{G}))^c \supseteq h_1^{-1}([\epsilon, \infty[) \supseteq h_2^{-1}([\epsilon, \infty[) \supseteq h_3^{-1}([\epsilon, \infty[) \supseteq \cdots$ is a decreasing sequence of closed sets because $h_1, h_2, h_3, \ldots$ is a decreasing sequence of functions. Moreover, $(R(\mathcal{G}))^c$ is bounded, and therefore compact (Theorem VII B 3). The fact that $\lim_{k \to \infty} h_k(x) = 0$ for every $x \in \mathbf{R}^n$ implies that $\bigcap \{h_k^{-1}([\epsilon, \infty[) : k \in \mathbf{N}\} = \varnothing$. Therefore, by Theorem VII A 5, the family $\{h_k^{-1}([\epsilon, \infty[) : k \in \mathbf{N}\}$ cannot have the Finite Intersection Property. Since $h_1^{-1}([\epsilon, \infty[) \cap \cdots \cap h_k^{-1}([\epsilon, \infty[) = h_k^{-1}([\epsilon, \infty[)$, $h_k^{-1}([\epsilon, \infty[)$ must be empty for some k. If q is the step function that takes the value 1 on $(R(\mathcal{G}))^c$ and the value 0 on $\mathbf{R}^n - (R(\mathcal{G}))^c$, then for each $j \geq k$, $h_j \leq h_k < \epsilon q$, so that $\int h_j \leq \epsilon \int q$. Since ϵ can be any positive number, $\lim_{k \to \infty} h_k = 0$.

Exercises

1. For each $k = 1, \ldots, N$, let L_k be the linear functional on $\mathbf{R}^N$ defined by L_k: $(x_1, \ldots, x_N) \mapsto x_k$. By applying Theorem 9(b) to each of the functionals $L_1, \ldots, L_N$, verify that the integral of a step function from $\mathbf{R}^n$ to $\mathbf{R}^N$ can be computed by integrating each of its components separately. (The systematic use of this device in infinite-dimensional vector spaces leads to the notion of *weak integral*. To define weak integrals, it must first be proved that for each nonzero x in a vector space V, there is a linear functional L on V such that $L(x) \neq \mathbf{0}$.)

*2. Show that the function $f \mapsto \int \|f\|$ from Step (n, V) to $\mathbf{R}$ has all the properties of a norm on Step (n, V) (Definition IV A 1) *except* that there are *nonzero* functions $f \in$ Step (n, V) such that $\int \|f\| = 0$. (See Remarks 8, and also compare this result with Examples III C 2(a) and IV A 3(d).) Such a function is called a **seminorm**.

3. Show that there is a sequence $k \mapsto f_k$ in Step $(n, \mathbf{R})$ such that for each $x \in \mathbf{R}^n$, $\lim_{k \to \infty} f_k(x) = 0$, but that $\int f_k \to +\infty$ as $k \to \infty$. Of course condition (a) of Theorem 12 must be violated by this sequence. *Hint:* Choose a sequence $k \mapsto R_k$ of mutually disjoint rectangles in $\mathbf{R}^n$, and let $\{x \in \mathbf{R}^n : f_k(x) \neq 0\} = R_k$.

B. The integral as a linear map—extension of the elementary integral

We shall now extend the integral to a class of functions larger than Step (n, V) by approximating more general functions by step functions. The construction is analogous to the geometric Method of Exhaustion. We must first be able to say when a function is closely approximated by a step function. This is the purpose of the following definition.

1.   DEFINITION. Let $\langle V, ||\ \ || \rangle$ be a normed space. A function $f \in \mathfrak{F}(\mathbf{R}^n, V)$ is **normable** if there exists a sequence $k \mapsto s_k$ in Step $(n, \mathbf{R})$ such that

(a)   for each $x \in \mathbf{R}^n$ and each $k \in \mathbf{N}$, $0 \leq s_k(x) \leq s_{k+1}(x)$,

(b)   for each $x \in \mathbf{R}^n$ such that $\{s_k(x) : k \in \mathbf{N}\}$ is bounded, $\lim_{k \to \infty} s_k(x) \geq ||f(x)||$
(see Theorem VI B 9),

(c) $\{\int s_k : k \in \mathbf{N}\}$ is bounded.

Such a sequence is a **bounding sequence** for f. According to Theorem A 9(d), $\int s_k \leq \int s_{k+1}$ for each $k \in \mathbf{N}$, so that $k \mapsto \int s_k$ has a limit by Theorem VI B 9. The set of all normable functions from $\mathbf{R}^n$ to V is $\mathfrak{N}(n, V)$. If $f \in \mathfrak{N}(n, V)$, $|f|_1 = \text{glb}\ \{\lim_{k \to \infty} \int s_k : k \mapsto s_k$ is a bounding sequence for $f\}$.

2.   PROPOSITION. $\mathfrak{N}(n, V)$ is a linear subspace of $\mathfrak{F}(\mathbf{R}^n, V)$ and $|\ |_1$ is a **seminorm** on $\mathfrak{N}(n, V)$. That is,

(a)   for each $f \in \mathfrak{N}(n, V)$, $|f|_1 \geq 0$,

(b)   for each $f \in \mathfrak{N}(n, V)$ and each $r \in \mathbf{R}$, $|rf|_1 = |r|\ |f|_1$,

(c)   if $f, g \in \mathfrak{N}(n, V)$, then $|f + g|_1 \leq |f|_1 + |g|_1$.

Proof. (a) is a consequence of Definition 1(a) and Theorem A 9(d).

(b)   If $r \in \mathbf{R}$, $f \in \mathfrak{N}(n, V)$, and $k \mapsto s_k$ is a bounding sequence for f, then $k \mapsto |r|s_k$ is a bounding sequence for rf. Thus $rf \in \mathfrak{N}(n, V)$, and $|rf|_1 \leq$ glb $\{\lim_{k \to \infty} \int |r|s_k : k \mapsto s_k$ is a bounding sequence for $f\} = |r|\ |f|_1$. If $r = 0$, then clearly $|rf|_1 = 0 = |r|\ |f|_1$. If $r \neq 0$, then by what we have just shown, $|f|_1 = |r^{-1}(rf)|_1 \leq |r|^{-1}|rf|_1 \Rightarrow |r|\ |f|_1 \leq |rf|_1 \Rightarrow |r|\ |f|_1 = |rf|_1$.

(c)   If $f, g \in \mathfrak{N}(n, V)$, $\epsilon > 0$, $k \mapsto s_k$ is a bounding sequence for f such that $\lim_{k \to \infty} \int s_k < |f|_1 + \epsilon/2$, and $k \mapsto s_k'$ is a bounding sequence for g such that $\lim_{k \to \infty} \int s_k' < |g|_1 + \epsilon/2$, then $k \mapsto s_k + s_k'$ is a bounding sequence for $f + g$. Thus $f + g \in \mathfrak{N}(n, V)$ and $|f + g|_1 < |f|_1 + \epsilon/2 + |g|_1 + \epsilon/2 = |f|_1 + |g|_1 + \epsilon$. Since ϵ can be as small as we please, (c) is proved.

3.   PROPOSITION. If $f \in$ Step (n, V), then $f \in \mathfrak{N}(n, V)$ and $|f|_1 = \int ||f||$.

Proof. The constant sequence $k \mapsto ||f||$ is a bounding sequence for f. Thus $f \in$ Step (n, V) and $|f|_1 \leq \int ||f||$. Now let $k \mapsto s_k$ be any bounding sequence for f. Then $k \mapsto s_k \wedge ||f||$ is also a bounding sequence for f. The sequence $k \mapsto ||f|| - s_k \wedge ||f||$ satisfies the hypotheses of Theorem A 12. Therefore $0 = \lim_{k \to \infty} \int (||f|| - s_k \wedge ||f||) = \int ||f|| - \lim_{k \to \infty} \int s_k \wedge ||f|| \geq \int ||f|| - \lim_{k \to \infty} \int s_k$. That is, $\int ||f|| \leq \lim_{k \to \infty} \int s_k$. Taking the greatest lower bound over all such bounding sequence, we conclude that $\int ||f|| \leq |f|_1$.

4.   REMARKS. (a) An example of a function in $\mathfrak{F}(\mathbf{R}^n, \mathbf{R})$ that is *not* normable is the constant function $f : x \mapsto 1$. Indeed, it is not hard to show that if

$k \mapsto s_k$ is any monotonically increasing sequence of step functions such that $\lim_{k \to \infty} s_k(x) \geq 1$ for all $x \in \mathbf{R}^n$, then $\int s_k \to +\infty$ as $k \to \infty$ (see Exercise 1).

(b) It follows from Proposition 3 and Exercise A 2 that $|\ |_1$ is *not* a norm on $\mathfrak{N}(n, V)$, because there exist nonzero functions f in Step $(n, V) \subseteq \mathfrak{N}(n, V)$ such that $|f|_1 = 0$. Nonetheless we can do most of the same things with the seminormed space $\langle \mathfrak{N}(n, V), |\ |_1 \rangle$ that we can do with a normed space, because *most of what we have done with normed spaces did not actually use the fact that if $\|\ \|$ is a norm and $\|x\| = 0$, then $x = \mathbf{0}$.* For example, we can define $\lim_{k \to \infty} f_k = f$ in the usual way to mean $\lim_{k \to \infty} |f - f_k|_1 = 0$. Likewise, we can define the notions of closed set, Cauchy sequence, closure, open set, and completeness for seminormed spaces in *exactly the same way* as we earlier defined these notions for normed spaces.

There are some differences, however. If $|f - g|_1 = 0$, then the constant sequence $k \mapsto f$ converges to g, even though f need not equal g. Thus $\{f\}^c = \{g \in \mathfrak{N}(n, V): |f - g|_1 = 0\} \neq \{f\}$. In particular, a sequence in a seminormed space may have more than one limit. For further details, see Exercise 3.

5.   PROPOSITION. Let $\langle V, \|\ \| \rangle$ be a Banach space (a *complete* normed space) and let $k \mapsto g_k$ be a sequence in Step (n, V) that converges to an element $f \in \mathfrak{N}(n, V)$ in the sense that $\lim_{k \to \infty} |f - g_k|_1 = 0$. Then the sequence $k \mapsto \int g_k$ has a limit in V. Moreover, if $k \mapsto h_k$ is another sequence in Step (n, V) that converges to f, then $\lim_{k \to \infty} \int h_k = \lim_{k \to \infty} \int g_k$.

Proof. Let $\epsilon > 0$ be given. Choose K so large that $k > K \Rightarrow |f - g_k|_1 < \epsilon/2$. Then if $k, j > K$, $|g_k - g_j|_1 \leq |g_k - f|_1 + |f - g_j|_1 < \epsilon$. It follows from Proposition 3 and Theorem A 9(c) that $\|\int g_k - \int g_j\| \leq \int \|g_k - g_j\| = |g_k - g_j|_1 < \epsilon$. That is, we have shown that $k \mapsto \int g_k$ is a Cauchy sequence in V. Since V is complete, this sequence has a limit. If $k \mapsto h_k$ is another sequence in Step (n, V) that converges to f, then for each $\epsilon > 0$, we can choose K so large that $k > K \Rightarrow |g_k - f|_1 < \epsilon/2$ and $|h_k - f|_1 < \epsilon/2$. Then if $k > K$, $\|\int g_k - \int h_k\| \leq \int \|g_k - h_k\| = |g_k - h_k|_1 \leq |g_k - f|_1 + |f - h_k|_1 < \epsilon$. Taking the limit as $k \to \infty$, we find that $\|\lim_{k \to \infty} \int g_k - \lim_{k \to \infty} \int h_k\| \leq \epsilon$. Since ϵ can be as small as we please, $\|\lim_{k \to \infty} \int g_k - \lim_{k \to \infty} \int h_k\| = 0 \Rightarrow \int g_k = \int h_k$ because $\|\ \|$ is a *norm* (not just a seminorm).

6.   DEFINITION. Let $\langle V, \|\ \| \rangle$ be a Banach space. $\mathfrak{L}^1(n, V)$, the space of **integrable functions** from $\mathbf{R}^n$ to V, is the closure of Step (n, V) in $\langle \mathfrak{N}(n, V), |\ |_1 \rangle$. That is, $f \in \mathfrak{L}^1(n, V)$ if, and only if, there is a sequence $k \mapsto s_k$ in Step (n, V) such that $\lim_{k \to \infty} |f - s_k|_1 = 0$. If $f \in \mathfrak{L}^1(n, V)$, then $\int f$ is the

common value assumed by all the limits $\lim_{k \to \infty} \int s_k$, where $\lim_{k \to \infty} | f - s_k |_1 = 0$, and $k \mapsto s_k$ is a sequence in Step (n, V). In other words, the relation $\int = \{\langle f, v \rangle \in \mathcal{L}^1(n, V) \times V: \text{there is a sequence } k \mapsto s_k \text{ in Step } (n, V) \text{ such that } \lim_{k \to \infty} | f - s_k |_1 = 0 \text{ and } \lim_{k \to \infty} \int s_k = v\}$ is a *function* (Proposition 5).

7.   REMARKS. (a) If $f \in$ Step (n, V), then the constant sequence $k \mapsto s_k = f$ satisfies $\lim_{k \to \infty} | f - s_k |_1 = 0$ and $\lim_{k \to \infty} \int s_k = \int f$. Thus the integral defined above is an extension of the integral defined earlier for step functions.

(b)   Using the Axiom of Choice, it can be shown that $\mathcal{L}^1(n, V)$ is not all of $\mathfrak{N}(n, V)$. However, the Axiom of Choice is actually needed to show this. Recently, it has been proved that if one is willing to abandon the Axiom of Choice, he can get a set theory (which is consistent if the usual set theory is consistent) in which $\mathcal{L}^1(n, V) = \mathfrak{N}(n, V)$.

8.   LEMMA. Let $\langle V, \| \ \| \rangle$ be a normed space, and let $g \in \mathfrak{F}(n, V)$.

(a) $g \in \mathfrak{N}(n, V) \Leftrightarrow \|g\| \in \mathfrak{N}(n, \mathbf{R})$.
(b)   If $g \in \mathfrak{N}(n, V)$, then $| g |_1 = | \|g\| |_1$.
(c)   If $\|f(x)\| \le \|g(x)\|$ for every $x \in \mathbf{R}^n$ and $g \in \mathfrak{N}(n, V)$, then $f \in \mathfrak{F}(n, V)$ and $| f |_1 \le | g |_1$.

Proof. (a) According to Definition 1, the sequence $k \mapsto s_k$ is a bounding sequence for g if, and only if, it is a bounding sequence for $\|g\|$. This also shows that

(b) $| g |_1 = | \|g\| |_1$.

(c)   Any bounding sequence $k \mapsto s_k$ for g is also a bounding sequence for f. It follows that $f \in \mathfrak{N}(n, V)$ and $| f | \le \lim_{k \to \infty} \int s_k$. (c) follows by taking the glb over all such bounding sequences.

9.   THEOREM. Let $\langle V, \| \ \| \rangle$ be a Banach space.

(a) $\mathcal{L}^1(n, V)$ is a linear subspace of $\mathfrak{N}(n, V)$.
(b)   If $f \in \mathcal{L}^1(n, V)$, then $\|f\| \in \mathcal{L}^1(n, \mathbf{R})$ and $\|\int f\| \le \int \|f\| = | f |_1$.
(c)   If $\langle W, \| \ \| \rangle$ is another Banach space and $L: V \to W$ is a bounded linear map, then for each $f \in \mathcal{L}^1(n, V)$, $L \circ f \in \mathcal{L}^1(n, W)$ and $L \int f = \int (L \circ f)$.
(d)   If $f, g \in \mathcal{L}^1(n, \mathbf{R})$, then $f \vee g$ and $f \wedge g$ are in $\mathcal{L}^1(n, \mathbf{R})$.
(e)   If $f, g \in \mathcal{L}^1(n, \mathbf{R})$ and $f \le g$, then $\int f \le \int g$.

Proof. Let $f, g \in \mathcal{L}^1(n, V)$. Then there are sequences $k \mapsto s_k$ and $k \mapsto s_k'$ in Step (n, V) such that $\lim_{k \to \infty} | f - s_k |_1 = \lim_{k \to \infty} | g - s_k' |_1 = 0$. There are also a bounding sequence $k \mapsto t_k$ for f and a bounding sequence $k \mapsto t_k'$ for g.

(a) $| (f + g) - (s_k + s_k') |_1 \le | f - s_k |_1 + | g - s_k' |_1 \to 0$ as $k \to \infty$ and for each $r \in \mathbf{R}$, $| rf - rs_k |_1 = |r| \, | f - s_k |_1 \to 0$ as $k \to \infty$. Thus $f + g$ and rf are in $\mathcal{L}^1(n, V)$.

(b)   By Lemma 8(a), $||f|| \in \mathfrak{N}(n, V)$. For each k, $|\ ||f|| - ||s_k||\ | \leq ||f - s_k||$, so by Lemma 8(c), $|\ ||f|| - ||s_k||\ |_1 \leq |f - s_k\ |_1 \to 0$ as $k \to \infty$. Thus $||f|| \in \mathcal{L}^1(n, \mathbf{R})$ and $\int ||f|| = \lim_{k \to \infty} \int ||s_k|| = \lim_{k \to \infty} |\ s_k\ |_1$ (Proposition 3 and Definition 6). But $|\ |f\ |_1 - |\ s_k\ |_1| \leq |f - s_k\ |_1 \to 0$ as $k \to \infty$, so $\lim_{k \to \infty} |\ s_k\ |_1 = |f\ |_1$. Finally, by Theorem A 9(b), for each $k \in \mathbf{N}$, $||\int s_k|| \leq \int ||s_k|| \Rightarrow ||\int f|| = \lim_{k \to \infty} ||\int s_k|| \leq \lim_{k \to \infty} \int ||s_k||$.

(c)   Evidently $k \mapsto |\ L\ |\ t_k$ is a bounding sequence for $L \circ f$, since $|||L \circ f(x)||| \leq |\ L\ |\ ||f(x)||$ for each $x \in \mathbf{R}^n$. Thus $L \circ f \in \mathfrak{N}(n, W)$. Moreover, $|||L \circ f(x) - L \circ s_k(x)||| \leq |\ L\ |\ ||f(x) - s_k(x)||$ for each $x \in \mathbf{R}^n$ and $k \in \mathbf{N}$. Therefore, by Lemma 8(b), $|\ L \circ f - L \circ s_k\ |_1 = |\ |||L \circ f - L \circ s_k|||\ |_1 \leq |\ L\ |\ |\ |f - s_k|\ |_1 = |\ L\ |\ |f - s_k\ |_1 \to 0$ as $k \to \infty$. It follows that $L \circ f \in \mathcal{L}^1(n, W)$. By Theorem A 9(b), $\int (L \circ f) = \lim_{k \to \infty} \int (L \circ s_k) = \lim_{k \to \infty} L \int s_k = L(\lim_{k \to \infty} \int s_k) = L \int f$.

(d)   According to Lemma A 10, for each $x \in \mathbf{R}^n$, $|f \wedge g(x)| = |f \wedge g(x) - 0| \leq |f(x) - 0| + |g(x) - 0| = |f(x)| + |g(x)|$. (b) shows that $|f|, |g| \in \mathcal{L}^1(n, \mathbf{R})$. Therefore, by Lemma 8(c), $f \wedge g \in \mathfrak{N}(n, \mathbf{R})$. Lemma A 10 then shows that for each $k \in \mathbf{N}$ and each $x \in \mathbf{R}^n$, $|f \wedge g(x) - s_k \wedge s_k'(x)| \leq |f(x) - s_k(x)| + |g(x) - s_k'(x)|$. By another application of Lemma 8(c), $|f \wedge g - s_k \wedge s_k'\ |_1 \leq |f - s_k\ |_1 + |\ g - s_k'\ |_1 \to 0$ as $k \to \infty$. It follows from Corollary A 5 that $f \wedge g \in \mathcal{L}^1(n, \mathbf{R})$. Since $f \vee g = -((-f) \wedge (-g))$, $f \vee g \in \mathcal{L}^1(n, \mathbf{R})$ as well.

(e)   If $f(x) \leq g(x)$ for all $x \in \mathbf{R}^n$, then for every x, $|f(x) - s_k \wedge s_k'(x)| = |f \wedge g(x) - s_k \wedge s_k'(x)| \leq |f(x) - s_k(x)| + |g(x) - s_k'(x)|$. It follows that $|f - s_k \wedge s_k'\ |_1 \leq |f - s_k\ |_1 + |\ g - s_k'\ |_1 \to 0$ as $k \to \infty$. Therefore $\int f = \lim_{k \to \infty} \int s_k \wedge s_k' \leq \lim_{k \to \infty} \int s_k' = \int g$, using Theorem A 9(d) and the fact that $s_k \wedge s_k'(x) \leq s_k'(x)$ for every $x \in \mathbf{R}^n$.

10. LEBESGUE'S DOMINATED CONVERGENCE THEOREM. Let $\langle V, ||\ ||\rangle$ be a Banach space and let $k \mapsto f_k$ be a sequence in $\mathcal{L}^1(n, V)$ with the following properties:

(a)   there is a function $f \in \mathfrak{F}(n, V)$ such that $\lim_{k \to \infty} f_k(x) = f(x)$ for every $x \in \mathbf{R}^n$,
   and

(b)   there is a function $b \in \mathfrak{N}(n, \mathbf{R})$ such that $||f_k(x)|| \leq b(x)$ for every $x \in \mathbf{R}^n$.

Then $f \in \mathcal{L}^1(n, \mathbf{R})$ and $\int f = \lim_{k \to \infty} \int f_k$.

Proof. $f \in \mathfrak{N}(n, V)$ by Lemma 8, since for every $x \in \mathbf{R}^n$, $||f(x)|| = \lim_{k \to \infty} ||f_k(x)|| \leq b(x)$.

Case 1. Suppose $V = \mathbf{R}$ and that for each $k \in \mathbf{N}$ and each $x \in \mathbf{R}^n$, $0 \leq f_k(x) \leq f_{k+1}(x)$. Put $g_1 = f_1$, $g_k = f_k - f_{k-1}$ for $k > 1$, so that $f_k = \sum_{j=1}^{k} g_j$ and $f(x) = \sum_{j=1}^{\infty} g_j(x)$ for every $x \in \mathbf{R}^n$. Since $g_j(x) \geq 0$ for every $j \in \mathbf{N}$, $|g_j(x)| = g_j(x)$. It follows then from Lemma 8(c) and Theorem 9(b) that

$| g |_1 + \cdots + | g_k |_1 = \int (|g_1| + \cdots + |g_k|) = \int (g_1 + \cdots + g_k) = \int f_k = \int |f_k| = |f|_1 \leq |b|_1$ for every $k \in \mathbf{N}$. By Corollary VI B 10, the series $| g_1 |_1 + | g_2 |_1 + | g_3 |_1 + \cdots$ converges to a limit $L \leq |b|_1$. Let $\epsilon > 0$ be given. For each $j \in \mathbf{N}$, we can choose a bounding sequence $k \mapsto s_{jk}$ for g_j such that $\lim_{k \to \infty} s_{jk} < | g_j |_1 + 2^{-j}\epsilon$. For each $m \in \mathbf{N}$ and each $k \geq m$, put $s'_{mk} = \sum_{j=m}^{k} s_{jk}$.

(i) For each $x \in \mathbf{R}^n$, $0 \leq s'_{mk}(x) \leq s'_{mk+1}(x)$.

(ii) For each $i \geq m$, $\lim_{k \to \infty} s'_{mk}(x) \geq \lim_{k \to \infty} \sum_{j=m}^{i} s_{jk}(x) \geq \sum_{j=m}^{i} g_j(x)$, so that $\lim_{k \to \infty} s'_{mk}(x) \geq \lim_{i \to \infty} \sum_{j=m}^{i} g_j(x) = \sum_{j=m}^{\infty} g_j(x) = f(x) - f_{m-1}(x)$, provided that $\{s'_{mk}(x) : k \in \mathbf{N}\}$ is bounded.

(iii) $\int s'_{mk} = \sum_{j=m}^{k} \int s_{jk} < \sum_{j=m}^{k} (| g_j |_1 + 2^{-i}\epsilon) < \sum_{j=m}^{\infty} (| g_j |_1 + 2^{-i}\epsilon) = (L - \sum_{j=1}^{m-1} | g_j |_1) + 2^{-m+1}\epsilon$.

(i), (ii), and (iii) show that $k \mapsto s'_{mk}$ is a bounding sequence for $f - f_{m-1}$, and that $| f - f_{m-1} |_1 < (L - \sum_{j=1}^{m-1} | g_j |_1) + 2^{-m+1}\epsilon \to 0$ as $m \to \infty$. Moreover, for each m, we can choose $h_m \in$ Step $(n, \mathbf{R})$ such that $|f_m - h_m|_1 < 1/m$, by Definition 6. Then $| f - h_m |_1 \leq | f - f_m |_1 + | f_m - h_m |_1 \to 0$ as $m \to \infty$. It follows that $f \in \mathcal{L}^1(n, \mathbf{R})$. $|\int f - \int f_m| \leq | f - f_m |_1 \to 0$ as $m \to \infty$, by Theorem 9(b). Thus $\int f = \lim_{m \to \infty} \int f_m$.

Case 2. Suppose $V = \mathbf{R}$ and that either (i) $f_k(x) \leq f_{k+1}(x)$ for every $x \in \mathbf{R}^n$ and every $k \in \mathbf{N}$, or (ii) $f_k(x) \geq f_{k+1}(x)$ for every $x \in \mathbf{R}^n$ and every $k \in \mathbf{N}$. Then if (i) is true, the sequence $k \mapsto f_k - f_1$ satisfies the hypotheses of Case 1 $(|f_k(x) - f_1(x)| \leq |f_k(x)| + |f_1(x)| \leq 2b(x))$. We conclude that $f - f_1 \in \mathcal{L}^1(n, \mathbf{R})$ and that $\int (f - f_1) = \lim_{k \to \infty} \int (f_k - f_1)$, from which it is apparent that $f = f_1 + (f - f_1) \in \mathcal{L}^1(n, \mathbf{R})$ and that $\int f = \lim_{k \to \infty} (\int f_1 + \int (f_k - f_1)) = \lim_{k \to \infty} \int f_k$. If (ii) is true, we can apply what we have just proved to the sequence $k \mapsto -f_k$.

Case 3. Suppose $V = \mathbf{R}$, without any further restriction on $k \mapsto f_k$. For each $m \in \mathbf{N}$, $|f_m \vee f_{m+1} \vee \cdots \vee f_{m+k}(x)| \leq b(x)$, and the sequence $k \mapsto f_m \vee f_{m+1} \vee \cdots \vee f_{m+k}(x)$ is increasing. By Theorem VI B 9 there is a $g_m(x) \in \mathbf{R}$ such that $\lim_{k \to \infty} f_m \vee f_{m+1} \vee \cdots \vee f_{m+k}(x) = g_m(x)$. Since the sequence $k \mapsto f_m \vee f_{m+1} \vee \cdots \vee f_{m+k}$ satisfies the hypotheses of Case 2(i), $g_m \in \mathcal{L}^1(n, \mathbf{R})$ and $\lim_{k \to \infty} \int f_m \vee f_{m+1} \vee \cdots \vee f_{m+k} = \int g_m$. For every $k \geq m + 1$, $f_m \vee f_{m+1} \vee \cdots \vee f_{m+k}(x) \geq f_{m+1} \vee \cdots \vee f_{m+k}(x)$ for all $x \in \mathbf{R}^n$. Taking the limit as $m \to \infty$, we find that $g_m(x) \geq g_{m+1}(x)$. Moreover $|g_m(x)| \leq b(x)$ for every $x \in \mathbf{R}^n$. Let us show that $\lim_{m \to \infty} g_m(x) = f(x)$ for every $x \in \mathbf{R}^n$. Indeed, for each $\epsilon > 0$, there is an $M \in \mathbf{N}$ so large that $k > M \Rightarrow |f(x) - f_k(x)| < \epsilon \Rightarrow |f(x) - f_m \vee \cdots \vee f_k(x)| < \epsilon$ whenever $k \geq m > M \Rightarrow |f(x) - g_m(x)| \leq \epsilon$ whenever $m > M$. Therefore, the sequence $m \mapsto g_m$ satisfies the hypotheses of Case 2(ii), so that $f \in \mathcal{L}^1(n, \mathbf{R})$, and $\int f = \lim_{m \to \infty} \int g_m$. Similarly, if $h_m = \lim_{k \to \infty} f_m$

$\wedge f_{m+1} \wedge \cdots \wedge f_k$, then $h_m \in \mathfrak{L}^1(n, \mathbf{R})$ for each $m \in \mathbf{N}$, and $\int f = \lim_{m \to \infty} \int h_m$. However, for every $m \in \mathbf{N}$ and every $x \in \mathbf{R}^n$, $h_m(x) \leq f_m(x) \leq g_m(x) \Rightarrow \int h_m \leq \int f_m \leq \int g_m$. Since $\lim_{m \to \infty} \int h_m = \lim_{m \to \infty} \int g_m = \int f$, $\lim_{m \to \infty} \int f_m = \int f$ as well.

We can now complete the proof of Theorem 10, without further restrictions on V or $k \mapsto f_k$. In general, if $k \mapsto f_k$ satisfies (a) and (b), then for each $m \in \mathbf{N}$ and each $x \in \mathbf{R}^n$, $\lim_{k \to \infty} ||f_k(x) - f_m(x)|| = ||f(x) - f_m(x)||$ and $||f_k(x) - f_m(x)|| \leq 2b(x)$. By Case 3, $||f - f_m|| \in \mathfrak{L}^1(n, \mathbf{R})$ and by Case 3, Lemma 8(b), and Theorem 9(b), $\mathbf{I} f - f_m \mathbf{l}_1 = \mathbf{I} \, ||f - f_m|| \, \mathbf{l}_1 = \int ||f - f_m|| = \lim_{k \to \infty} \int ||f_k - f_m||$. For each $x \in \mathbf{R}^n$, $\lim_{m \to \infty} ||f(x) - f_m(x)|| = 0$. By another application of Case 3, $\lim_{m \to \infty} \int ||f - f_m|| = 0$, and thus $\lim_{m \to \infty} \mathbf{I} f - f_m \mathbf{l}_1 = 0$. For each m, we can choose $g_m \in$ Step (n, V) such that $\mathbf{I} f_m - g_m \mathbf{l}_1 < 1/m$. Then $\mathbf{I} f - g_m \mathbf{l}_1 \leq \mathbf{I} f - f_m \mathbf{l}_1 + \mathbf{I} f_m - g_m \mathbf{l}_1 \to 0$ as $m \to \infty$, so that $f \in \mathfrak{L}^1(n, V)$. $||\int f - \int f_m|| \leq \mathbf{I} f - f_m \mathbf{l}_1 \to 0$ as $m \to \infty \Rightarrow \int f = \lim_{m \to \infty} \int f_m$. ●

11. COROLLARY. $\mathfrak{L}^1(n, V)$ is closed in $\mathfrak{N}(n, V)$.

Proof. This was established twice in the proof of Theorem 10, once at the end of the proof of Case 1, and again at the end of the proof of the Theorem. It also follows from Definition IV C 7, which says that a closure is closed.

12. COROLLARY (BEPPO LEVI'S THEOREM). Let $k \mapsto f_k$ be a sequence in $\mathfrak{L}^1(n, \mathbf{R})$ with the following properties:

(a) there is a function $f \in \mathfrak{F}(n, \mathbf{R})$ such that $\lim_{k \to \infty} f_k(x) = f(x)$ for every $x \in \mathbf{R}^n$,

(b) for every $x \in \mathbf{R}^n$ and every $k \in \mathbf{N}$, $f_k(x) \leq f_{k+1}(x)$, and

(c) $\{\int f_k : k \in \mathbf{N}\}$ is bounded.

Then $f \in \mathfrak{L}^1(n, \mathbf{R})$ and $\int f = \lim_{k \to \infty} \int f_k$.

Proof. This is really what was established in Case 1 of the proof of Theorem 10.

13. THEOREM. If $\langle V, || \; || \rangle$ is a Banach space, the seminormed space $\langle \mathfrak{L}^1(n, V), \mathbf{I} \; \mathbf{l}_1 \rangle$ is complete.

Proof. Let $k \mapsto f_k$ be a Cauchy sequence in $\langle \mathfrak{L}^1(n, V), \mathbf{I} \; \mathbf{l}_1 \rangle$, so that $\mathbf{I} f_j - f_k \mathbf{l}_1 \to 0$ as $j, k \to \infty$. For each $m \in \mathbf{N}$, choose $N(m) > N(m-1)$ so that $\mathbf{I} f_j - f_k \mathbf{l}_1 < 2^{-m}$ whenever $j, k \geq N(m)$. Put $g_1 = f_{N(1)}$, and for $m > 1$, put $g_m = f_{N(m)} - f_{N(m-1)}$, so that $\mathbf{I} g_m \mathbf{l}_1 < 2^{-m}$. Then the series $\sum_{m=1}^{\infty} \mathbf{I} g_m \mathbf{l}_1$ converges. Suppose that for every $x \in \mathbf{R}^n$, the series $\sum_{m=1}^{\infty} ||g_m(x)||$ converged. Then by Proposition VI B 2 the series $\sum_{m=1}^{\infty} g_m(x)$ would converge to a limit $f(x)$ in V. It would then be an easy matter to show, using Theorem 10, that $f \in \mathfrak{L}^1(n, V)$ and that $\lim_{m \to \infty} \mathbf{I} f_m - f \mathbf{l}_1 = 0$. What makes the proof of Theorem 13 a bit troublesome is that $S = \{x \in \mathbf{R}^n : \sum_{m=1}^{\infty} ||g_m(x)||$ does not converge} need not be empty. The strategy of proof is to show that in any case, S is so small that we can alter g_1, g_2, g_3, ... on S so as to obtain a new sequence h_1, h_2, h_3, ... that does sum to a function $f \in \mathcal{L}^1(n, V)$, and still have $\lim_{m \to \infty} I\, f_m - f\, l_1 = 0$.

For each $m \in \mathbf{N}$, let $k \mapsto s_{mk}$ be a bounding sequence for $||g_m||$ such that $\lim_{k \to \infty} \int s_{mk} < I\, g_m\, l_1 + 2^{-m}$, and for each $k \in \mathbf{N}$ let $s'_k = \sum_{m=1}^k s_{mk}$. As in Case 1 of the proof of Theorem 10, we have the following. (i) $s'_k \leq s'_{k+1}$. (ii) If $x \in \mathbf{R}^n$ and $\{s'_k(x): k \in \mathbf{N}\}$ is bounded, then for each $i \in \mathbf{N}$, $\lim_{k \to \infty} s'_k(x) \geq$
$$\lim_{k \to \infty} \sum_{m=1}^i s_{mk}(x) \geq \sum_{m=1}^i ||g_m(x)|| \Rightarrow \sum_{m=1}^\infty ||g_m(x)|| \leq \lim_{k \to \infty} s'_k(x) \Rightarrow x \notin S.$$
(iii) $\int s'_k(x) = \sum_{m=1}^k \int s_{mk}(x) < \sum_{m=1}^k (I\, g_m\, l_1 + 2^{-m}) < \sum_{m=1}^\infty (I\, g_m\, l_1 + 2^{-m}) = 1 + \sum_{m=1}^\infty I\, g_m\, l_1 = B$. Now put $h_m(x) = g_m(x)$ if $x \in \mathbf{R}^n - S$, and $h_m(x) = 0$ if $x \in S$, for each $m \in \mathbf{N}$. If $g_m(x) - h_m(x) \neq 0$, then $x \in S$, so that $\{s'_k(x): k \in \mathbf{N}\}$ is *not* bounded according to (ii). It follows from Definition 1 that for *every* $N \in \mathbf{N}$, $k \mapsto (1/N)s'_k$ is a bounding sequence for $h_m - g_m$. Then $I\, h_m - g_m\, l_1 = 0$. According to Corollary 11, $h_m \in \mathcal{L}^1(n, V)$.

Put $b(x) = \sum_{m=1}^\infty ||h_m(x)||$ for every $x \in \mathbf{R}^n$. Then if $x \in S$, $b(x) = \sum_{m=1}^\infty 0 = 0$. If $x \in \mathbf{R}^n - S$, $b(x) = \sum_{m=1}^\infty ||g_m(x)||$. (i), (ii), and (iii) above show that $k \mapsto s'_k$ is a bounding sequence for b. Therefore $b \in \mathfrak{R}(n, V)$. As we have already remarked, $\sum_{m=1}^\infty h_m(x)$ converges absolutely to an element $f(x) \in V$ for each $x \in \mathbf{R}^n$. According to Theorem 10, $f \in \mathcal{L}^1(n, V)$. For any $m \in \mathbf{N}$,

$$I\, f - f_{N(m)}\, l_1 = I\, f - \sum_{j=1}^m g_j\, l_1 = I\, f - \sum_{j=1}^m h_j + \sum_{j=1}^m (h_j - g_j)\, l_1$$

$$\leq I\, f - \sum_{j=1}^m h_j\, l_1 + \sum_{j=1}^m I\, h_j - g_j\, l_1 = I\, f - \sum_{j=1}^m h_j\, l_1 = \int ||f - \sum_{j=1}^m h_j||.$$

By another application of Theorem 10 (to the sequence $m \mapsto ||f - \sum_{j=1}^m h_j||$), we conclude that $\lim_{m \to \infty} \int ||f - \sum_{m=1}^\infty h_j|| = 0$. It follows that $\lim_{m \to \infty} I\, f - f_{N(m)}\, l_1 = 0$, and then, using Corollary VII A 9, that $\lim_{k \to \infty} I\, f - f_k\, l_1 = 0$.

14. DEFINITION. A subset S of $\mathbf{R}^n$ is a **λ-null** set if for every normed space $\langle V, ||\ ||\rangle$ and every function $f \in \mathfrak{F}(\mathbf{R}^n, V)$ that vanishes on $\mathbf{R}^n - S$, $f \in \mathfrak{R}(n, V)$ and $I\, f\, l_1 = 0$. A property P of points in $\mathbf{R}^n$ holds **λ-almost everywhere** if $\{x \in \mathbf{R}^n: x \text{ does not have the property } P\}$ is a λ-null set. Thus a sequence $k \mapsto f_k$ in $\mathfrak{F}(\mathbf{R}^n, V)$ converges λ-almost everywhere if $k \mapsto f_k(x)$ converges except for those x in a λ-null set. Similarly, two functions in $\mathfrak{F}(\mathbf{R}^n, V)$ are **λ-almost equal** if they are equal except on a λ-null set.

15. COROLLARY. Theorem 10 and Corollary 12 remain true if their hypotheses are weakened by requiring only that $\lim_{k \to \infty} f_k(x) = f(x)$ and that $||f_k(x)|| \leq b(x)$ λ-almost everywhere.

Proof. Let S be the λ-null set on which $k \mapsto f_k(x)$ does *not* converge to $f(x)$ or on which $||f_k(x)|| > b(x)$ for at least one k. Replace each f_k by the function f_k' defined by $f_k'(x) = 0$ if $x \in S$, $f_k'(x) = f_k(x)$ if $x \in \mathbf{R}^n - S$, and let $f'(x) = \lim_{k \to \infty} f_k'(x)$ for each $x \in \mathbf{R}^n$. Then Theorem 10 or Corollary 12 shows that $f' \in \mathcal{L}^1(n, V)$ or $f' \in \mathcal{L}^1(n, \mathbf{R})$ and that $\lim_{k \to \infty} \int f_k' = \int f'$. The argument of Theorem 13 shows that $f \in \mathcal{L}^1(n, V)$ or $f \in \mathcal{L}^1(n, \mathbf{R})$, that $\int f = \int f'$, and that for each $k \in \mathbf{N}$, $\int f_k' = \int f_k$. $\bullet$

16. COROLLARY. Let $\langle V, || \ \, || \rangle$ be a Banach space, let $f \in \mathcal{L}^1(n, V)$, and let $k \mapsto f_k$ be a sequence in $\mathcal{L}^1(n, V)$ such that $\lim_{k \to \infty} | f - f_k |_1 = 0$. Then there is a subsequence of $k \mapsto f_k$ that converges to f λ-almost everywhere.

Proof. This was proved in the course of proving Theorem 13.

17. PROPOSITION. Let $E \subseteq \mathbf{R}^n$. Suppose there are a normed space $\langle W, ||| \ \, ||| \rangle$ and a function $f \in \mathfrak{N}(n, W)$ such that $x \in E \Rightarrow f(x) \neq 0$ and that $| f |_1 = 0$. Then E is a λ-null set. In particular, let χ_E be the function that takes the value 1 on E and the value 0 on $\mathbf{R}^n - E$. Then E is a λ-null set if, and only if, $\chi_E \in \mathfrak{N}(n, \mathbf{R})$ and $| \chi_E |_1 = 0$.

Proof. Let $\langle V, || \ \, || \rangle$ be a normed space, and let g be any function from $\mathbf{R}^n$ to V that vanishes outside E. By Lemma 8, for every $k \in \mathbf{N}$, $||g|| \wedge k|||f||| \in \mathfrak{N}(n, \mathbf{R})$ and $| \ ||g|| \wedge k|||f||| \ |_1 = 0$. But $\lim_{k \to \infty} ||g|| \wedge k|||f|||(x) = ||g(x)||$ for every $x \in \mathbf{R}^n$. Therefore, by Beppo Levi's Theorem, $||g|| \in \mathcal{L}^1(n, \mathbf{R})$ and $\int ||g|| = 0$. It follows that $g \in \mathfrak{N}(n, V)$ and $| g |_1 = \int ||g|| = 0$, by Lemma 8.

18. COROLLARY. If $E(k)$ is a λ-null set in $\mathbf{R}^n$ for every $k \in \mathbf{N}$, then $\mathbf{U}_{k=1}^\infty E(k)$ is a λ-null set.

Proof. Let $E = \mathbf{U}_{k=1}^\infty E_k$. Then $\chi_E(x) = \lim_{k \to \infty} \chi_{E(1)} \vee \cdots \vee \chi_{E(k)}(x) \leq \lim_{k \to \infty} \chi_{E(1)}(x) + \cdots + \chi_{E(k)}(x)$ for every $x \in \mathbf{R}^n$. Therefore $\chi_E \in \mathcal{L}^1(n, \mathbf{R})$ and $| \chi_E |_1 = 0$, by Beppo Levi's Theorem.

19. DEFINITION. Let S be a subset of $\mathbf{R}^n$, and let $\langle V, || \ \, || \rangle$ be a Banach space. A function $f \in \mathfrak{F}(S, V)$ is **integrable** if, and only if, the function $f^*: \mathbf{R}^n \to V$ defined by $f^*(x) = f(x)$ if $x \in S$ and $f^*(x) = 0$ if $x \in \mathbf{R}^n - S$ belongs to $\mathcal{L}^1(n, V)$. If f is integrable, we define $\int f$ to be $\int f^*$ and define $| f |_1$ to be $| f^* |_1$. The set of all integrable functions in $\mathfrak{F}(S, V)$ is $\mathcal{L}^1(S, V)$.

20. PROPOSITION. Let $\langle V, || \ \, || \rangle$ be a Banach space, let S be a compact subset of $\mathbf{R}^n$, and let $f: S \to V$ be continuous. Then $f \in \mathcal{L}^1(S, V)$.

Proof. The function $||f||: S \to \mathbf{R}$ is continuous, so there is a bound B for

$\{||f(x)||: x \in S\}$ (Theorems VII B 1 and VII B 3). Moreover, f is uniformly continuous (Corollary VII C 3), so for each $m \in M$ there is a $\delta_m > 0$ such that $||f(x) - f(y)|| < 1/m$ whenever $x, y \in S$ and $||x - y||_\infty < 1/m$. Since S is bounded (Theorem VII B 3), there is an open rectangle R in $\mathbf{R}^n$ that contains S. For each $m \in \mathbf{N}$, let $\mathcal{G}_m$ be a grid in $\mathbf{R}^n$ such that $R(\mathcal{G}_m) = R$, and such that if x and y belong to the same cell of $\mathcal{G}_m$, then $||x - y||_\infty < \delta_m$. Then define a step function g_m as follows. Let R' be a cell of $\mathcal{G}_m$. If $S \cap R' \neq \varnothing$, choose $y \in S \cap R'$ and put $g_m(x) = g_m(y)$ for each $x \in R'$. If $S \cap R' = \varnothing$, put $g_m(x) = \mathbf{0}$ for each $x \in R'$. If x is not in any cell of $\mathcal{G}_m$, put $g_m(x) = f^*(x)$. Also let b be the step function from $\mathbf{R}^n$ to $\mathbf{R}$ defined as follows. If $x \in R^c$, $b(x) = B$. If $x \in \mathbf{R}^n - R^c$, $b(x) = 0$. Clearly $b \in \mathrm{Step}\,(n, \mathbf{R})$ and $||g_m(x)|| \leq b(x)$ for every $x \in \mathbf{R}^n$. If x does not belong to any cell of $\mathcal{G}_m$, or if x belongs to a cell of $\mathcal{G}_m$ that does not intersect S, then $g_m(x) = f^*(x)$. If $x \in S$, x is in the cell R' of $\mathcal{G}_m$, and $g_m(R') = f(y)$, where $y \in R' \cap S$, then $||x - y||_\infty < \delta_m \Rightarrow 1/m > ||f(x) - f(y)|| = ||f^*(x) - g_m(x)||$. If $x \notin S$, then because S is compact and the function $y \mapsto ||x - y||_\infty$ is continuous, there is an $\epsilon > 0$ such that $||x - y||_\infty \geq \epsilon$ for every $y \in S$. In particular, if x and y belong to the same cell of $\mathcal{G}_m$ and if $1/m < \epsilon$, then $y \notin S$. Therefore $m > 1/\epsilon \Rightarrow f^*(x) = g_m(x) = 0$. It follows that $\lim_{m \to \infty} g_m(x) = f^*(x)$ for every $x \in \mathbf{R}^n$. By Lebesgue's Dominated Convergence Theorem, $f^* \in \mathcal{L}^1(n, V)$, so that $f \in \mathcal{L}^1(S, V)$.

21. COROLLARY. Let $\langle V, ||\ || \rangle$ be a Banach space, let S be a closed subset of $\mathbf{R}^n$, and let $f: S \to V$ be continuous. If there is a nonnegative function $b \in \mathfrak{N}(n, \mathbf{R})$ such that $||f(x)|| \leq b(x)$ for every $x \in S$, then $f \in \mathcal{L}^1(S, V)$.

Proof. For each $m \in \mathbf{N}$, let $f_m^*(x) = f(x)$ if $x \in S \cap B_m^c(\mathbf{0})$, and let $f_m^*(x) = \mathbf{0}$ if $x \in \mathbf{R}^n - S \cap B_m^c(\mathbf{0})$. Since $S \cap B_m^c(\mathbf{0})$ is closed and bounded, it is compact (Theorem VII B 3). Therefore $f_m^* \in \mathcal{L}^1(n, V)$ by Proposition 20. Moreover, $||f_m^*(x)|| \leq b(x)$ for every $x \in \mathbf{R}^n$ and every $m \in \mathbf{N}$, and $\lim_{m \to \infty} f_m^*(x) = f^*(x)$ for every $x \in \mathbf{R}^n$. Theorem 10 implies that $f^* \in \mathcal{L}^1(n, V)$, so that $f \in \mathcal{L}^1(S, V)$.

22. COROLLARY. Let $\langle V, ||\ || \rangle$ be a Banach space, let S be an open subset of $\mathbf{R}^n$, and let $f: S \to V$ be continuous. If there is a nonnegative $b \in \mathfrak{N}(n, \mathbf{R})$ such that $||f(x)|| \leq b(x)$ for every $x \in S$, then $f \in \mathcal{L}^1(S, V)$.

Proof. For each $m \in \mathbf{N}$, let $S_m = \{x \in S: B_{1/m}(x) \subseteq S\}$. If $\{x_k\} \subseteq S_m$, $\lim_{k \to \infty} x_k = x$ and $||y - x||_\infty < 1/m$, then for some $k \in \mathbf{N}$, $||y - x_k||_\infty < 1/m$, so that $y \in B_{1/m}(x_k) \subseteq S$. It follows that $B_{1/m}(x) \subseteq S$. This shows that S_m is closed. Clearly $\bigcup \{S_m: m \in \mathbf{N}\} = S$, because S is open. For each $m \in \mathbf{N}$ and $x \in \mathbf{R}^n$, put $f_m^*(x) = f(x)$ if $x \in S_m$, $f_m^*(x) = \mathbf{0}$ if $x \in \mathbf{R}^n - S_m$. According to Corollary 21, $f_m^* \in \mathcal{L}^1(n, V)$. For each $x \in \mathbf{R}^n$, $\lim_{m \to \infty} f_m^*(x) = f^*(x)$ and $||f_m^*(x)|| \leq b(x)$ for each $m \in \mathbf{N}$. Therefore, by another application of Theorem 10, $f^* \in \mathcal{L}^1(n, V)$, so $f \in \mathcal{L}^1(S, V)$.

23. REMARKS ON MEASURES. For each subset S of $\mathbf{R}^n$, let χ_S be the function in $\mathcal{F}(\mathbf{R}^n, \mathbf{R})$ that takes the value 1 on S and the value 0 on $\mathbf{R}^n - S$. If R is an open rectangle, $|\chi_R|_1 = \lambda(R) = $ volume (R). It is natural then to define the **Lebesgue outer measure** of a set S as follows. If $\chi_S \in \mathfrak{N}(n, \mathbf{R})$, $\lambda^*(S) = |\chi_S|_1$. If $\chi_S \notin \mathfrak{N}(n, \mathbf{R})$, $\lambda^*(S) = +\infty$. If for every open rectangle R in $\mathbf{R}^n$, $\chi_S \wedge \chi_R = \chi_{S \cap R}$ belongs to $\mathcal{L}^1(n, \mathbf{R})$, we say that S is *measurable*, and put $\lambda(S) = \lambda^*(S)$. We thus extend the n-dimensional volume defined above to the class of measurable sets. It is not hard to show, using what we have proved above, that the Lebesgue outer measure has the following properties.

(a) $\lambda^*(\varnothing) = 0$.
(b) If $S \subseteq T$, $\lambda^*(S) \leq \lambda^*(T)$.
(c) $\lambda^*(\bigcup_{k=1}^{\infty} S_k) \leq \sum_{k=1}^{\infty} \lambda^*(S_k)$.
(d) If $S_1, S_2, S_3, \ldots$ are *mutually disjoint measurable* sets, then $\lambda^*(\bigcup_{k=1}^{\infty} S_k) = \lambda(\bigcup_{k=1}^{\infty} S_k) = \sum_{k=1}^{\infty} \lambda(S_k)$.

There are other outer measures on $\mathbf{R}^n$ having the properties (a) through (c). For example, if $f \in \mathcal{L}^1(n, \mathbf{R})$ is nonnegative, the function $S \mapsto |f\chi_S|_1$ is an outer measure on $\mathbf{R}^n$. The study of such things is *measure theory*. It is the mature form of the Method of Exhaustion.

In Lebesgue's thesis, in which the type of integral presented here was first developed, the exposition began by defining an outer measure and proceeded from that to the definition of an integral. The "integrals first" approach was developed later by Daniell. The present approach is similar to that of Bourbaki. Vector-valued integrals appeared long ago, but vector-valued Lebesgue integrals are a comparatively recent development.

24. REMARKS ON RIEMANN INTEGRATION. An apparent simplification of Definition 1 leads to the theory of *Riemann integration*, which is much inferior to that of Lebesgue. Let $\langle V, \| \ \| \rangle$ be a normed space. Let us call a function $f \in \mathcal{F}(\mathbf{R}^n, V)$ **R-normable** if there exists a single function $s \in$ Step $(n, \mathbf{R})$ such that $s(x) \geq \|f(x)\|$ for each $x \in \mathbf{R}^n$. Then put $|f|_R = $ glb $\{\int s : s \in$ Step $(n, \mathbf{R})$ and $s(x) \geq \|f(x)\|$ for every $x \in \mathbf{R}^n\}$, and let $\mathfrak{N}_R(n, V)$ be the class of R-normable functions. We note at once that $\mathfrak{N}_R(n, V) \subseteq \mathfrak{N}(n, V)$ and that $|f|_R \geq |f|_1$, since any such **bounding function** s for a function $f \in \mathfrak{N}_R(n, V)$ gives rise to a *constant* bounding sequence for f, $k \mapsto s_k = s$.

The development of the Riemann integral then proceeds like that of the Lebesgue integral through Propositions 2, 3, and 5, and Definition 6, although as might be expected, the work is somewhat easier because all bounding sequences for the Riemann theory are constant sequences. The appropriate form of Proposition 8 remains true in Riemann's theory, as does Theorem 9. It is only when we come to the crucial Dominated Convergence Theorem that Riemann's theory fails. The Dominated Convergence Theorem is *false* in the Riemannian theory. It should not be hard to find where nonconstant bounding sequences were needed in its proof. The best theorem that is true in the Riemannian theory *assumes* that the limit function f is R-integrable as part of the hypothesis. The only honest proofs I know of this theorem use at least a part of the Lebesgue theory. Corollary 12 and Theorem 13 are also false in the Riemannian theory.

Because $\mathfrak{N}_R(n, V) \subseteq \mathfrak{N}(n, V)$ and $| f |_R \geq | f |_1$ for any $f \in \mathfrak{N}_R(n, V)$, it follows that if $f \in \mathfrak{N}_R(n, V)$ and $k \mapsto f_k$ is a sequence in Step (n, V) such that $\lim_{k \to \infty} | f - f_k |_R = 0$, then $\lim_{k \to \infty} | f - f_k |_1 = 0$, so that $f \in \mathcal{L}^1(n, V)$. For a similar reason, if $f \in R(n, V)$ (the closure of Step (n, V) in $\langle \mathfrak{N}_R(n, V), | \ |_R \rangle$), then the Riemann integral of f equals the Lebesgue integral of f.

A simple example points up the difference. Let $k \mapsto r_k$ be an enumeration of the set of rational numbers, and let $S(k) = \{r_1, \ldots, r_k\}$ for each $k \in \mathbf{N}$. For each $k \in \mathbf{N}$, $\chi_{S(k)}$ is a step function from $\mathbf{R}$ to $\mathbf{R}$, and $| \chi_{S(k)} |_R = | \chi_{S(k)} |_1 = \int \chi_{S(k)} = 0$. According to Beppo Levi's Theorem 12, $\chi_{\mathbf{Q}} = \lim_{k \to \infty} \chi_{S(k)} \in \mathcal{L}^1(1, \mathbf{R})$ and $\int \chi_{\mathbf{Q}} = 0$. However, $\chi_{\mathbf{Q}}$ does not even belong to $\mathfrak{N}_R(1, \mathbf{R})$. A fortiori it does not have a Riemann integral.

Exercises

1. Prove that the constant function $f : x \mapsto 1$ is not normable. *Hint:* If $k \mapsto s_k$ were a bounding sequence for f and R is any open rectangle, then $k \mapsto s_k$ would be a bounding sequence for χ_R. Hence $\lim_{k \to \infty} \int s_k \geq | R |_1$.

2. Show that if C is any countable subset of $\mathbf{R}^n$, then C is a λ-null set.

*3. List the definitions and theorems in Chapters IV, VII, and VIII that apply to seminormed spaces and pseudometric spaces without any significant change. (A **pseudometric space** is a pair $\langle M, d \rangle$ consisting of a set M and a function $d : M \times M \to \mathbf{R}$ satisfying all the axioms for a metric space except that $d(x, y)$ may be 0 even if $x \neq y$.)

*4. Let $\langle V, \| \ \| \rangle$ be a Banach space, and let $L^1(n, V) = \{\{f\}^c : f \in \mathcal{L}^1(n, V)\}$, where $\{f\}^c = \{g \in \mathcal{L}^1(n, V) : | g - f |_1 = 0\}$.

    (a) Show that each of the following relations is a function:

$$+ = \{\langle\langle \{f\}^c, \{g\}^c \rangle, \{h\}^c \rangle \in (L^1(n, V) \times L^1(n, V)) \times L^1(n, V) : f + g = h\},$$

$$\cdot = \{\langle\langle r, \{f\}^c \rangle, \{g\}^c \rangle \in (\mathbf{R} \times L^1(n, V)) \times L^1(n, V) : g = rf\},$$

$$| \ |_1 = \{\langle \{f\}^c, r \rangle \in L^1(n, V) \times \mathbf{R} : r = | f |_1\}.$$

    Hint: You have to show, for example, that if $\{f'\}^c = \{f\}^c$ and $\{g'\}^c = \{g\}^c$, then $\{f' + g'\}^c = \{f + g\}^c$.

    (b) Show that with these definitions of addition, scalar product, and norm, that $\langle\langle L^1(n, V), +, \cdot \rangle, | \ |_1 \rangle$ is a Banach space. *Hint:* It is routine to prove that $\langle L^1(n, V), +, \cdot \rangle$ is a vector space, that $| \ |_1$ is a seminorm on $L^1(n, V)$, and that $| \ |_1$ is actually a norm on $L^1(n, V)$, since $| \{f\}^c |_1 = 0 \Rightarrow \{f\}^c = \{0\}^c$. The completeness of this normed space follows from Theorem 13.

5. Show that hypothesis (b) of Lebesgue's Dominated Convergence Theorem cannot be dropped. *Hint:* Let $f_k = \chi_{[0,k]} \in \mathcal{L}^1(1, \mathbf{R})$ for each $k \in \mathbf{N}$.

6. Show that hypothesis (b) of Beppo Levi's Theorem cannot be dropped. *Hint:* Let $f_k = \chi_{[k,k+1]}$ for each $k \in \mathbf{N}$.

7. Show that even though $k \mapsto f_k$ is a sequence in $\mathcal{L}^1(1, \mathbf{R})$ such that $\lim_{k \to \infty} |f_k|_1 = 0$, the sequence $k \mapsto f_k(x)$ need not converge for any $x \in [0, 1]$. (See Corollary 16, which refers to *subsequences*.) *Hint:* Let $f_1 = 0$, $f_2 = \chi_{[0,1]}$, and in general, if $k = 2^m + j$ ($j \in \{1, \ldots, 2^m\}$), let f_k be the function that is 1 on $[2^{-m}(j-1), 2^{-m}j]$ and 0 outside this interval.

8. Let $f: \,]0, 1] \to \mathbf{R}$ be defined by $f: x \mapsto x^{-1/2}$.
   (a) Explicitly construct a bounding sequence for f^*.
   (b) Show that $f \in \mathcal{L}^1(]0, 1], \mathbf{R})$. *Hint:* Use Corollary 22 and the fact that $\chi_{\{1\}} \in \mathcal{L}^1(1, \mathbf{R})$.
   (c) Show that $f^* \in \mathcal{L}^1(1, \mathbf{R})$ by applying Beppo Levi's Theorem with $f_k = f^* \times \chi_{[1/k,1]}$.

*9. Prove that λ^* has the properties (a) through (d) set forth in Remark 23.

*10. Let $f \in \mathcal{L}^1(1, \mathbf{R})$ be nonnegative, and let $S = \{(x, y) \in \mathbf{R}^2 : 0 \leq y \leq f(x)\}$. Prove that $\chi_S \in \mathcal{L}^1(2, \mathbf{R})$ and that $\lambda(S) = \int f$. This justifies the usual application of integration to the calculation of areas. *Hint:* First prove the assertion in case f is a step function.

*11. Work out the details of the theory of Riemann integration up to and including the appropriate variant of Theorem 10, in which it is part of the hypothesis that f is Riemann-integrable. You may use the theory of Lebesgue integration at any stage in the proof.

*12. Show that a function $f \in \mathcal{F}(\mathbf{R}^n, \mathbf{R})$ is R-integrable if, and only if, for every $\epsilon > 0$ there are functions g and h in Step $(n, \mathbf{R}^n)$ such that $g(x) \leq f(x) \leq h(x)$ for every $x \in \mathbf{R}^n$ and that $\int (h - g) < \epsilon$. *Hint:* To prove the necessity of this condition, suppose f is R-integrable. Choose $s \in$ Step (n, V) such that $|f - s|_R < \epsilon/2$, and choose a bounding function t for $f - s$ such that $\int t < \epsilon/2$. Put $g = s - t$, $h = s + t$.

C. Interchange of limiting operations

We now apply the results of the preceding section to prove several theorems of calculus.

1. PROPOSITION. If $\langle V, \| \, \| \rangle$ is a Banach space, $a \leq b \leq c \leq d$, and $f \in \mathcal{L}^1([a, d], V)$, then the restriction g of f to $[b, c]$ lies in $\mathcal{L}^1([b, c], V)$.

Proof. Let $k \mapsto s_k$ be a sequence in Step $(1, V)$ such that $\lim |f^* - s_k|_1 = 0$. Let $\chi = \chi_{[b,c]}$. Then $\chi s_k \in$ Step $(1, V)$, and $|g^* - \chi s_k|_1 \leq |f^* - s_k|_1$, by Lemma B 8(c). It follows that $g^* \in \mathcal{L}^1(1, V) \Rightarrow g \in \mathcal{L}^1([b, c], V)$.

2. DEFINITION. Let $\langle V, \| \, \| \rangle$ be a Banach space and let $f \in \mathcal{L}^1([a, b], V)$. Then $\int_a^b f = \int f^* = \int f = -\int_b^a f$.

3. PROPOSITION. If $\langle V, \| \, \| \rangle$ is a Banach space, $a \leq b \leq c$, and $f \in \mathcal{L}^1([a, c], V)$, then $\int_a^b f + \int_b^c f = \int_a^c f$.

Proof. Let g be the restriction of f to $[a, b]$ and let h be the restriction of f to $[b, c]$. Then $g^* + h^*$ differs from f^* only at b, where $g^*(b) + h^*(b) = 2f(b)$. Thus $\int_a^c f = \int_a^c (g^* + h^*) = \int_a^c g^* + \int_a^c h^* = \int_a^b g^* + \int_b^c h^* = \int_a^b f + \int_b^c f$.

4.   FUNDAMENTAL THEOREM OF CALCULUS. Let $\langle V, \| \ \| \rangle$ be a Banach space and let $f \in \mathcal{L}^1([a, b], V)$. Put $F(x) = \int_a^x f$ for $x \in [a, b]$. Then F is continuous on $[a, b]$. If f is continuous at $x_0 \in [a, b]$, F is differentiable at x_0 and $F'(x_0) = f(x_0)$.

Proof. If $x_0 \in \ ]a, b[$ and $b - x_0 \geq h > 0$, $F(x_0 + h) - F(x_0) = \int_{x_0}^{x_0+h} f$ by Proposition 3. In particular, if $g_k(x) = 0$ when $x \in \mathbf{R} - [x_0, x_0 + 1/k]$ and $g_k(x) = \|f(x)\|$ when $x \in [x_0, x_0 + 1/k]$, and if $0 < h \leq 1/k$, then $\|F(x_0 + h) - F(x_0)\| \leq \int_{x_0}^{x_0+h} \|f\| = \int_{x_0}^{x_0+h} g_k \leq \int g_k^*$. As $k \to \infty$, $g_k \to 0$ λ-almost everywhere on $\mathbf{R}$. Therefore, by Lebesgue's Dominated Convergence Theorem, $(g_k(x) \leq \|f(x)\|$ for all $x \in \mathbf{R})$, $\lim\limits_{k \to \infty} \int g_k^* = 0 \Rightarrow F(x_0 + h) - F(x_0) \to 0$ as $h \to 0$ through positive values. Similarly, $F(x_0 + h) - F(x_0) \to 0$ as $h \to 0$ through negative values $\left(\text{if } h < 0, F(x_0 + h) - F(x_0) = -\int_{x_0+h}^{x_0} f\right)$. That is, $\lim\limits_{h \to \infty} F(x_0 + h) - F(x_0) = 0$, so F is continuous at x_0. If $x_0 = a$, we use only positive h. If $x_0 = b$, we use only negative h.

Now suppose f is continuous at $x_0 \in \ ]a, b[$. For any $\epsilon > 0$, choose $\delta > 0$ so small that $\delta < b - x_0$ and $0 < h < \delta \Rightarrow \|f(x_0 + h) - f(x_0)\| < \epsilon$. Then if $0 < h < \delta$,

$$\|(F(x_0 + h) - F(x_0))/h - f(x_0)\| = \left\| \int_{x_0}^{x_0+h} \frac{f}{h} - \int_{x_0}^{x_0+h} \frac{f(x_0)}{h} \right\|$$

$$= \left\| \int_{x_0}^{x_0+h} \frac{f - f(x_0)}{h} \right\| \leq \int_{x_0}^{x_0+h} \left\| \frac{f - f(x_0)}{h} \right\| \leq \int_{x_0}^{x_0+h} \frac{\epsilon}{h} = \epsilon.$$

It follows that $(F(x_0 + h) - F(x_0))/h \to f(x_0)$ as $h \to 0$ through positive values. Similarly, $(F(x_0 + h) - F(x_0))/h \to f(x_0)$ as $h \to 0$ through negative values. That is, $F'(x_0) = f(x_0)$. For the end points, we use only the one-sided limits.   ●

It is this theorem, of course, that allows us to calculate integrals in the usual way. Note that $F : x \mapsto \int_a^x f$ is continuous so long as f is integrable. It is *not* necessary to assume that f is continuous or bounded.

5.   COROLLARY. Let $\langle V, \| \ \| \rangle$ be a Banach space, let F be a differentiable function from $[a, b]$ to V, and let F' be continuous. Then $F(b) - F(a) = \int_a^b F'$.

Proof. Put $G(x) = F(x) - F(a) - \int_a^x F'$. Then $G(a) = \mathbf{0}$ and, according to the Fundamental Theorem of Calculus, G is differentiable on $[a, b]$ and $G' = \mathbf{0}$. If $V = \mathbf{R}$, we can simply apply the Mean Value Theorem to conclude that $G(b) - G(a) = (b - a)G'(\zeta) = 0$ for some $\zeta \in \,]a, b[$. If we knew enough about Banach spaces (if we knew the Hahn-Banach Theorem in particular) we could use this fact to deal with the general case. Here is another argument that will do.

Observe that if $[c, d] \subseteq [a, b]$, then either $\|G(d) - G((c + d)/2)\| \geq \frac{1}{2}\|G(d) - G(c)\|$ or else $\|G(c + d)/2) - G(c)\| \geq \frac{1}{2}\|G(d) - G(c)\|$. Otherwise we could conclude that $\|G(d) - G(c)\| < \|G(d) - G(c)\|$ by using the triangle inequality. Now choose a nested sequence of intervals $k \mapsto [a_k, b_k]$ using the Method of Bisection. Put $a_1 = a$, $b_1 = b$. Having chosen $[a_k, b_k]$, let $[a_{k+1}, b_{k+1}]$ be one of its halves, so chosen that $\|G(b_{k+1}) - G(a_{k+1})\| \geq \frac{1}{2}\|G(b_k) - G(a_k)\|$. By induction, using the relation $b_{k+1} - a_{k+1} = \frac{1}{2}(b_k - a_k)$, we see that $\|(G(b_k) - G(a_k))/(b_k - a_k)\| \geq \|(G(b) - G(a))/(b - a)\|$ for each k. The sequences $k \mapsto a_k$ and $k \mapsto b_k$ satisfy the Cauchy criterion, and converge to a common limit $c \in [a, b]$. Since $k \mapsto a_k$ is an increasing sequence and $k \mapsto b_k$ is a decreasing sequence, $a_k \leq c \leq b_k$ for every k, so that $|a_k - c| \leq b_k - a_k$ and $|b_k - c| \leq b_k - a_k$. Since $G'(c) = \mathbf{0}$, we know that $G(d) = G(c) + \epsilon(d)$, where $\lim_{d \to c} \epsilon(d)/(d - c) = \mathbf{0}$. Thus

$$(G(b_k) - G(a_k))/(b_k - a_k) = (\epsilon(b_k) - \epsilon(a_k))/(b_k - a_k) \leq \|\epsilon(b_k)/(b_k - a_k)\|$$
$$+ \|\epsilon(a_k)/(b_k - a_k)\| \leq \|\epsilon(b_k)/(b_k - c)\| + \|\epsilon(a_k)/(a_k - c)\| \to 0$$

as $k \to \infty$. This shows that $\|(G(b) - G(a))/(b - a)\| = 0 \Rightarrow G(b) = G(a) = \mathbf{0} \Rightarrow F(b) - F(a) = \int_a^b F'$. $\bullet$

Lebesgue's Dominated Convergence Theorem and Beppo Levi's Theorem can be interpreted as affirming the possibility of integrating a sequence or a series term-by-term. Using the Fundamental Theorem of Calculus, they can be employed to prove theorems on differentiation term-by-term. Here is one that is simple and serviceable.

6. THEOREM ON DIFFERENTIATION TERM-BY-TERM. Let $\langle V, \| \,\, \| \rangle$ be a Banach space, let $a \leq b \leq c$, and let $k \mapsto f_k$ be a sequence of functions from $[a, c]$ into V with the following properties.

(a) For each $k \in \mathbf{N}$, f_k is differentiable and f_k' is continuous.
(b) The series $\sum_{k=1}^{\infty} f_k'$ converges uniformly to a function $g: [a, c] \to V$.
(c) $\sum_{k=1}^{\infty} f_k(b)$ converges.

Then the series $\sum_{k=1}^{\infty} f_k$ converges uniformly to a function f on $[a, c]$, f is differentiable, and $f' = g$.

Proof. If $b \le x \le c$, $f_k(x) - f_k(b) = \int_b^x f_k'$ by Corollary 5, so that

$$\sum_{k=1}^{\infty} (f_k(x) - f_k(b)) = \sum_{k=1}^{\infty} \int_b^x f_k' = \int_b^x \sum_{k=1}^{\infty} f_k' = \int_b^x g.$$

The integration term-by-term is justified by Lebesgue's Dominated Convergence Theorem and the fact that a uniformly convergent series of continuous functions has uniformly bounded partial sums. Thus the series $\sum_{k=1}^{\infty} f_k(x) = \sum_{k=1}^{\infty} f_k(b) + \sum_{k=1}^{\infty} (f_k(x) - f_k(b))$ converges to an $f(x) \in V$. According to the Fundamental Theorem of Calculus, f is continuous at x and differentiable at x, and $f'(x) = g(x)$ ($f(x) = f(x) - f(b) + f(b) = f(b) + \int_b^x g$).

We deal with the case $a \le x \le b$ by a change of variable: look at the function $h \colon x \mapsto f(-x)$ on the interval $[-b, -a]$, and note that $a \le x \le b \Leftrightarrow -b \le -x \le -a$.

7.   DEFINITION. Let S, T, and U be sets, and let $f \colon S \times T \to U$. For each $x \in S$, $f(x, \cdot) \colon T \to U$ is the function defined by $f(x, \cdot) \colon y \mapsto f(x, y)$ and for each $y \in T$, $f(\cdot, y) \colon S \to U$ is the function defined by $f(\cdot, y) \colon x \mapsto f(x, y)$.

8.   THEOREM ON CONTINUOUS DEPENDENCE ON A PARAMETER. Let $\langle V, \| \ \| \rangle$ be a Banach space and let $\langle M, d \rangle$ be a metric space. Let f be a function from $M \times \mathbf{R}^n$ to V with the following properties.

(a)   For each $x \in M$, $f(x, \cdot) \in \mathcal{L}^1(n, V)$.

(b)   There is a $b \in \mathfrak{N}(n, \mathbf{R})$ such that $\|f(x, y)\| \le b(y)$ for every $x \in M$ and λ-almost all $y \in \mathbf{R}^n$.

(c) $\lim_{x \to x_0} f(x, y) = f(x_0, y)$ for λ-almost all $y \in \mathbf{R}^n$. Then $\lim_{x \to x_0} \int f(x, \cdot) = \int f(x_0, \cdot)$.

Proof. If the theorem were false, there would have to be an $\epsilon > 0$ with the following property: for every $\delta > 0$, there is an $x \in M$ such that $d(x, x_0) < \delta$ but $\| \int f(x, \cdot) - \int f(x_0, \cdot) \| \ge \epsilon$. There would then be a sequence $k \mapsto x_k$ in M such that $d(x_0, x_k) < 1/k$ for each $k \in \mathbf{N}$ but $\| \int f(x_k, \cdot) - \int f(x_0, \cdot) \| \ge \epsilon$ for each $k \in \mathbf{N}$. Let S be the λ-null set of $y \in \mathbf{R}^n$ such that $\lim_{x \to x_0} f(x, y) \ne f(x_0, y)$ or that for some $x \in M$, $\|f(x, y)\| > b(y)$. Put $g_k = f(x_k, \cdot) - f(x_0, \cdot)$. Then for each $y \in \mathbf{R}^n - S$, $\|g_k(y)\| \le 2b(y)$ and $\lim_{k \to \infty} g_k(y) = 0$. Therefore, by Lebesgue's Dominated Convergence Theorem, $0 = \lim_{k \to \infty} \int g_k = \lim_{k \to \infty} \int f(x_k, \cdot) - \int f(x_0, \cdot)$, a contradiction.

9.   THEOREM ON DIFFERENTIATION UNDER AN INTEGRAL SIGN. Let $\langle V, \| \ \| \rangle$ be a Banach space, $a < b$, and f be a function from $[a, b] \times \mathbf{R}^n$ into V with the following properties.

(a)   For each $x \in [a, b]$, $f(x, \cdot) \in \mathcal{L}^1(n, V)$.

(b)   There is a $b \in \mathfrak{N}(n, \mathbf{R})$ such that for λ-almost all $y \in \mathbf{R}^n$, $||(f(x, y) - f(x_0, y))/(x - x_0)|| \leq b(y)$ for all $x \in [a, b] - \{x_0\}$.

(c) $\dfrac{\partial f}{\partial x}(x_0, y)$ exists for λ-almost all $y \in \mathbf{R}^n$.

Put $g(x) = \int f(x, \cdot)$ for each $x \in [a, b]$, and let $D \in \mathfrak{F}(\mathbf{R}^n, V)$ be equal to $\dfrac{\partial f}{\partial x}(x_0, y)$ λ-almost everywhere. Then g is differentiable at x_0, $D \in \mathfrak{L}^1(n, V)$, and $g'(x_0) = \int D$.

Proof. Let S be the λ-null set of $y \in \mathbf{R}^n$ such that $||(f(x, y) - f(x_0, y)/(x - x_0)|| > b(y)$ for some $x \in [a, b] - \{x_0\}$, that f does not have a partial derivative $\dfrac{\partial f}{\partial x}(x_0, y)$, or that

$$D(y) \neq \frac{\partial f}{\partial x}(x_0, y).$$

Define a function $h \colon [a, b] \times \mathbf{R}^n \to V$ as follows. If $y \in \mathbf{R}^n - S$, then

$$h(x_0, y) = \frac{\partial f}{\partial x}(x_0, y)$$

and $h(x, y) = (f(x, y) - f(x_0, y))/(x - x_0)$ if $x \neq x_0$. If $y \in S$, $h(x, y) = 0$. Choose a sequence $k \mapsto x_k$ in $[a, b]$ such that $\lim_{k \to \infty} x_k = x_0$. Then $D = \lim_{k \to \infty} h(x_k, \cdot)$ λ-almost everywhere. By Lebesgue's Dominated Convergence Theorem, it follows that $D \in \mathfrak{L}^1(n, V)$. h then satisfies the hypotheses of Theorem 8, so that $g'(x_0) = \lim_{x \to x_0} \int h(x, \cdot) = \int h(x_0, \cdot) = \int D$. ●

Let $f \in \mathfrak{L}^1(n + m, V)$, and suppose that for each $x \in \mathbf{R}^n$ and each $y \in \mathbf{R}^m$, $f(x, \cdot) \in \mathfrak{L}^1(m, V)$ and $f(\cdot, y) \in \mathfrak{L}^1(n, V)$. Following ancient usage, we let $\int f \, dy$ be the *function* $x \mapsto \int f(x, \cdot)$ and we let $\int f \, dx$ be the *function* $y \mapsto \int f(\cdot, y)$. In case $m = 0$, $\int f \, dx$ is simply $\int f$. This notation is imprecise, but it is familiar and suggestive. It should help make sense of the following theorem.

The familiar formula for the area of a rectangle $R = R((a_1, a_2), (b_1, b_2))$ has an analogue in integration theory: $\lambda(R) = (b_2 - a_2)(b_1 - a_1) = \int [\int \chi_R \, dx] \, dy$, where χ_R is the function which takes the value 1 on R and the value 0 on $\mathbf{R}^2 - R$. This is, of course, a special case of the usual formula for the evaluation of a double integral by repeated integrals.

Even if $f(x, \cdot) \in \mathfrak{L}^1(m, V)$ for every $x \in \mathbf{R}^n$ and $f(\cdot, y) \in \mathfrak{L}^1(n, V)$ for every $y \in \mathbf{R}^m$, f need not be in $\mathfrak{L}^1(n + m, V)$ (see Exercise 7). And if $f \in \mathfrak{L}^1(n + m, V)$, $f(x, \cdot)$ need not be in $\mathfrak{L}^1(m, V)$ for every $x \in \mathbf{R}^n$. For example, define $f \in \mathfrak{L}^1(2, \mathbf{R})$ by putting $f(0, 0) = 0$, $f(x, y) = (x^2 + y^2)^{-1/2}$ if $0 < x^2 + y^2 \leq 1$, and $f(x, y) = 0$ if $x^2 + y^2 > 1$. Then $f(x, \cdot) \in \mathfrak{L}^1(1, \mathbf{R})$ *except when $x = 0$.*

The next theorem gives a precise analysis of this situation. Together with Lebesgue's Dominated Convergence Theorem, it is one of the cornerstones of the modern theory of integration.

10. FUBINI'S THEOREM ON REPEATED INTEGRALS. Let $\langle V, \| \ \| \rangle$ be a Banach space and let $f \in \mathcal{L}^1(n + m, V)$. Then $\int [\int f \, dx] \, dy = \int f = \int [\int f \, dy] \, dx$ in the following sense: $f(x, \cdot) \in \mathcal{L}^1(m, V)$ for λ-almost all $x \in \mathbf{R}^n$, and $f(\cdot, y) \in \mathcal{L}^1(n, V)$ for λ-almost all $y \in \mathbf{R}^m$. If $g \in \mathcal{F}(\mathbf{R}^n, V)$, $h \in \mathcal{F}(\mathbf{R}^m, V)$, $g(x) = \int f(x, \cdot)$ for λ-almost all $x \in \mathbf{R}^n$, and $h(y) = \int f(\cdot, y)$ for λ-almost all $y \in \mathbf{R}^m$, then $g \in \mathcal{L}^1(n, V)$, $h \in \mathcal{L}^1(m, V)$, and $\int h = \int f = \int g$.

Proof. We break the proof up into a series of lemmas. The first shows that the theorem is true if $f \in \text{Step}\,(n + m, V)$.

11. LEMMA. Let $f \in \text{Step}\,(n + m, V)$ and let $\mathcal{G}$ be a presentation of f. Choose $v \in V$, and let $f' \in \text{Step}\,(n + m, V)$ be defined as follows: if $x \in R \in C(\mathcal{G})$, then $f'(x) = f(x)$; if $x \in R(\mathcal{G})^c - \bigcup C(\mathcal{G})$, then $f'(x) = v$; if $x \in \mathbf{R}^{n+m} - R(\mathcal{G})^c$, then $f'(x) = 0$. Then $\mathbf{I} f - f' \mathbf{I}_1 = 0$. For each $x \in \mathbf{R}^n$, $f'(x, \cdot) \in \text{Step}\,(m, V)$. If $g \colon x \mapsto \int f'(x, \cdot)$, then $g \in \text{Step}\,(n, V)$ and $\int g = \int f' = \int f$.

Proof. Since $\mathcal{G}$ is a presentation of f and also of f', and since $f(x) - f'(x) = 0$ if $x \in R \in C(\mathcal{G})$, $\mathbf{I} f - f' \mathbf{I}_1 = \int \|f - f'\| = \sum \{\|f(R) - f'(R)\| \colon R \in C(\mathcal{G})\} = 0$. Thus $\|\int f - \int f'\| \le \mathbf{I} f - f' \mathbf{I}_1 = 0$. For each $x \in \mathbf{R}^n$, $(E_{n+1}(\mathcal{G}), \ldots, E_{n+m}(\mathcal{G}))$ is a grid that presents $f'(x, \cdot)$, and $(E_1(\mathcal{G}), \ldots, E_n(\mathcal{G}))$ is a grid on $\mathbf{R}^n$ that presents g (see Definition A 1). It remains to prove the formula $\int g = \int f'$.

If $R = R(a, b)$ is a rectangle in $\mathbf{R}^{n+m}$, let

$$S = R((a_1, \ldots, a_n), (b_1, \ldots, b_n))$$

be the projection of R on $\mathbf{R}^n$ and let

$$T = R((a_{n+1}, \ldots, a_{n+m}), (b_{n+1}, \ldots, b_{n+m}))$$

be the projection of R on $\mathbf{R}^m$. If $x \in \mathbf{R}^n$, $y \in \mathbf{R}^m$, then $\chi_R(x, y) = \chi_S(x)\chi_T(y)$. Let $\psi_R \colon x \mapsto \int \chi_R(x, \cdot) = \int \chi_S(x)\chi_T = \chi_S(x) \int \chi_T = (b_{n+1} - a_{n+1}) \cdots (b_{n+m} - a_{n+m})\chi_S(x)$. Then

$$\int \chi_R = (b_1 - a_1) \cdots (b_n - a_n)(b_{n+1} - a_{n+1}) \cdots (b_{n+m} - a_{n+m})$$

$$= (b_{n+1} - a_{n+1}) \cdots (b_{n+1} - a_{n+m}) \int \chi_S = \int \psi_R.$$

$$\int f' = \sum \left(f'(R) \int \chi_R \colon R \in C(\mathcal{G}) \right) = \sum \left(f'(R) \int \psi_R \colon R \in C(\mathcal{G}) \right)$$

$$= \int \sum \{f'(R)\psi_R \colon R \in C(\mathcal{G})\} = \int g.$$

12. LEMMA. For each $k \in \mathbf{N}$, let $g_k \in \mathcal{L}^1(n, \mathbf{R})$ and let $g_k(x) \ge 0$ for all $x \in \mathbf{R}^n$. Suppose further that $\sum_{k=1}^{\infty} \int g_k$ converges. Then $\sum_{k=1}^{\infty} g_k$ converges λ-almost everywhere.

Proof. This was part of the proof of Theorem B 13. We can deduce the Lemma from Theorem B 13 as follows: $| g_k |_1 = \int |g_k| = \int g_k$. Thus the series $\sum_{k=1}^{\infty} g_k$ is an absolutely Cauchy series in $\langle \mathcal{L}^1(n, \mathbf{R}), | \ |_1 \rangle$, so by Theorem B 13, there is a $g \in \mathcal{L}^1(n, \mathbf{R})$ such that $\lim_{k \to \infty} | g - \sum_{j=1}^{k} g_j |_1 = 0$. According to Corollary B 16, some subsequence of the sequence $k \mapsto \sum_{j=1}^{k} g_j$ converges to g λ-almost everywhere. But by Theorem VI B 9, a subsequence of a monotonically increasing sequence converges if, and only if, the original sequence converges.

13. **LEMMA.** For each $k \in \mathbf{N}$, let $f_k \in \mathcal{L}^1(m, V)$ and suppose that $\sum_{k=1}^{\infty} \int ||f_k||$ converges. Then $\sum_{k=1}^{\infty} f_k$ converges λ-almost everywhere to a function $f \in \mathcal{L}^1(m, V)$, and $\lim_{N \to \infty} | f - \sum_{k=1}^{\infty} f_k |_1 = 0$.

Proof. By Lemma 12, $S = \{ x \in \mathbf{R}^m : \sum_{k=1}^{\infty} ||f_k(x)|| \text{ does } not \text{ converge} \}$ is a λ-null set. If $x \in \mathbf{R}^m - S$, then $\sum_{k=1}^{\infty} f_k(x)$ is an absolutely Cauchy series in $\langle V, || \ || \rangle$, so it converges to a limit $f(x)$ in V. Put $f(x) = \mathbf{0}$ if $x \in S$. Put $f_k'(x) = f_k(x)$ if $x \in \mathbf{R}^m - S$, and $f_k'(x) = \mathbf{0}$ if $x \in S$. Then by Beppo Levi's Theorem, $b = \sum_{k=1}^{\infty} ||f_k'|| \in \mathcal{L}^1(m, \mathbf{R})$. Since $||\sum_{k=1}^{N} f_k'(x)|| \leq b(x)$ for every $x \in \mathbf{R}^m$, it follows from Lebesgue's dominated convergence theorem that $f \in \mathcal{L}^1(m, V)$.

$$\left| f - \sum_{k=1}^{N} f_k \right|_1 = \left| f - \sum_{k=1}^{N} f_k' \right|_1 = \left| \sum_{k=N+1}^{\infty} f_k' \right|_1 = \int \left\| \sum_{k=N+1}^{\infty} f_k' \right\|$$
$$\leq \sum_{k=N+1}^{\infty} \int ||f_k'|| = \sum_{k=N+1}^{\infty} \int ||f_k|| \to 0 \quad \text{as } N \to \infty. \quad \bullet$$

We now prove another special case of Fubini's Theorem, that in which $f = \chi_E$, where E is a λ-null subset of $\mathbf{R}^{n+m}$.

14. **LEMMA.** Let $E \subseteq \mathbf{R}^{n+m}$ be a λ-null set. For each $x \in \mathbf{R}^n$, let $E_x = \{ y \in \mathbf{R}^m : (x, y) \in E \}$. Then E_x is a λ-null subset of $\mathbf{R}^m$ for λ-almost all $x \in \mathbf{R}^n$.

Proof. For each $j \in \mathbf{N}$, let $k \mapsto s_{jk}$ be a bounding sequence for χ_E such that $\lim_{k \to \infty} \int s_{jk} < 2^{-j}$. Put $s_k' = \sum_{j=1}^{k} s_{jk}$. Then $\int s_k' = \sum_{j=1}^{k} \int s_{jk} < \sum_{j=1}^{k} 2^{-j} < 1$; $s_k' \leq s_{k+1}'$; and if $x \in E$, $\{ s_k'(x) : k \in \mathbf{N} \}$ is unbounded. Therefore $k \mapsto s_k'$ is a bounding sequence for χ_E. According to Lemma 11 (with $v = 1 \in \mathbf{R} = V$), we can modify s_{jk} if necessary (for each $j, k \in \mathbf{N}$) so that (i) $k \mapsto s_k'$ remains a bounding sequence for χ_E that is unbounded at each $x \in E$, (ii) $s_k'(x, \cdot) \in$ Step $(m, \mathbf{R})$ for each $x \in \mathbf{R}^n$, (iii) $t_k' : x \mapsto \int s_k'(x, \cdot) \in$ Step $(n, \mathbf{R})$, and (iv) $\int s_k' = \int t_k'$. By Lemma 12, $T = \{ x \in \mathbf{R}^n : \sum_{k=1}^{\infty} (t_{k+1}'(x) - t_k'(x)) \text{ does } not \text{ converge} \}$ is a λ-null set. By another application of Lemma 12, if $x \in \mathbf{R}^n - T$, $\sum_{k=1}^{\infty} (s_{k+1}'(x, \cdot) - s_k'(x, \cdot))$ converges λ-almost everywhere on $\mathbf{R}^m$. Since $\{ s_k'(x, y) : k \in \mathbf{N} \}$ is unbounded if $y \in E_x$ and since $\sum_{k=1}^{N} (s_{k+1}'(x, \cdot) - s_k'(x, \cdot)) = s_{N+1}'(x, \cdot) - s_1'(x, \cdot)$, it follows that E_x is a λ-null set if $x \in \mathbf{R}^n - T$. $\quad \bullet$

We can now prove Fubini's Theorem. According to Definition B 6, there is a sequence $k \mapsto s_k$ in Step $(n + m, V)$ such that $\lim\limits_{k \to \infty} |f - s_k|_1 = 0$. According to Lemma 11, we may assume that this sequence is so chosen that for each $k \in \mathsf{N}$, $s_k(x, \cdot) \in$ Step (m, V) for every $x \in \mathsf{R}^n$, $t_k \colon x \mapsto \int s_k(x, \cdot) \in$ Step (n, V), and $\int s_k = \int t_k$. Furthermore, Corollary B 16 assures us that by replacing $k \mapsto s_k$ by a subsequence if necessary, we can arrange that $\lim\limits_{k \to \infty} s_k(x, y) = f(x, y)$ unless (x, y) belongs to a λ-null set $E \subseteq \mathsf{R}^{n+m}$.

For each $i \in \mathsf{N}$, choose $N(i)$ so large that $N(i) > N(i - 1)$ and that $j, k \geq N(i) \Rightarrow |s_j - s_k|_1 < 2^{-i}$. Put $p_1 = s_{N(1)}$ and, for $k > 1$, put $p_k = s_{N(k)} - s_{N(k-1)}$. Then $\sum_{k=1}^{\infty} |p_k|_1$ converges, and $\sum_{k=1}^{M} p_k = s_{N(M)}$ converges to f as $M \to \infty$ except on E. $\|p_k\| \in$ Step $(n + m, \mathsf{R})$, and $\|p_k(x, \cdot)\| \in$ Step (m, R) for each $x \in \mathsf{R}^n$. Let $q_k \colon x \mapsto \int \|p_k(x, \cdot)\|$, so that $q_k \in$ Step (n, R) and $\int q_k = \int \|p_k\| = |p_k|_1$ by Lemma 11. According to Lemma 12, $\sum_{k=1}^{\infty} q_k$ converges except on a λ-null set $S' \subseteq \mathsf{R}^n$. By Lemma 14, $S'' = \{x \in \mathsf{R}^n \colon E_x$ is not a λ-null set$\}$ is a λ-null set in R^n. Put $S = S' \cup S''$. Then S is a λ-null set by Proposition B 18.

Let us show that if $x \in \mathsf{R}^n - S$, then $f(x, \cdot) \in \mathcal{L}^1(m, V)$. Since $x \notin S'$, $\sum_{k=1}^{\infty} q_k(x) = \sum_{k=1}^{\infty} \int \|p_k(x, \cdot)\|$ converges. By Lemma 13, there is a function $\phi \in \mathcal{L}^1(m, V)$ such that $A = \{y \in \mathsf{R}^m \colon \sum_{k=1}^{\infty} p_k(x, y)$ does not converge to $\phi(y)\}$ is a λ-null set and that $\lim\limits_{N \to \infty} |\phi - \sum_{k=1}^{N} p_k(x, \cdot)|_1 = 0$. Since $x \notin S''$, E_x is a λ-null set. If $y \notin E_x \cup A$, then $f(x, y) = \sum_{k=1}^{\infty} p_k(x, y) = \phi(y)$. Therefore $f(x, \cdot)$ is λ-almost equal to ϕ. Since $\phi \in \mathcal{L}^1(n, V)$, $f(x, \cdot) \in \mathcal{L}^1(n, V)$ as well, $\int f(x, \cdot) = \int \phi = \sum_{k=1}^{\infty} \int p_k(x, \cdot)$, and $\lim\limits_{N \to \infty} |f(x, \cdot) - \sum_{k=1}^{N} p_k(x, \cdot)|_1 = 0$.

Finally, let us show that if $g \in \mathfrak{F}(\mathsf{R}^n, V)$ and $g(x) = \int f(x, \cdot)$ for λ-almost all $x \in \mathsf{R}^n$, then $g \in \mathcal{L}^1(n, V)$ and $\int f = \int g$. Let $u_k \colon x \mapsto \int p_k(x, \cdot)$ for each $k \in \mathsf{N}$. Then $u_k \in$ Step (n, V) and $\int u_k = \int p_k$ by Lemma 11. Moreover, for each $x \in \mathsf{R}^n$, $\|u_k(x)\| = \|\int p_k(x, \cdot)\| \leq \int \|p_k(x, \cdot)\| = q_k(x) \Rightarrow \int \|u_k\| \leq \int q_k = |p_k|_1$. By another application of Lemma 13, there is a function $g' \in \mathcal{L}^1(n, V)$ such that $\sum_{k=1}^{\infty} u_k$ converges to g' except on a λ-null set $T \subseteq \mathsf{R}^n$, and that $\lim\limits_{N \to \infty} |g' - \sum_{k=1}^{N} u_k|_1 = 0$. But if $x \in \mathsf{R}^n - (S \cup T)$, $\lim\limits_{N \to \infty} |f(x, \cdot) - \sum_{k=1}^{N} p_k(x, \cdot)|_1 = 0 \Rightarrow 0 = \lim\limits_{N \to \infty} \|\int f(x, \cdot) - \sum_{k=1}^{N} \int p_k(x, \cdot)\| = \lim\limits_{N \to \infty} \|\int f(x, \cdot) - \sum_{k=1}^{N} u_k(x)\| = \|\int f(x, \cdot) - g'(x)\|$. Since $S \cup T$ is a λ-null set, g is λ-almost equal to g'. It follows that $g \in \mathcal{L}^1(n, V)$, and that $\int g = \int g' = \lim\limits_{N \to \infty} \int \sum_{k=1}^{N} u_k = \lim\limits_{N \to \infty} \sum_{k=1}^{N} \int u_k = \lim\limits_{N \to \infty} \sum_{k=1}^{N} \int p_k = \int f$. The remainder of the Theorem (the part about $\int f(\cdot, y)$ and h) is proved in exactly the same way. ●

15. COROLLARY. *Let S be a compact subset of R^{n+m}, let $\langle V, \| \ \| \rangle$ be a Banach space, and let $f \colon S \to V$ be continuous. Then for each $x \in \mathsf{R}^n$ and each $y \in \mathsf{R}^m$, $f^*(x, \cdot) \in \mathcal{L}^1(m, V)$ and $f^*(\cdot, y) \in \mathcal{L}^1(n, V)$. $\int [\int f^* \, dx] \, dy = \int f = \int [\int f^* \, dy] \, dx$.*

Proof. If $x \in \mathbf{R}^n$, $S_x = \{y \in \mathbf{R}^m : (x, y) \in S\}$ is a compact subset of $\mathbf{R}^m$, and $f(x, \cdot) : S_x \to V$ is continuous. Thus $f^*(x, \cdot) \in \mathcal{L}^1(m, V)$ by Proposition B 20. Similarly, $f^*(\cdot, y) \in \mathcal{L}^1(n, V)$ for each $y \in \mathbf{R}^m$, and $f^* \in \mathcal{L}^1(n + m, V)$. The rest follows from Fubini's Theorem.

16. THEOREM ON INTERCHANGE OF PARTIAL DIFFERENTIATION OPER-
ATORS. Let U be an open subset of $\mathbf{R}^2$, let $\langle V, \|\ \| \rangle$ be a Banach space, and let $f : U \to V$ have the following properties.

(a) f is continuous, and has continuous partial derivatives $\partial f/\partial x$ and $\partial f/\partial y$ on U.
(b) The function $\partial f/\partial x$ has a continuous partial derivative with respect to y on U; or
(c) the function $\partial f/\partial y$ has a continuous partial derivative with respect to x on U.

Then f has both partial derivatives $(\partial/\partial y)(\partial f/\partial x)$ and $(\partial/\partial x)(\partial f/\partial y)$ throughout U, and they are equal.

Proof. Suppose (b) is true. Let $g = (\partial/\partial y)(\partial f/\partial x)$. By two applications of the Fundamental Theorem of the Calculus (and its Corollary) and one application of the Corollary to Fubini's Theorem,

$$f(x, y) - f(a, y) - f(x, b) + f(a, b) = \int_a^x \left[\frac{\partial f}{\partial x}(\cdot, y) - \frac{\partial f}{\partial x}(\cdot, b) \right]$$

$$= \int_a^x \left[\int_b^y \frac{\partial}{\partial y} \frac{\partial f}{\partial x} \, dy \right] dx = \int_a^x \left[\int_b^y g \, dy \right] dx = \int_b^y \left[\int_a^x g \, dx \right] dy,$$

provided that $(R((a, b), (x, y)))^c \subseteq U$. By the Theorem on Continuous Dependence on a Parameter, $\int_a^x g \, dx$ is a continuous function of y for each fixed x, so by the Fundamental Theorem of the Calculus, $f(x, y) - f(a, y) - f(x, b) - f(a, b)$ is differentiable with respect to y at b and its derivative is

$$\frac{\partial f}{\partial y}(x, b) - \frac{\partial f}{\partial y}(a, b) = \int_a^x g(\cdot, b).$$

Another application of the Fundamental Theorem of the Calculus shows that $\frac{\partial f}{\partial y}(x, b) - \frac{\partial f}{\partial y}(a, b)$ is differentiable with respect to x at a, and its derivative is

$$\frac{\partial}{\partial x} \frac{\partial f}{\partial y}(a, b) = g(a, b) = \frac{\partial}{\partial y} \frac{\partial f}{\partial x}(a, b).$$

The other case, (c), is treated similarly.

Exercises

*1. Let I be an interval in $\mathbf{R}$ and let f be a continuous function from I to $\mathbf{R}^+ \cup \{0\}$. For each $k \in \mathbf{N}$, let $I_k = [a_k, b_k]$ be a compact interval. Suppose that $I_1 \subseteq I_2 \subseteq I_3 \subseteq \cdots$ and that $\bigcup_{k=1}^\infty I_k = I$. Show that $f \in \mathcal{L}^1(I, \mathbf{R})$ if, and only if, $\left\{ \int_{a_k}^{b_k} f : k \in \mathbf{N} \right\}$ is bounded. *Hint:* Let f_k be the function defined by $f_k(x) = 0$ for $x \in \mathbf{R} - I_k$ and $f_k(x) = f(x)$ for $x \in I_k$. Then $0 \le f_k \le f_{k+1} \le f^*$ for each $k \in \mathbf{N}$, and $\lim_{k \to \infty} f_k(x) = f^*(x)$ for every $x \in \mathbf{R}$.

2. Let $f: \mathbf{R} \to \mathbf{R}$ be defined as follows: $f(x) = 0$ if $x \le 0$ or $x > 1$, $f(x) = x^{-1/2}$ if $0 < x \le 1$. Show that $f \in \mathcal{L}^1(1, \mathbf{R})$ (use Exercise 1). Verify by direct calculation that $F: x \mapsto \int_{-1}^x f$ is continuous on $[-1, \infty[$ even though f is not bounded, and that F is *not* differentiable at the points of discontinuity of f.

3. Let $f: \mathbf{R} \to \mathbf{R}$ be defined as follows: $f(x) = 0$ if $x \le 0$, $f(x) = 1/x$ if $x > 0$. Is $f \in \mathcal{L}^1(1, \mathbf{R})$?

4. Prove the following **Mean Value Inequality** (see VI G 6(b)). If $\langle V, \| \ \| \rangle$ is a Banach space and F is a continuously differentiable function from $[a, b]$ to V, then $\|F(b) - F(a)\| \le (b - a) \operatorname{lub} \{\|F'(x)\| : x \in [a, b]\}$. *Hint:* Use Corollary 5.

5. Show that in the situation of Exercise 4 there need not be any $x \in [a, b]$ such that $F(b) - F(a) = (b - a)F'(x)$, or even that $\|F(b) - F(a)\| = (b - a)\|F'(x)\|$. *Hint:* Let $a = 0$, $b = 2\pi$, $\langle V, \| \ \| \rangle = \langle \mathbf{C}, | \ | \rangle$, and $F: x \mapsto e^{ix}$. Nonetheless, *when* $V = \mathbf{R}$, the Mean Value Theorem assures us that there is an $x \in [a, b]$ such that $F(b) - F(a) = (b - a)F'(x)$.

6. Define the function $f: \mathbf{R} \to \mathbf{R}$ as follows: $f(x) = 0$ if $x \le 0$, $f(x) = (\sin x)/x$ if $x > 0$. Show that $f \notin \mathfrak{N}(1, \mathbf{R})$, although $\lim_{n \to \infty} \int_0^{n\pi} f$ exists. Why does this not contradict the result of Exercise 1? *Hint:* Show that $|f| \notin \mathfrak{N}(1, \mathbf{R})$. Let $t_n = \int_{(n-1)\pi}^{n\pi} f$. Show that for every $n \in \mathbf{N}$, $|t_{n+1}| < |t_n|$, that $2/(n\pi) \le |t_n|$, and that the terms $t_1, t_2, t_3, \ldots$ alternate in sign. Then apply standard tests to the series $t_1 + t_2 + t_3 + \cdots$ and $|t_1| + |t_2| + |t_3| + \cdots$.

7. Put $f(x, y) = 2xye^{-xy} - x^2y^2e^{-xy}$ for every $(x, y) \in [0, 1] \times [0, \infty[$.
   (a) Show that f is continuous on $[0, 1] \times [0, \infty[$ and that
   $$f(x, y) = \frac{\partial}{\partial x} (x^2ye^{-xy}) = \frac{\partial}{\partial y} (xy^2e^{-xy}).$$
   (b) Show that for every $x \in [0, 1]$, $f(x, \cdot) \in \mathcal{L}^1([0, \infty[, \mathbf{R})$, and that for every $y \in [0, \infty[$, $f(\cdot, y) \in \mathcal{L}^1([0, 1], \mathbf{R})$. (Use Exercise 1.)
   (c) Show, using (a), that $\int_0^1 \left[\int_0^\infty f \, dy \right] dx = 0$, but that $\int_0^\infty \left[\int_0^1 f \, dx \right] dy = 1$. Why does this *not* contradict Fubini's Theorem?

8. Let $g: [0, 1] \times [0, \infty[\to \mathbf{R}$ be defined by $g(x, y) = x^2ye^{-xy}$.
   (a) Show that $g(x, \cdot) \in \mathcal{L}^1([0, \infty[)$ for each $x \in [0, 1]$.
   (b) Show that $\int_0^\infty g(x, \cdot) = 1$ if $0 < x \le 1$, but that $\int_0^\infty g(0, \cdot) = 0$. Why does this *not* contradict Theorem 8?

9. Let $h: [0, 1] \times [0, \infty[\to \mathbf{R}$ be defined by $h(x, y) = x^3ye^{-xy}$.
   (a) Show that $h(x, \cdot) \in \mathcal{L}^1([0, \infty[)$ for each $x \in [0, 1]$.

(b) Show that $\int_0^\infty h(x, \cdot) = x$ for each $x \in [0, 1]$. *Hint:* You can use 8(b).

(c) Let

$$p(x, y) = \frac{\partial h}{\partial x}(x, y) = 3x^2 y e^{-xy} - x^3 y^2 e^{-xy}.$$

Show that $\int_0^\infty p(0, y) = 0$. Why does this *not* contradict Theorem 9?

D.   Some applications to functions of a complex variable

1.   PROPOSITION. Let S, D, P be the functions from $\mathsf{C} \times \mathsf{C}$ to C defined by $S(z, w) = z + w$, $D(z, w) = z - w$, $P(z, w) = zw$, and let Q be the function from $\mathsf{C} \times (\mathsf{C} - \{0\})$ to C defined by $Q(z, w) = z/w$. Then S, D, P, and Q are continuous.

Proof. The proof is isomorphic to the proof of Theorem IV D 7.

2.   DEFINITION. Let U be an open subset of C, and let $f \colon U \to \mathsf{C}$. f has a **complex derivative** at $z \in U$ if $\lim_{w \to z} \dfrac{f(w) - f(z)}{w - z}$ exists. If f has a complex derivative at z, we put

$$f'(z) = \lim_{w \to z} \frac{f(w) - f(z)}{w - z}.$$

If f has a continuous complex derivative on a neighborhood of z, then f is **analytic** at z. If f is analytic at every point of U, then f is **analytic on U.** ●

If f has a complex derivative at z, then f is evidently differentiable at z in the sense of Definition VI G 1. In fact, $df_z \colon h \mapsto f'(z) \cdot h$. In particular, the Chain Rule holds for complex derivatives, in the form $(f \circ g)'(z) = f'(g(z)) \cdot g'(z)$. Moreover, the usual formulas for differentiation are valid as well.

3.   PROPOSITION. If f and g have complex derivatives at z, so do $f + g$, $f - g$, and fg, and $(f + g)'(z) = f'(z) + g'(z)$, $(f - g)'(z) = f'(z) - g'(z)$, $(f \cdot g)'(z) = f(z)g'(z) + f'(z)g(z)$. If $g(z) \neq 0$, then f/g also has a complex derivative at z, and

$$(f/g)'(z) = (f'(z)g(z) - f(z)g'(z))/(g(z))^2.$$

Proof. The proofs of these formulas are isomorphic to the corresponding formulas for real functions.

4.   PROPOSITION. Let U be a *connected* open subset of C, and let f be an analytic function from U to C. Then $f' = \mathbf{0}$ if, and only if, f is a constant function.

Proof. If f is a constant function, then $(f(w) - f(z))/(w - z) = 0$ for every $w \in U - \{z\}$. Thus $f'(z) = 0$. Conversely, suppose $f' = 0$. By the Mean Value Inequality (VI G 6(b)), f is constant on every convex subset of U. In particular, f is constant on every open ball in U. Choose $z \in U$. Then $f^{-1}\{f(z)\}$ is closed in U. However, if $w \in f^{-1}\{f(z)\}$, there is an open ball $B_\epsilon(w)$ contained in U since U is open. Since f is constant on $B_\epsilon(w)$, $B_\epsilon(w) \subseteq f^{-1}\{f(z)\}$. It follows that $f^{-1}\{f(z)\}$ is both open and closed in U. Since U is connected, $f^{-1}\{f(z)\} = U$.

5.   LEMMA. Suppose $R_0 \le R_1$ and that f is an analytic complex-valued function on an open set containing the annulus $\{z \in \mathbf{C}: R_0 \le |z - a| \le R_1\}$ Then $g: r \mapsto \int_0^{2\pi} f(a + re^{i\theta})re^{i\theta}\, d\theta$ is a constant function on $[R_0, R_1]$ (see Figure 35).

Proof. We shall show that $g'(r) = 0$ for every $r \in [R_0, R_1]$. According to Theorem C 9, the Chain Rule, and the Fundamental Theorem of Calculus, if $r \ne 0$,

$$g'(r) = \int_0^{2\pi} [f'(a + re^{i\theta})e^{i\theta}re^{i\theta} + f(a + re^{i\theta})e^{i\theta}]\, d\theta$$

$$= \int_0^{2\pi} (1/ir)d/d\theta[f(a + re^{i\theta})re^{i\theta}]\, d\theta$$

$$= (1/ir)[f(a + re^{2\pi i})re^{2\pi i} - f(a + re^0)re^0]$$

$$= (1/ir)[f(a + r)r - f(a + r)r] = 0.$$

If $R_0 = 0$, we can conclude that $g'(0) = 0$ by taking the limit of $g'(r)$ as $r \to 0$.

6.   PROPOSITION (CAUCHY'S INTEGRAL THEOREM FOR CIRCLES). Let U be an open set containing the closed disk $B_R^c(a)$ in $\mathbf{C}$, and let E be a *finite* subset of U. Let $f: U \to \mathbf{C}$ be a bounded function that is analytic on $U - E$. Then for every $r \in [0, R]$, $g(r) = \int_0^{2\pi} f(a + re^{i\theta})re^{i\theta}\, d\theta = 0$ (see Figure 35).

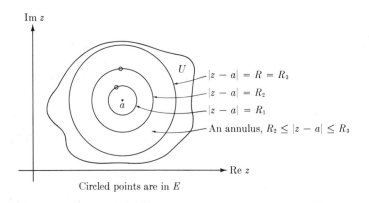

Figure 35. Illustration of Lemma 5 and Proposition 6.

Proof. Let $0 = R_0 < R_1 < \cdots < R_n = R$ be an enumeration of the set $\{0, R\} \cup \{|z - a| : z \in E \cap B_R^c(a)\}$. For each $i = 0, \ldots, n - 1$, if $2\epsilon < R_{i+1} - R_i$, then Lemma 5 shows that g is constant on the interval $[R_i + \epsilon, R_{i+1} - \epsilon]$. Since g is continuous by Theorem C 8, we can conclude that g is constant on $[R_i, R_{i+1}]$ by taking the limit as $\epsilon \to 0$. It follows that g is constant on $[0, R]$. Obviously $g(0) = 0$.

7. **THEOREM (CAUCHY'S INTEGRAL FORMULA FOR A DISK).** Let U be an open set containing the closed disk $B_R^c(a)$ in C, and let E be a finite subset of $B_R(a)$. Let $f : U \to \mathsf{C}$ be a bounded function that is analytic on $U - E$. Then for each $z \in B_R(a) - E$,

$$f(z) = \frac{1}{2\pi} \int_0^{2\pi} \frac{f(a + Re^{i\theta})Re^{i\theta}}{a + Re^{i\theta} - z} \, d\theta.$$

Proof. Given $z \in B_R(a) - E$, consider the function $g : U \to \mathsf{C}$, defined as follows: if $w \neq z$, $g(w) = (f(w) - f(z))/(w - z)$; $g(z) = f'(z)$. Since f is analytic at z, g is actually continuous at z. In particular, g is bounded, and g is analytic on $U - (E \cup \{z\})$. By Proposition 6, $\int_0^{2\pi} g(a + Re^{i\theta})Re^{i\theta} \, d\theta = 0$. That is,

$$\frac{1}{2\pi} \int_0^{2\pi} \frac{f(a + Re^{i\theta})Re^{i\theta}}{a + Re^{i\theta} - z} \, d\theta = \frac{1}{2\pi} \int_0^{2\pi} \frac{f(z)Re^{i\theta}}{a + Re^{i\theta} - z} \, d\theta$$

$$= \frac{f(z)}{2\pi} \int_0^{2\pi} \frac{Re^{i\theta}}{a + Re^{i\theta} - z} \, d\theta.$$

Since $|z - a| < R$, we can expand the function

$$\frac{Re^{i\theta}}{a + Re^{i\theta} - z} = \frac{1}{1 - \dfrac{z - a}{Re^{i\theta}}}$$

in a geometric series $\sum_{k=0}^{\infty} \left(\dfrac{z - a}{Re^{i\theta}}\right)^k$ that converges uniformly on $[0, 2\pi]$. Thus

$$\int_0^{2\pi} \frac{Re^{i\theta}}{a + Re^{i\theta} - z} \, d\theta = \sum_{k=0}^{\infty} \int_0^{2\pi} \left(\frac{z - a}{R}\right)^k e^{-ik\theta} \, d\theta.$$

The term with $k = 0$ is $\int_0^{2\pi} 1 \, d\theta = 2\pi$. If $k > 0$,

$$\int_0^{2\pi} \left(\frac{z - a}{R}\right)^k e^{-ik\theta} = \left(\frac{z - a}{R}\right)^k \int_0^{2\pi} \frac{d}{d\theta} [(-ik)^{-1} e^{-ik\theta}] \, d\theta$$

$$= \left(\frac{z - a}{R}\right)^k (-ik)^{-1} [e^{-2\pi k i} - e^0] = \left(\frac{z - a}{R}\right)^k (-ik)^{-1} [1 - 1] = 0.$$

Thus

$$\int_0^{2\pi} \frac{Re^{i\theta}}{a + Re^{i\theta} - z} \, d\theta = 2\pi. \quad \bullet$$

8.   COROLLARY (EXPANSION IN TAYLOR'S SERIES). Under the hypotheses of Theorem 7, if

$$c_k = \int_0^{2\pi} \frac{f(a + Re^{i\theta})}{(Re^{i\theta})^k} \, d\theta$$

for $k \geq 0$, then the power series $\sum_{k=0}^{\infty} c_k(z - a)^k$ converges to $f(z)$ for each $z \in B_R(a) - E$, and the convergence is uniform on $B_r^c(a) - E$ for each $r \in [0, R[$.

Proof. As in the proof of Theorem 7, expand the integrand $\dfrac{f(a + Re^{i\theta})Re^{i\theta}}{a + Re^{i\theta} - z}$

in a power series $\sum_{k=0}^{\infty} f(a + Re^{i\theta}) \left(\dfrac{z - a}{Re^{i\theta}}\right)^k$, and integrate term-by-term. The fact that the power series converges uniformly on $B_r(a)$ follows from Theorem VI C 4(b).

REMARK. According to Theorems VI C 4(d) and IX C 6, the series of Corollary 8 can be differentiated term by term any number of times. The *nth* derivative is

$$f^{(n)}(z) = \sum_{k=n}^{\infty} k(k - 1) \cdots (k - n + 1)c_k(z - a)^{k-n}.$$

In particular,

☛ $$c_n = \frac{f^{(n)}(a)}{n!}$$

9.   COROLLARY. Under the hypotheses of Theorem 7, f' is also analytic on $B_R(a) - E$, and for each $z \in B_R(a) - E$,

$$f'(z) = \frac{1}{2\pi} \int_0^{2\pi} \frac{f(a + Re^{i\theta})Re^{i\theta}}{(a + Re^{i\theta} - z)^2} \, d\theta.$$

Proof. Exactly as in the proof of Theorem C 9, we can show that the complex derivative of f with respect to z can be calculated by differentiation with respect to z under the integral sign.

10.   COROLLARY. If f is analytic on an open set $U \subseteq \mathbf{C}$, so is f', and indeed f has analytic complex derivatives of all orders on U.

Proof. Each $z \in U$ is contained in a closed disk $B_R^c(z) \subseteq U$. Corollary 9 (with $E = \varnothing$) shows that f' can be given by an integral

$$\frac{1}{2\pi} \int_0^{2\pi} \frac{f(a + Re^{i\theta})Re^{i\theta}}{(a + Re^{i\theta} - z)^2} \, d\theta.$$

Since this integral in turn can be differentiated under the integral sign with respect to z, f' has a complex derivative on U,

$$\frac{1}{\pi} \int_0^{2\pi} \frac{f(a + Re^{i\theta})Re^{i\theta}}{(a + Re^{i\theta} - z)^3} \, d\theta.$$

Evidently this integral is a continuous function of z, so f' is analytic. The rest follows by induction.

11. LIOUVILLE'S THEOREM. A *bounded* analytic function from $\mathbf{C}$ to $\mathbf{C}$ must be a constant function.

Proof. Suppose f is an analytic function from $\mathbf{C}$ to $\mathbf{C}$ such that $|f(z)| \leq B$ for every $z \in \mathbf{C}$. Then $f(z + Re^{i\theta})/Re^{i\theta} \to 0$ uniformly on $[0, 2\pi]$ as $R \to \infty$. By applying Lebesgue's Dominated Convergence Theorem to the formula

$$f'(a) = \frac{1}{2\pi} \int_0^{2\pi} \frac{f(a + Re^{i\theta})}{Re^{i\theta}} \, d\theta$$

(Corollary 9), we conclude that $f'(a) = 0$ for every $a \in \mathbf{C}$. According to Proposition 4, f is a constant function.

12. FUNDAMENTAL THEOREM OF ALGEBRA. Every nonconstant polynomial with complex coefficients has a complex root.

Proof. Let $P(z) = a_n z^n + a_{n-1} z^{n-1} + \cdots + a_0$ be a complex polynomial with $n \geq 1$ and $a_n \neq 0$. Suppose that $P(z) \neq 0$ for every $z \in \mathbf{C}$. Then $f = 1/P$ would be an analytic function from $\mathbf{C}$ to $\mathbf{C}$. Since

$$f(z) = \frac{1}{z^n(a_n + a_{n-1}z^{-1} + \cdots + a_0 z^{-n})},$$

for every $\epsilon > 0$ there is an $R \in \mathbf{R}$ such that $|f(z)| < \epsilon$ whenever $|z| \geq R$. In particular, f is bounded. By Liouville's Theorem, f is constant, so P is constant, a contradiction.

13. RIEMANN'S THEOREM ON REMOVABLE SINGULARITIES. Under the hypotheses of Theorem 7, for each $z \in E$, $\lim_{w \to z} f(w)$ exists. If we define $g : U \to \mathbf{C}$ by putting $g(z) = f(z)$ if $z \in U - E$, $g(z) = \lim_{w \to z} f(w)$ if $z \in E$, then g is analytic on U.

Proof. Put

$$h(z) = \frac{1}{2\pi} \int_0^{2\pi} \frac{f(a + Re^{i\theta})Re^{i\theta}}{a + Re^{i\theta} - z} \, d\theta$$

for $z \in B_R(a)$. Then h is analytic on $B_R(a)$, as we see by differentiating under the integral sign. Moreover, $h(z) = f(z)$ for each $z \in B_R(a) - E$. It follows that $\lim_{w \to z} f(w) = \lim_{w \to z} h(w) = h(z)$ for each $z \in B_R(a)$—in particular, for each $z \in E$. If we define g as above, then $g(z) = h(z)$ for each $z \in B_R(a)$. Thus g is analytic on $B_R(a)$. g is also analytic on $U - B_R(a)$, since this set contains no point of E. ●

There were two reasons for introducing the set E of removable singularities in Proposition 6 and Theorem 7. One was to make possible the proof of Theorem 7, since the function g introduced there is only discovered to be analytic at z *after Theorem 7 has been proved.* The other is to prove Riemann's Theorem on Removable Singularities. For applications of this theorem, see the Exercises.

14. IDENTITY THEOREM. Let U be a *connected* open subset of **C**, and let f and g be analytic functions from U to **C**. Suppose that $S = \{z \in U : f(z) = g(z)\}$ has an accumulation point in U. Then $f(z) = g(z)$ for every $z \in U$.

Proof. Put $h = f - g$. Let A be the set of accumulation points of S in U. If z belongs to the closure of A in U and V is an open neighborhood of z in U, then V contains a point $w \in A$. Since V is also a neighborhood of w, V contains infinitely many points of S. Therefore z is also an accumulation point of S in U. Thus A is closed in U. If we can show that A is open in U, it will follow that $A = U$, since U is connected. Since h is continuous, $z \in A \Rightarrow h(z) = 0 \Rightarrow z \in S$. It will then follow that $S = U$. Suppose then that $a \in A$. According to Corollary 8, we can expand h in a power series $\sum_{k=0}^{\infty} c_k(z - a)^k$ around a. If $c_k = 0$ for every $k \geq 0$, then $h(z) = 0$ for every z in a neighborhood $B_\epsilon(a)$ of a. It follows that $B_\epsilon(a) \subseteq A$. If this were true for every $a \in A$, A would be open. On the other hand, if for some k, $c_k \neq 0$, then there would be a smallest such k, say K. Then we should have $h(z) = (z - a)^K \sum_{k=K}^{\infty} c_k(z - a)^{k-K}$. There is then an $\epsilon > 0$ such that $0 < |z - a| < \epsilon \Rightarrow |\sum_{k=K+1}^{\infty} c_k(z - a)^{k-K}| < |c_K| \Rightarrow \sum_{k=K}^{\infty} c_k(z - a)^{k-K} \neq 0 \Rightarrow h(z) \neq 0$. But then $B_\epsilon(a)$ would contain only *one* point of S, namely a, so $a \notin A$. This would be a contradiction. ●

15. REMARKS. (a) Corollary 8 illuminates the phenomenon of the radius of convergence of a power series (see Remark VI C 5(c)). Let f be an analytic function from an open set $U \subseteq$ **C** into **C**. Let $a \in U$, and let R be the distance from a to **C** $- U$ ($R = +\infty$ if $U =$ **C**) (see Figure 36). Corollary 8 shows that f can be expanded in a power series $\sum_{k=0}^{\infty} c_k(z - a)^k$ that converges to f on $B_R(a)$. On the other hand, if f has such a power series expansion on $B_R(a)$, then by Theorem C 6, this series can be differentiated term-by-term, so that f is analytic on $B_R(a)$.

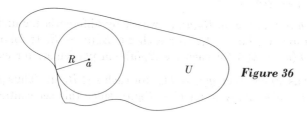

Figure 36

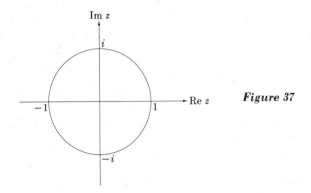

Figure 37

Now consider a series such as $1 - x^2 + x^4 - x^6 + \cdots = 1/(1 + x^2)$. If we think of x as a real variable, then the function $1/(1 + x^2)$ is well-behaved for all $x \in \mathbf{R}$. From this point of view, nothing goes wrong with the function at $x = \pm 1$, so it is not apparent why the power series expansion should fail at these points. However, if we think of x as a complex variable, we see immediately that $1/(1 + x^2) \to \infty$ as $x \to \pm i$. That is why the circle of convergence is the circle of radius 1 about 0. The intersection of this circle with $\mathbf{R}$ is the *interval of convergence* of the real series, $[-1, 1]$. (See Figure 37.)

(b) It can be shown that if U is an open subset of $\mathbf{C}$ and $f: U \to \mathbf{C}$ has a complex derivative at every point of U, then f' is automatically continuous. Thus the condition for analyticity in Definition 1 can be weakened to mere differentiability. (See Ahlfors, *Complex Analysis*, 2nd Edition, McGraw-Hill, New York, 1966.)

(c) Almost all of what has been done in this section works just as well, with the same definitions and proofs, for functions from an open subset U of $\mathbf{C}$ into an arbitrary Banach space *over the complex field*. The sole exceptions are the results on multiplication and division, since we cannot generally multiply and divide vectors, although we can multiply and divide vectors by scalars. Indeed, this whole section is an application of the theory of Banach-valued integrals, since $\langle \mathbf{C}, |\ | \rangle$ is a Banach space of dimension 2 over the field $\mathbf{R}$. Had we merely dealt with real-valued functions in the preceding three sections, we would not be prepared to deal here with integrals of complex-valued functions without doing some additional work.

Exercises

1. Show that the function $f: \mathbf{C} \to \mathbf{C}$ defined by $f(z) = (e^z - 1)/z$ if $z \neq 0$, $f(0) = 0$, has a removable singularity at 0. Remove it.

2. More generally, show that if U is open in $\mathbf{C}$, $f: U \to \mathbf{C}$ is analytic, and $z \in U$, then we define an analytic function $g: U \to \mathbf{C}$ by putting $g(w) = (f(w) - f(z))/(w - z)$ for $w \neq z$, and $g(z) = f'(z)$.

3. Let U be open in $\mathbf{C}$ and suppose $f: U - \{z\} \to \mathbf{C}$ is analytic, and that there is a positive number $p \in \mathbf{R}^+$ such that $\{|w - z|^p f(w): w \in U - \{z\}\}$ is bounded. Let n be an integer such that $n > p$. Show that there is an analytic function $g: U \to \mathbf{C}$ such that $f(w) = g(w)/(w - z)^n$ for $w \in U - \{z\}$. A singularity of this type that is not removable is called a *pole*. *Hint:* Put $g(w) = (w - z)^n f(w)$ if $w \in U - \{z\}$, $g(z) = 0$, and apply Riemann's Theorem on Removable Singularities.

4. Let $f: \mathbf{C} \to \mathbf{C}$ be an analytic function with the following property: for every $z \in \mathbf{C}$, there is an integer n such that the nth derivative of f at z, $f^{(n)}(z)$, vanishes. Prove that f is a polynomial. *Hint:* Let $E_n = \{z \in \mathbf{C}: f^{(n)}(z) = 0\}$. Apply the Baire Category Theorem and the Identity Theorem.

5. Show that an analytic function $f: \mathbf{C} \to \mathbf{R}$ must be a constant function. *Hint:* Show that the function $g: z \mapsto 1/(i - f(z))$ must be bounded, and use Liouville's Theorem. Alternatively, expand f in a power series about a, $f(z) = \sum_{k=0}^{\infty} c_k(z - a)^k$, and show that $c_1 = f'(a) = 0$.

6. Let U be a connected open subset of $\mathbf{C}$ and let $f: U \to \mathbf{C}$ be nonconstant and analytic. Show that if $K \subseteq U$ is compact, then $f^{-1}\{0\} \cap K$ is finite. *Hint:* Use the Identity Theorem and the Bolzano-Weierstrass Theorem (VII A 8(c)).

7. Give an example of a *disconnected* open subset U of $\mathbf{C}$ and an analytic function $f: U \to \mathbf{C}$ such that (a) $f^{-1}\{0\}$ contains a nonempty open subset of U, but (b) f is not a constant function on all of U.

Bibliography

Bibliographical remarks

Most of the matters discussed in this book are extensively treated in other texts. The following remarks are for the guidance of readers who may want to pursue some of them in detail. I have tried to list texts that I like to use, but in some cases I have listed widely used texts that I don't know firsthand. A few texts are listed because I couldn't think of better ones. The bracketed numerals refer to the list of references at the end of this bibliography. On any bibliographical question, you would do well to consult the journal *Mathematical Reviews* and the reviews in the *American Mathematical Monthly*.

General references. A reader of this book is expected to come to it with a solid background in elementary calculus. Most textbooks on calculus are either bad prose or bad mathematics. Here are a few that I like: Spivak's book [1] is an honest, rigorous, and witty modern text. Courant and John [2] is an updated version of a masterwork, with a strong emphasis on mathematical physics. Fadell's book [3] is recommended for the sake of its excellent intuitive discussions of mathematical concepts. The books of Apostol [4] and Dieudonné [5] are more advanced, and cover much of the material in this book, as well as much that is not in it. Neither could be called a text on *elementary* calculus.

Dunford and Schwartz [6] is a three- (or perhaps four-) volume encyclopedia of the theories of Banach spaces, linear maps, integration, differential equations, and so on. Individual sections and chapters are quite readable, despite the forbidding size of the whole. The twenty-plus volumes of N. Bourbaki's *Éléments de Mathématique* [7] cover the whole range of topics I have mentioned. Every mathematician should eventually become familiar with Bourbaki's work, which dogmatically presents a "correct" approach to the whole body of mathematics, in reaction against the former ossification of mathematical education in France.

References on rigor, the foundations of mathematics, and the nature of mathematical proof. There seems to be a dearth of serious treatments of mathematical proof as it is actually practiced by mathematicians—as convincing argument, ritual, game, or element of culture. Most theories of mathematical proof bear a relation to real mathematics similar to that which mathematical physics bears to the world; they present a simplified and idealized theoretical model of reality, which is close enough to reality to be useful. One important difference is that theories of mathematical proof may be *normative*. The theory may influence the way mathematicians go about their business, in a way in which physical theories cannot influence the physical world.

Among the few books on mathematical proof as it is practiced are the books of Polya [8], [9]. Wilder's book [10], though primarily devoted to theories of proof, is sensitive to the relation between theory and practice. The introductory chapters of Kleene [11] give an account of troublesome questions about the foundations of mathematics that arose around the turn of the century. Quine's treatise [12] is a very literate exposition of the view that mathematics can be reduced to logic. One of the most interesting things about it is pointed out in its appendix: all but a handful of the proofs in [12] are informal proofs that there are formal proofs for some theorems. This is a fine illustration of the point made in Section I C, that the Formalist program does not call upon mathematicians to make formal proofs, but rather to convince each other that they could do so if they had the time. There are interesting accounts of controversy among mathematicians about the foundations of their discipline in Fraenkel and Bar-Hillel [13]. This book on the foundations of set theory explicitly makes the point that mathematicians use, or refuse to use, axioms of set theory according to personal prejudice and emotional bent. Körner's book [14] compares and criticizes some theories of mathematical proof on a nontechnical level.

References on set theory. Halmos' book [15] is an informal introduction to formalizable set theory. There is an almost formal set of axioms for set theory in the back of Kelley [16]. Quine's book [12] also presents a theory of sets— one that is not used by working mathematicians. Fraenkel's book [17] on abstract set theory develops that part of the subject that mathematicians use. In contrast, the concern of Fraenkel and Bar-Hillel's book [13] is not to work within set theory, but to investigate questions about it.

References on the real number system. The algebraic properties of ordered fields are discussed in detail in texts on algebra (see below). The consequences of the completeness axiom are discussed in adequate detail in this book. There remain two questions about the axioms for the real number system that deserve further attention.

(1) Is there any complete ordered field?
(2) Are any two complete ordered fields isomorphic?

In one sense, the answers to these questions involve deep problems in set theory, and are affected by such developments as the recent work of Paul Cohen, Skolem's Paradox, and the Intuitionist critique of classical analysis. Fraenkel and Bar-Hillel [13] and Kleene [11] can be consulted on these last two developments. In another sense, it is known that both questions have affirmative answers.

(1) If we grant that there is a system of positive integers, and a little elementary set theory, then a complete ordered field can be constructed out of the positive integers using set-theoretical methods. Thus there is no more danger of inconsistency in the axioms for a complete ordered field than there is danger of inconsistency in Peano's postulates for the positive integers and in the elements of set theory.

(2) Any two complete ordered fields within a given set theory are isomorphic.

These answers can be found in Cohen and Ehrlich [18]. There is also an appendix on question (1) in Spivak's book [1], and there is some information on these matters in the references on algebra.

References on algebra. Bourbaki [19] is always a strong candidate. Van der Waerden [20] is out of date, but still very useful. Birkhoff and MacLane [21] is a widely used text. There is a new version [22], which is very *au courant*. Another new book is Lang [23]. Most of Lang's books are elegant, but betray his impatience with textbook writing. Jacobson [24] is another widely used text. All these books are surveys. Some more specialized texts are Hall [25] on group theory, Artin [26] on fields, Jacobson [27] on rings, and Halmos [28] and Hoffman and Kunze [29] on linear algebra (vector spaces, matrices, determinants, and so on). Faddeeva [30] may be consulted for a discussion of *practical* schemes for computation with linear equations, determinants, eigenvalues, and the like. Another text on these matters is Varga [31].

References on metric spaces, normed spaces, and topological spaces. The treatment of metric spaces in this book is fairly thorough. The standard reference on the next step up the ladder of abstraction is Kelley's *General Topology* [16], on questions of convergence and continuity. There is another side to topology, having to do with connectedness and the classification of topological spaces, for which Spanier's *Algebraic Topology* [32] is recommended. A more elementary book, which has one leg in point-set topology and the other in algebraic topology, and contains some fascinating old theorems, is Hocking and Young [33].

Riesz and Sz.-Nagy [34] is a beautifully written introduction to the applications of Banach spaces and integration to classical analysis. The treatment in Dunford and Schwartz [6] is much more thorough, and the parts on Banach spaces are not hard to read. Bourbaki's volumes on topological vector spaces [35] are still standard. Newer texts, which go still deeper into the theory of Banach spaces, are Kelley and Namioka [36] and Schaefer [37]. At some point I have to mention the lovely, difficult text of Loomis [38]. Now is as good as another. This book says most of the important things about Banach spaces in sixteen pages, and does integration in another nineteen. Another beautiful book on a related matter is Rickart [39].

References on differential equations. Ince [40] is an old favorite, and Birkhoff and Rota [41] is a more modern introductory text. Beyond mentioning these two, I must claim ignorance of the greater part of a vast literature. In particular, I won't stick my neck out by recommending a text on partial differential equations.

References on differential calculus in several variables. The texts of Apostol [4] and Dieudonné [5], as well as Dunford and Schwartz [6], may be consulted for more information about the differential calculus of functions of several variables of the kind presented here. Apostol [4] also has some material on vector analysis, which points in the direction of the really serious further development of these matters: differential geometry. Spivak [42] is a good beginning. O'Neill [43] deals with classical matters. Buck [44] makes the transition between the kind of thing done in this book and more sophisticated theories. Getting on to modern differential geometry, we have Bishop and Crittenden [45] and Sternberg [46]. Lang [47] is very brief, and high-powered. Recently a number of books on differential geometry have appeared. I have mentioned just a few that I know or know of. For more information, you may want to consult the journal *Mathematical Reviews*.

References on integration. The method of introducing the Lebesgue integral used here is closest to that of Bourbaki [48]. Dunford and Schwartz [6] has an extensive treatment of vector-valued integrals and measures, plus an extensive bibliography. Riesz and Sz.-Nagy [34] and Loomis [38] have elegant treatments of integration. Halmos [49] has long been a standard reference work on measure and integration. An attractive new elementary text is Asplund and Bungart [50]. Bartle [51] goes a bit beyond the material in Chapter IX, and the approach is measures first, which you might find interesting. There are too many good and satisfactory books on integration available to make it worthwhile to list more than a representative sample here. You should find Chapter IX a good start.

If, for some reason, you should want to know more about the less satisfactory theory of Riemann integration, there is a thorough treatment in

Apostol [4]. As you can observe, once the initial development is over, it is easier to do almost anything with Lebesgue's integral than to do it with Riemann's.

References on complex analysis. Section IX D only scratches the surface of the very beautiful theory of analytic functions of a complex variable. My favorite text is Ahlfors [52], in the much improved second edition. Cartan [53] is also excellent, as is the older book of Knopp [54]. I also like what I have seen of Markushevich [55], which does approximately what Ahlfors does but uses three times as many pages. Dienes [56] offers another point of view. Heins [57] covers some special topics.

The theory of analytic functions of several complex variables (my own specialty) has a flavor quite different from the theories of one complex variable and several real variables. The standard text now is Gunning and Rossi [58]. Narasimhan [59] does some of the same things more elegantly.

References on Fourier series. Much of the modern theory of integration was motivated by the theory of trigonometrical series, which has strong connections with complex analysis as well. Originally, I had intended to include a section on Fourier series in this book, but abandoned the project to keep Chapter IX from being too long. Zygmund [60] is very thorough. There are also treatments of Fourier series in Apostol [4], Asplund and Bungart [50], Sz.-Nagy [61], and Natanson [62]. Of these, the first is an old-fashioned approach, the second and third make strong use of the Lebesgue integral, and the fourth is concerned with Fourier series as a means of approximating functions. In a sense, Loomis [38] is devoted to a theory of abstract Fourier series. Hoffman [63] has much to say about the relation between Fourier series and complex analysis; de Branges and Rovnyak [64] and Helson [65] also discuss it. See also the supplementary problems for Section IX B.

References

The references are listed in the order in which they are mentioned in the foregoing remarks, under the same subject headings. There are cross-references to texts that are mentioned under more than one heading.

General references

1. M. Spivak, *Calculus*, Benjamin, New York, 1967.
2. R. Courant and F. John, *Introduction to Calculus and Analysis* (2 volumes), Interscience, New York, 1965.
3. A. Fadell, *Calculus with Analytic Geometry*, Van Nostrand, Princeton, 1964.
4. T. Apostol, *Mathematical Analysis*, Addison-Wesley, Reading, 1957.
5. J. Dieudonné, *Foundations of Modern Analysis*, Academic Press, New York, 1960.

6. N. Dunford and J. Schwartz, *Linear Operators*, Interscience, New York. Part I, 1958; Part II, 1963; Part III, to appear; Part IV, to appear (perhaps).

7. N. Bourbaki, *Éléments de Mathématique*, Hermann, Paris. Various chapters of this multivolume work have appeared from time to time since 1939. New chapters continue to appear, and old ones are revised. Some are listed separately below. Among those that have appeared are Book I, *Théorie des Ensembles*, four chapters: (1 & 2), 3, 4 (1 and 2 are bound together, 3 and 4 are separate volumes). Book II, *Algèbre*, nine chapters: 1, 2, 3, (4 & 5), (6 & 7), 8, 9. Book III, *Topologie Générale*, ten chapters: (1 & 2), (3 & 4), (5, 6, 7, & 8), 9, 10. Book IV, *Fonctions d'une Variable Réelle*, seven chapters: (1, 2, & 3), (4, 5, 6, & 7). Book V, *Espaces Vectoriels Topologiques*, five chapters: (1 & 2), (3, 4, & 5). Book VI, *Intégration*, eight chapters: (1, 2, 3, & 4), 5, 6, (7 & 8).

References on rigor, the foundations of mathematics and the nature of mathematical proof

8. G. Polya, *Induction and Analogy in Mathematics*, Princeton University Press, Princeton, 1954.

9. G. Polya, *Patterns of Plausible Inference*, Princeton University Press, Princeton, 1954.

10. R. Wilder, *The Foundations of Mathematics* (2nd Edition), Wiley, New York, 1952.

11. S. Kleene, *Introduction to Metamathematics*, Van Nostrand, Princeton, 1952.

12. W. Quine, *Mathematical Logic* (Revised Edition), Harvard University Press, Cambridge, 1951.

13. A. Fraenkel and Y. Bar-Hillel, *Foundations of Set Theory*, North-Holland, Amsterdam, 1958.

14. S. Körner, *The Philosophy of Mathematics*, Harper, 1960.

References on set theory

15. P. Halmos, *Naive Set Theory*, Van Nostrand, Princeton, 1960.

16. J. Kelley, *General Topology*, Van Nostrand, Princeton, 1955.

17. A. Fraenkel, *Abstract Set Theory*, North-Holland, Amsterdam, 1961.
Also see A. Fraenkel and Y. Bar-Hillel [13] and W. Quine [12].

References on the real number system

18. L. Cohen and G. Ehrlich, *The Structure of the Real Number System*, Van Nostrand, Princeton, 1963.
Also see M. Spivak [1], S. Kleene [11], A. Fraenkel and Y. Bar-Hillel [13], and the following texts on algebra.

References on algebra

19. N. Bourbaki, *Éléments de Mathématique, Livre II, Algèbre*, Hermann, Paris. Chapter 1, *Structures Algébriques*, 1958. Chapter 2, *Algèbre Linéaire* (3rd edition), 1962.

Chapter 3, *Algèbre Multilinéaire*, 1958. Chapter 4, *Polynomes et Fractions Rationelles* and Chapter 5, *Corps Commutatifs*, 1959. Chapter 6, *Groupes et Corps Ordonnés* and Chapter 7, *Modules sur les Anneaux Principaux*, 1952. Chapter 8, *Modules et Anneaux Semi-Simples*, 1958. (See Reference [7].)

20. B. van der Waerden, *Modern Algebra*, Ungar, New York. Volume I, 1953; Volume II, 1950.
21. G. Birkhoff and S. MacLane, *A Survey of Modern Algebra* (Revised Edition), Macmillan, New York, 1953.
22. S. MacLane and G. Birkhoff, *Algebra*, Macmillan, 1967.
23. S. Lang, *Algebra*, Addison-Wesley, Reading, 1965.
24. N. Jacobson, *Lectures in Abstract Algebra*, Van Nostrand, New York. Volumes 1 & 2, 1951; Volume 3, 1964.
25. M. Hall, *The Theory of Groups*, Macmillan, New York, 1959.
26. E. Artin, *Galois Theory*, University of Notre Dame Press, Notre Dame, 1964.
27. N. Jacobson, *Structure of Rings*, American Mathematical Society, Providence, 1964.
28. P. Halmos, *Finite-dimensional Vector Spaces*, Van Nostrand, Princeton, 1958.
29. K. Hoffman and R. Kunze, *Linear Algebra*, Prentice-Hall, Englewood Cliffs, 1961.
30. V. Faddeeva, *Computational Methods of Linear Algebra*, Dover, New York, 1959.
31. R. Varga, *Matrix Iterative Analysis*, Prentice-Hall, Englewood Cliffs, 1962.

References on metric spaces, normed spaces, and topological spaces

32. E. Spanier, *Algebraic Topology*, McGraw-Hill, New York, 1966.
33. J. Hocking and G. Young, *Topology*, Addison-Wesley, Reading, 1961.
34. F. Riesz and B. Sz.-Nagy, *Functional Analysis*, Ungar, New York, 1955.
35. N. Bourbaki, *Espaces Vectoriels Topologiques*, Hermann, Paris. Chapters 1 and 2, 1953. Chapters 3, 4, and 5, 1955. (See Reference [7].)
36. J. Kelley and I. Namioka, *Linear Topological Spaces*, Van Nostrand, Princeton, 1963.
37. H. Schaefer, *Topological Vector Spaces*, Macmillan, New York, 1966.
38. L. Loomis, *An Introduction to Abstract Harmonic Analysis*, Van Nostrand, New York, 1953.
39. C. Rickart, *General Theory of Banach Algebras*, Van Nostrand, Princeton, 1960.
   ☞ Also see J. Kelley [16] and N. Dunford and J. Schwartz [6], which are especially valuable.

References on differential equations

40. E. Ince, *Ordinary Differential Equations*, Dover, New York, 1944.
41. G. Birkhoff and G.-C. Rota, *Ordinary Differential Equations*, Ginn, New York, 1962.

References on differential calculus in several variables

42. M. Spivak, *Calculus on Manifolds*, Benjamin, New York, 1965.

43. B. O'Neill, *Elementary Differential Geometry*, Academic Press, New York, 1966.
44. R. Buck, *Advanced Calculus*, McGraw-Hill, New York, 1956.
45. R. Bishop and R. Crittenden, *Geometry of Manifolds*, Academic Press, New York, 1964.
46. S. Sternberg, *Lectures on Differential Geometry*, Prentice-Hall, Englewood Cliffs, 1964.
47. S. Lang, *Introduction to Differentiable Manifolds*, Interscience, New York, 1962. Also see T. Apostol [4], J. Dieudonné [5], N. Dunford and J. Schwartz [6].

References on integration

48. N. Bourbaki, *Intégration*, Hermann, Paris. Chapters 1, 2, 3, and 4, 1952. Chapter 5, 1956. Chapter 6, 1959. Chapters 7 & 8, 1963. (See Reference [7].)
49. P. Halmos, *Measure Theory*, Van Nostrand, New York, 1950.
50. E. Asplund and L. Bungart, *A First Course in Integration*, Holt, Rinehart and Winston, New York, 1966.
51. R. Bartle, *The Elements of Integration*, Wiley, New York, 1966.
☞ Also see T. Apostol [4], N. Dunford and J. Schwartz [6], F. Riesz and B. Sz.-Nagy [34], L. Loomis [38].

References on complex analysis

52. L. Ahlfors, *Complex Analysis* (2nd Edition), McGraw-Hill, New York, 1966.
53. H. Cartan, *Elementary Theory of Analytic Functions of One or Several Complex Variables*, Addison-Wesley, New York, 1963.
54. K. Knopp, *Theory of Functions*, Dover, New York, 1952.
55. A. Markushevich, *Theory of Functions of a Complex Variable* (translated by R. Silverman), Prentice-Hall, Englewood Cliffs. Volumes 1 and 2, 1965; Volume 3, to appear.
56. P. Dienes, *The Taylor Series*, Dover, New York, 1957.
57. M. Heins, *Selected Topics in the Classical Theory of Functions of a Complex Variable*, Holt, Rinehart and Winston, New York, 1962.
58. R. Gunning and H. Rossi, *Analytic Functions of Several Complex Variables*, Prentice-Hall, Englewood Cliffs, 1965.
59. R. Narasimhan, *Introduction to the Theory of Analytic Spaces*, Springer, New York, 1966.

References on Fourier series

60. A. Zygmund, *Trigonometric Series*, Cambridge University Press, Cambridge, 1959.
61. B. Sz.-Nagy, *Introduction to Real Functions and Orthogonal Expansions*, Oxford University Press, New York, 1965.
62. I. Natanson, *Constructive Function Theory* (translated by J. Schulenberger), Ungar, New York, 1965.
63. K. Hoffman, *Banach Spaces of Analytic Functions*, Prentice-Hall, Englewood Cliffs, 1962.

64. L. de Branges and J. Rovnyak, *Square Summable Power Series*, Holt, Rinehart and Winston, New York, 1966.
65. H. Helson, *Lectures on Invariant Subspaces*, Academic Press, New York, 1964. Also see T. Apostol [4], L. Loomis [38], and E. Asplund and L. Bungart [50].

Answers and solutions to starred exercises

Exercises in Section I D

1. The difference and quotient of rational numbers are rational. Therefore, if r were rational and $r + Hitler$ were rational, $Hitler = (r + Hitler) - r$ would be rational. Similarly, if $r \neq 0$ and $r \times Hitler$ were rational, then $Hitler = (r \times Hitler)/r$ would be rational.

5. Let $x = 0.d_1d_2d_3d_4 \ldots$ be any decimal expansion of a real number between 0 and 1. We produce a new decimal expansion $f(x)$ according to the following rule: $f(x) = 0.\delta_1\delta_2\delta_3 \ldots$, where $\delta_{n^2} = d_n$, $\delta_{n^2+1} = 1$, for each $n \in \mathbf{N}$, and if neither k nor $k-1$ is a square, $\delta_k = 0$. Then $f(x) \neq 0$ and the digits of $f(x)$ do not repeat, so that $f(x)$ is irrational. The function $x \mapsto f(x)$ is a one-to-one mapping of the set of all decimal expansions into the set of all irrational real numbers, so there are as many irrational real numbers in the range of f as there are decimal expansions. But there are uncountably many decimal expansions by Corollary B 16.

Exercises in Section II B

3. The identity is the permutation $I : x \mapsto x$, which leaves everything in S fixed. The inverse of a permutation f is simply the inverse function f^{-1}. The associative law is trivial.

   If S contains only one element; $\pi(S)$ contains only the identity permutation. If S contains two elements a, b, there is also the permutation $f : a \mapsto b$, $b \mapsto a$, with $f \circ f = I$. f commutes with itself and with I, the only two possibilities. If S contains at least three elements, a, b, c, then $\pi(S)$ contains permutations f and g that do the following: f and g leave fixed everything but a, b, and c, and $f : a \mapsto b$, $b \mapsto a$, $c \mapsto c$, $g : a \mapsto a$, $b \mapsto c$, $c \mapsto b$. Then $f \circ g : a \mapsto a \mapsto b$ but $g \circ f : a \mapsto b \mapsto c$. Thus $f \circ g \neq g \circ f$.

Exercises in Section II C

1. In case $n = 1$,

$$(1 + a)^n = 1 + a = 1 + na + \frac{n(n-1)}{2} a^2.$$

Suppose $n \geq 1$, and that

$$(1 + a)^n \geq 1 + na + \frac{n(n-1)}{2} a^2.$$

Since $n(n-1) \geq 0$,

$$(1+a)^{n+1} = (1+a)(1+a)^n \geq (1+a)\left(1 + na + \frac{n(n-1)}{2} a^2\right)$$

$$= 1 + na + \frac{n(n-1)}{2} a^2 + a + na^2 + \frac{n(n-1)}{2} a^3$$

$$\geq 1 + na + a + \frac{n(n-1)}{2} a^2 + a^2$$

$$= 1 + (n+1)a + \frac{n(n-1) + 2}{2} a^2$$

$$= 1 + (n+1)a + \frac{(n+1)((n+1)-1)}{2} a^2.$$

2. Given any $\epsilon > 0$, there is, by the Archimedean Property, an integer $N > 0$ such that $N\epsilon > |a|$. Then if $n > N$, $n\epsilon > N\epsilon > |a|$, so $|a/n - 0| = |a|/n < \epsilon$.

3. Put $r = 1 + a$. Then

$$r^n = (1+a)^n \geq 1 + na + \frac{n(n-1)}{2} a^2.$$

For any b, there is, by the Archimedean Property, an integer N such that $r^N > Na > |b| \geq b$, and there is an integer M such that $\frac{1}{2}(M-1)a^2 > |b| \geq b$, since $\frac{1}{2}(M-1)a^2 > |b| \Leftrightarrow Ma^2 > 2|b| + a^2$. Now suppose that $0 < s < 1$. Put $r = 1/s > 1$. If $\epsilon > 0$, we can put $b = 1/\epsilon$, and choose N and M as above. Then $n > N \Rightarrow r^n > r^N > b \Rightarrow 0 < s^n = 1/r^n < 1/b = \epsilon$, and $m > M \Rightarrow r^m/m > ((m-1)/2)a^2 > ((M-1)/2)a^2 > b \Rightarrow 0 < ms^m = m/r^m < 1/b = \epsilon$.

4. If $S = \{n + m\sqrt{2} : n, m \in \mathbf{Z}\}$, it is clear that if $x, y \in S$ and $k \in \mathbf{Z}$, then $x + y$, $x - y$, $kx \in S$. Now for each integer m, there is a largest integer $n(m)$ that is $\leq m\sqrt{2}$. $0 \leq m\sqrt{2} - n(m) < 1$, for $n(m) \leq m\sqrt{2}$, and if $1 + n(m) \leq m\sqrt{2}$, then $n(m)$ would not be the *largest* integer $\leq m\sqrt{2}$. If $n + m\sqrt{2} = n' + m'\sqrt{2}$, then $(m - m')\sqrt{2} = n' - n$. If $m \neq m'$, then $\sqrt{2}$ would be a rational number $(n' - n)/(m - m')$. This is impossible, so $m = m'$ and therefore $n = n'$. In particular, if $m \neq m'$, then $m\sqrt{2} - n(m) \neq m'\sqrt{2} - n(m')$. Put $s_m = m\sqrt{2} - n(m)$. Then $s_m \in S$, $0 \leq s_m < 1$, and $m \neq m' \Rightarrow s_m \neq s'_m$. Thus $\{s_m : m \in \mathbf{Z}\}$ is an infinite subset of $S \cap \{x \in \mathbf{R} : 0 \leq x < 1\}$. Given any $\epsilon > 0$, we can partition the interval $[0, 1[= \{x \in \mathbf{R} : 0 \leq x < 1\}$ into finitely many subintervals of length less than ϵ, by choosing an integer $k > 1/\epsilon$ and using the intervals $[i/k, (i+1)/k[$, $i = 0, \ldots, k - 1$. At least one of these intervals must contain two elements of $S \cap \{x \in \mathbf{R} : 0 \leq x < 1\}$ (by the Pigeonhole Principle—otherwise there would be at most k elements in $\{s_m : m \in \mathbf{Z}\}$). Thus there exist m and m' such that s_m and s'_m are distinct points in some interval of length less than $\epsilon : 0 < s_m - s'_m < \epsilon$. But $s_m - s'_m \in S$. This shows that for every $\epsilon > 0$, there is an $s \in S$ such that $0 < s < \epsilon$. By the Archimedean property, if $\epsilon = (b - a)/2$, there is an integer n such that $a < ns < b$, and $ns \in S$.

Alternatively, we could use the Bolzano-Weierstrass Property to show that the infinite set $S \cap [0, 1[$ has a point of accumulation x. Then there must be infinitely many points of S within a distance $\epsilon/3$ of x. If s and t are two distinct ones, $0 < |s - t| \leq |s - x| + |x - t| \leq 2\epsilon/3 < \epsilon$. Then we proceed as before.

5. (a) $\mathbf{Z}$ is bounded neither above nor below.

   (b) $\mathbf{N}$ is not bounded above, and glb $\mathbf{N} = 1$.

   (c) $0 = \text{glb} \{1/n : n \in \mathbf{N}\}$, $1 = \text{lub} \{1/n : n \in \mathbf{N}\}$.

   (d) Let $s = \{x \in \mathbf{R} : x > .3 + .3/10 + \cdots + .3/10^n \text{ for every } n\}$. Then S is not bounded above, and glb $S = \frac{1}{3}$.

   (e) glb $\{r \in \mathbf{Q} : r^2 < b\} = -\sqrt{b}$. lub $\{r \in \mathbf{Q} : r^2 < b\} = \sqrt{b}$.

   (f) glb $\{r \in \mathbf{R} : r^2 < b\} = -\sqrt{b}$. lub $\{r \in \mathbf{R} : r^2 < b\} = \sqrt{b}$.

   (g) glb $\{x \in \mathbf{R}^+ : r^2 < b\} = 0$. lub $\{x \in \mathbf{R}^+ : r^2 < b\} = \sqrt{b}$.

   (h) $\{x \in \mathbf{R} : x < \pi\}$ is not bounded below. lub $\{x \in \mathbf{R} : x < \pi\} = \pi$.

6. Suppose $\lim_{n \to \infty} x_n = L$. Choose N so large that $n > N \Rightarrow |x_n - L| < 1$. Let $B = \max \{|x_n| : n \leq N\}$, which exists because $\{|x_n| : n \leq N\}$ is a *finite* set. Then $|L| + B + 1$ is a bound for $\{x_n\}$, for if $n \leq N$, $|x_n| \leq B < |L| + B + 1$, and if $n > N$, $|x_n| \leq |L| + |L - x_n| < |L| + 1 \leq |L| + 1 + B$.

9. Suppose the ordered field $\langle F, F^+, +, x \rangle$ has the Bolzano-Weierstrass Property. Let ϵ be any element of F^+, and let f be any element of F. Then there is an integer n such that

$$n\epsilon = \underbrace{\epsilon + \cdots + \epsilon}_{n \text{ terms}} > f.$$

Proof. If $f \leq 0$, we can take $n = 1$. If $f > 0$ and if the assertion were false, $S = \{n\epsilon : n \in \mathbf{N}\}$ would be bounded below by 0 and above by f, and would be infinite. Therefore, by B–W, S would have an accumulation point x. Infinitely many elements of S would lie between $x - \epsilon/2$ and $x + \epsilon/2$. Choose two distinct ones, say $x - \epsilon/2 < n\epsilon < m\epsilon < x + \epsilon/2$. Then $(m - n)\epsilon < \epsilon \Rightarrow m - n < 1$. This is impossible, since $m - n \in \mathbf{N}$. ●

Now let us show that F has the Archimedean Property. If $b - a > \epsilon > 0$, there is an integer $N \in \mathbf{N}$ such that $N\epsilon > -a$ or $-N\epsilon < a$, by what we have just shown. There is also an $M \in \mathbf{N}$ such that $M\epsilon > b$. Among the finitely many integers $-N, -N + 1, \ldots, M$, we can choose a smallest one, k, such that $k\epsilon > a$ (note that $M\epsilon > b > a > -N\epsilon$). Then $(k - 1)\epsilon \leq a \Rightarrow a < k\epsilon = (k - 1)\epsilon + \epsilon \leq a + \epsilon < b$. ●

Finally, let us show that every Cauchy sequence in F converges. Let $n \mapsto s_n$ be a Cauchy sequence in F. If $\{s_n : n \in \mathbf{N}\}$ is a finite set, then there must be some $N \in \mathbf{N}$ such that $\{n \in \mathbf{N} : s_n = s_N\}$ is infinite. Then $\lim_{n \to \infty} s_n = s_N$, for if $\epsilon > 0$, we can choose K so large that $n, m > K \Rightarrow |s_n - s_m| < \epsilon$, and then choose $m > K$ such that $s_m = s_N$. Then $n > K \Rightarrow |s_n - s_N| = |s_n - s_m| < \epsilon$. On the other hand, suppose $\{s_n : n \in \mathbf{N}\}$ is infinite. Choose N so large that $n, m \geq N \Rightarrow |s_n - s_m| < 1$. Then $\max \{|s_n| : n < N\} + |s_N| + 1$ is a bound for $\{|s_n| : n \in N\}$ (see Exercise 6). Therefore, by B–W, $\{s_n : n \in \mathbf{N}\}$ has a cluster point x. In fact, $\lim_{n \to \infty} s_n = x$, for if $\epsilon > 0$, we can choose M so large that $n, m > M \Rightarrow |s_n - s_m| < \epsilon/2$. Since x is a cluster point of $\{s_n : n \in \mathbf{N}\}$, we can choose $m > M$ such that $|x - s_m| < \epsilon/2$. Then if $n > M$, $|s_n - x| \leq |s_n - s_m| + |s_m - x| < \epsilon/2 + \epsilon/2 = \epsilon$.

10. Suppose that $\langle F, F^+, +, x \rangle$ has the Archimedean Property, and that every Cauchy sequence in F converges. Let us show that every nonempty subset S of F that is bounded above has a least upper bound.

Since S is nonempty, there is an $x \in S$. Put $b = x - 1 < x$. Then b is not an upper bound for S. Let B be an upper bound for S. Then choose sequences $n \mapsto l_n$ and $n \mapsto u_n$ as follows.

(i) $l_0 = b$, $u_0 = B$.

(ii) Once l_n and u_n have been chosen, choose their successors as follows. If $(l_n + u_n)/2$ is an upper bound for S, put $l_{n+1} = l_n$, $u_{n+1} = (l_n + u_n)/2$. If $(l_n + u_n)/2$ is not an upper bound for S, put $l_{n+1} = (l_n + u_n)/2$, $u_{n+1} = u_n$.

Note that $2 = 1 + 1 = 1^2 + 1^2 > 1^2 > 0$. The choice of $n \mapsto l_n$ and $n \mapsto u_n$ renders trivial a proof by induction that for each $n \in \mathbf{N}$,

(a) u_n is an upper bound for S but l_n is not,

(b) $u_n - l_n = (B - b) \times 2^{-n}$,

(c) $l_n \le l_{n+1} < u_{n+1} \le u_n$.

Let $\epsilon > 0$ be given. According to Exercise 1, $2^n = (1 + 1)^n > n \cdot 1$. By the Archimedean Property, we can choose $n \in \mathbf{N}$ such that $2^n > n \cdot 1 > (B - b)/\epsilon \Rightarrow (B - b)2^{-n} < \epsilon$. If $k \ge j \ge n$, $l_n \le l_{n+1} \le \cdots \le l_k < u_k \le u_{k-1} \le \cdots \le u_j \le u_{j-1} \le \cdots \le u_n \Rightarrow 0 \le u_j - u_k \le u_n - l_n = (B - b)2^{-n} < \epsilon$. This shows that $n \mapsto u_n$ is a Cauchy sequence, which has a limit L.

L is an upper bound for S, for if it were not, we could choose $y \in S$ such that $y > L$. Put $\epsilon = y - L$. If $|u_n - L| < \epsilon$, then $y = L + \epsilon > u_n$, so that u_n would not be an upper bound for S. But there is an n such that $|u_n - L| < \epsilon$, since $\lim_{n \to \infty} u_n = L$.

There can be no smaller upper bound for S than L, for if $z < L$, we can choose n so large that $(B - b) \cdot 2^{-n} + |L - u_n| < L - z$. Then $l_n = u_n - (B - b)2^{-n} = L - [(L - u_n) + (B - b)2^{-n}] > L - (L - z) = z$. Since l_n is not an upper bound for S, z cannot be an upper bound for S. Thus $L = \text{lub } S$.

Exercises in Section III A

1. Let e_j be the vector $(0, \ldots, 0, \overset{j\text{th place}}{1}, 0, \ldots, 0)$ in F^n. It is clear that a vector $(x_1, \ldots, x_n) \in F^n$ can be *uniquely* expressed as a linear combination of $e_1, \ldots, e_n$ as follows: $(x_1, \ldots, x_n) = x_1 e_1 + \cdots + x_n e_n$. According to Theorem 11, $\{e_1, \ldots, e_n\}$ is a basis for F^n, so dim $F^n = n$.

3. $(4, 5, 6) = (1, 2, 3) + 3(1, 1, 1)$. There are, of course, other possibilities, such as $(1, 2, 3) = (4, 5, 6) - 3(1, 1, 1)$.

4. Let V be the subspace of $\mathbf{Q}^3$ spanned by $(1, 2, 3)$, $(1, 3, 3)$, and $(1, 2, 4)$. By performing row operations (see Section III B),

$$\begin{pmatrix} 1 & 2 & 3 \\ 1 & 3 & 3 \\ 1 & 2 & 4 \end{pmatrix} \to \begin{pmatrix} 1 & 2 & 3 \\ 0 & 1 & 0 \\ 0 & 0 & 1 \end{pmatrix} \to \begin{pmatrix} 1 & 0 & 0 \\ 0 & 1 & 0 \\ 0 & 0 & 1 \end{pmatrix},$$

we see that $(0, 1, 0) = (1, 3, 3) - (1, 2, 3) \in V$, $(0, 0, 1) = (1, 2, 4) - (1, 2, 3) \in V$, and $(1, 0, 0) = (1, 2, 3) - 2(0, 1, 0) - 3(0, 0, 1) \in V$. But $(1, 0, 0)$, $(0, 1, 0)$, and $(0, 0, 1)$ span $\mathbf{Q}^3$ (Exercise 1). Thus $V = \mathbf{Q}^3$ by Proposition 6. Since $(1, 2, 3)$, $(1, 3, 3)$, and $(1, 2, 4)$ are three vectors that span a three-dimensional space $\mathbf{Q}^3$, they are independent by Corollary 16. Therefore they are a basis for $\mathbf{Q}^3$.

Exercises in Section III C

1. $f = g/\|g\|$, where $\|g\|^2 = \int_a^b (g(x))^2 \, dx$. Assume $g \neq 0$.

2. Put $w_1 = v_1/\|v_1\|$. Then w_1 is normal. Now for an induction on k, suppose we have found $w_1, \ldots, w_k$ that form an orthonormal basis for the subspace of V spanned by $v_1, \ldots, v_k$. If $u_{k+1} = v_{k+1} - c_1 w_1 - \cdots - c_k w_k$ is to be orthogonal to $w_1, \ldots,$ w_k, we must have $0 = [u_{k+1}, w_i] = [v_{k+1}, w_i] - c_1[w_1, w_i] - \cdots - c_k[w_k, w_i] = [v_{k+1}, w_i] - c_i[w_i, w_i] = [v_{k+1}, w_i] - c_i$ for each $i = 1, \ldots, k$; conversely, if we choose $c_i = [v_{k+1}, w_i]$ for $i = 1, \ldots, k$, then the same calculation shows that $u_{k+1} = v_{k+1} - c_1 w_1 - \cdots - c_k w_k$ is indeed orthogonal to $w_1, \ldots, w_k$. Then u_{k+1} is clearly in the subspace of V spanned by $v_{k+1}, w_k, \ldots, w_1$. Since each of $w_k, \ldots,$ w_1 is (by hypothesis) dependent on $\{v_k, \ldots, v_1\}$, u_{k+1} is dependent on $v_{k+1}, \ldots, v_1$. Also, since $v_{k+1} = u_{k+1} + c_1 w_1 + \cdots + c_k w_k$, $\{w_1, \ldots, w_k, u_{k+1}\}$ spans the subspace of V spanned by $v_1, \ldots, v_{k+1}$. This subspace has dimension $k + 1$. Therefore the $k + 1$ vectors $w_1, \ldots, w_k, u_{k+1}$ are a basis for it by Corollary A 16. Thus we have an orthogonal basis for the subspace of V spanned by $v_1, \ldots, v_{k+1}$. To normalize it, we have only to put $w_{k+1} = u_{k+1}/\|u_{k+1}\|$. $\|u_{k+1}\|$ cannot be 0, since the vectors $w_1, \ldots, w_k, u_{k+1}$ are independent.

3. $w_1(x) \equiv 1$, $w_2(x) \equiv 2\sqrt{3}(x - 1/2)$, $w_3(x) \equiv 6\sqrt{5}(x^2 - x + 1/6)$.

Exercises in Section III E

4. It is clear that the map $(y_1, \ldots, y_n) \mapsto x_1 y_1 + \cdots + x_n y_n$ is a linear functional on F^n. Conversely, if $\{e_1, \ldots, e_n\}$ is the usual basis for F^n (answer to Exercise III A 1) and L is a linear functional on F^n, $L(y_1, \ldots, y_n) = y_1 L(e_1) + \cdots + y_n L(e_n) = x_1 y_1 + \cdots + x_n y_n$, where $x_i = L(e_i)$ for $i = 1, \ldots, n$.

5. (a) Trivial.   (b) $\begin{pmatrix} 1 & 1 \\ -2 & -1 \end{pmatrix}$.

6. $\begin{pmatrix} 5 & -1 \\ -2 & 0 \end{pmatrix}$.

Exercises in Section III G

2. The eigenvalues and corresponding eigenvectors are:
$0 \leftrightarrow (1, 0, 1)$,
$1 \leftrightarrow (1, 2, 2)$,
$2 \leftrightarrow (0, 1, 1)$.

Exercises in Section IV D

2. (a) Routine.

(b) Suppose $\lim_{k \to \infty} d(n_k, \infty) = 0$. Let N be any positive integer. Then there is a $K \in \mathbf{N}$ such that $k > K \Rightarrow 1/n_k = d(n_k, \infty) < 1/N \Rightarrow n_k > N$. Conversely, if for every $N \in \mathbf{N}$ there is a $K \in \mathbf{N}$ such that $n_k > N$ whenever $k > K$, then for any $\epsilon > 0$ we can choose $N \in \mathbf{N}$ such that $1/N < \epsilon$, and then choose $k \in \mathbf{N}$ such that $k > K \Rightarrow n_k > N \Rightarrow d(n_k, \infty) = 1/n_k < 1/N < \epsilon$. That is, $\lim_{k \to \infty} d(n_k, \infty) = 0$.

(c) If $m \in \mathbf{N}$, then for every $x \in J - \{m\}$, $d(m, x) \geq d(m, m + 1) = 1/m(m + 1)$. Thus if $\epsilon = 1/m(m + 1)$, $B_\epsilon(m) = \{m\}$, which shows that $\{m\}$ is *open* in J. Now let $f: J \to M$ be defined by $f(n) = x_n$ if $n \in \mathbf{N}$,

$f(\infty) = L$. If V is a neighborhood of x_n in M, then $\{n\} \subseteq f^{-1}(V)$ is a neighborhood of n in J. That is, f is continuous at n for each $n \in \mathbf{N}$.

Suppose f is continuous at ∞. Then $\lim\limits_{n \to \infty} x_n = L$ by Theorem C 3, taking into account that $n \mapsto n$ is a sequence converging to ∞ in J. Conversely, suppose $\lim\limits_{n \to \infty} x_n = L$. Let V be any neighborhood of L in M. Then there is a $K \in \mathbf{N}$ such that $x_n \in V$ when $n > K$. This shows that $f^{-1}(V)$ contains $\{\infty\} \cup \{n \in \mathbf{N}: n > K\} = \{x \in J: d(x, \infty) < 1/K\} = B_{1/K}(\infty)$. This shows that f is continuous at ∞. That is, $f: J \to M$ is continuous.

(d) If $\lim\limits_{n \to \infty} x_n = X$ and $\lim\limits_{n \to \infty} y_n = Y$, we can define continuous functions f: $J \to \mathbf{R}$ and $g: J \to \mathbf{R}$ by putting $f(n) = x_n$, $g(n) = y_n$ for $n \in \mathbf{N}$, and $f(\infty) = X$, $g(\infty) = Y$. By Corollary 8, $\lim\limits_{n \to \infty} (x_n + y_n) = \lim\limits_{n \to \infty} (f(n_n) + g(n_n)) = f(\infty) + g(\infty)$, and $\lim\limits_{n \to \infty} (x_n y_n) = \lim\limits_{n \to \infty} f(n_n)g(n_n) = f(\infty)g(\infty) = XY$.

Exercises in Section VI A

2. $\int_0^1 g_n(x)\, dx = (n+1)(n+2) \int_0^1 (x^n - x^{n+1})\, dx = (n+1)(n+2)[1/(n+1) - 1/(n+2)] = (n+2) - (n+1) = 1$, so $\lim\limits_{n \to \infty} \int_0^1 g_n(x)\, dx = 1$. On the other hand, according to Example 2(c), $\lim\limits_{n \to \infty} g_n(x) = 0$ for each $x \in [0, 1]$. Thus $\int_0^1 \left(\lim\limits_{n \to \infty} g_n(x) \right) dx = 0$.

A theorem giving a very weak condition that is sufficient to guarantee that $\lim\limits_{n \to \infty} \int_0^1 g_n(x)\, dx = \int_0^1 \left(\lim\limits_{n \to \infty} g_n(x) \right) dx$ is given in Chapter IX. At this point, you should not be trying to use the Lebesgue theory of integration, but should formulate an elementary sufficient condition appropriate to this section. You need no further help from me than that.

Exercises in Section VI B

4. (a) Suppose $\sum x_n$ converges absolutely to L. Then for every $\epsilon > 0$, there is an $N \in \mathbf{N}$ such that $\sum_{n=N+1}^{\infty} \|x_n\| < \epsilon/2$. Let $E(\epsilon) = \{1, \ldots, N\}$. If H is a finite subset of $\mathbf{N}$ that includes $E(\epsilon)$ and if K is the largest integer in H, then

$$\left\| \sum_{n=1}^{\infty} x_n - \sum_{n \in H} x_n \right\| \le \left\| \sum_{n=1}^{\infty} x_n - \sum_{n \in E(\epsilon)} x_n \right\| + \left\| \sum_{n \in H} x_n - \sum_{n \in E(\epsilon)} x_n \right\|$$

$$= \left\| \sum_{n=N+1}^{\infty} x_n \right\| + \left\| \sum_{n \in H - E(\epsilon)} x_n \right\| \le \sum_{n=N+1}^{\infty} \|x_n\| + \sum_{n \in H - E(\epsilon)} \|x_n\|$$

$$< \epsilon/2 + \sum_{N=N+1}^{K} \|x_n\| \le \epsilon/2 + \sum_{n=N+1}^{\infty} \|x_n\| < \epsilon/2 + \epsilon/2 = \epsilon.$$

Thus L is the unordered sum of the function $n \mapsto x_n$.

(b) Conversely, suppose that the function $f: \mathbf{N} \to \mathbf{R}$ has an unordered sum, L. For each $k \in \mathbf{N}$, let $P(k) = \{n \in \mathbf{N}: n \le k \text{ and } f(n) \ge 0\}$, and let $Q(k) = \{n \in \mathbf{N}: n \le k \text{ and } f(n) < 0\}$. Let $A = \{\sum_{n \in P(k)} f(n): k \in \mathbf{N}\}$ and let $B = \{\sum_{n \in Q(k)} f(n): k \in \mathbf{N}\}$. Since $\sum_{n=1}^{k} |f(n)| = \sum_{n \in P(k)} f(n) - \sum_{n \in Q(k)} f(n)$,

$\sum_{n=1}^{\infty} |f(n)|$ can diverge only if at least one of A and B is unbounded. Let us show that, in fact, both must be bounded. Choose a finite subset $E \subseteq \mathbf{N}$ such that $|\sum_{n \in H} f(n) - L| < 1$ whenever $E \subseteq H \subseteq \mathbf{N}$ and H is finite. Then, in particular, $E \subseteq E \cup P(k) \subseteq \mathbf{N}$ for each k, so that $\sum_{n \in P(k)} f(n) = \sum_{n \in E \cup P(k)} f(n) - \sum_{n \in E - P(k)} f(n) < (L + 1) + \sum_{n \in E} |f(n)|$. Thus A is bounded. Similarly, $\sum_{n \in Q(k)} f(n) > -(L + 1) - \sum_{n \in E} |f(n)|$, so B is bounded. ●

5. For each $\epsilon > 0$, we can choose a finite set $E(\epsilon) \subseteq S$ such that $\|\sum_{n \in H} f(n) - L\| < \epsilon$ whenever $E(\epsilon) \subseteq H \subseteq S$ and H is finite. Let $N(\epsilon)$ be the largest integer in $\sigma^{-1}(E(\epsilon))$. If $n \geq N(\epsilon)$, then $\{\sigma(1), \ldots, \sigma(n)\} \supseteq \{\sigma(1), \ldots, \sigma(N(\epsilon))\} \supseteq E(\epsilon)$. Therefore $\|\sum_{k=1}^{n} f(\sigma(k)) - L\| < \epsilon$. This shows that $\sum_{n=1}^{\infty} f(\sigma(n)) = L$. Now if $\sum x_n$ is an absolutely Cauchy series, the function $n \mapsto x_n$ has an unordered sum L by Exercise 4(a), and what we have just proved shows that $\sum_{n=1}^{\infty} x_{\sigma(n)} = L$, whenever $\sigma: \mathbf{N} \to \mathbf{N}$ is one-to-one and onto. Thus the series $\sum x_n$ can be rearranged without changing its sum.

Remark. The fact that σ is onto was used tacitly above to show that $\{\sigma(1), \ldots, \sigma(N(\epsilon))\} \supseteq E(\epsilon)$, and the fact that σ is one-to-one was used to show that $\sum_{k=1}^{n} f(\sigma(k)) = \sum_{k \in E} f(k)$, where $E = \{\sigma(1), \ldots, \sigma(n)\}$.

6. Let A, B, $P(k)$, and $Q(k)$ be the sets defined in the answer to Exercise 4(b). Since $\sum |a_n|$ diverges, at least one of A and B must be unbounded. Since $\sum_{n=1}^{N} a_n = \sum_{n \in P(N)} a_n + \sum_{n \in Q(N)} a_n$ converges to a limit as $N \to \infty$, neither A nor B can be bounded. Let $P = \{n \in \mathbf{N}: a_n \geq 0\}$ and $Q = \{n \in \mathbf{N}: a_n < 0\}$. Since A and B are both unbounded, P and Q must be infinite. Now define a function $\sigma: \mathbf{N} \to \mathbf{N}$ by the following rule: Put $\sigma(1) = 1$. Suppose $\sigma(1), \ldots, \sigma(n - 1)$ have been defined.

(a) If $\sum_{k=1}^{n-1} a_{\sigma(k)} \leq L$, let $\sigma(n)$ be the smallest integer in $P - \{\sigma(1), \ldots, \sigma(n - 1)\}$.

(b) If $\sum_{k=1}^{n-1} a_{\sigma(k)} > L$, let $\sigma(n)$ be the smallest integer in $Q - \{\sigma(1), \ldots, \sigma(n - 1)\}$.

The choice is always possible because P and Q are infinite. σ is one-to-one because we always choose $\sigma(n)$ to be distinct from $\sigma(1), \ldots, \sigma(n - 1)$.

If there were an $N \in \mathbf{N}$ such that $\sigma(n) \in P$ whenever $n > N$, then $\sigma(N + 1)$, $\sigma(N + 2)$, $\sigma(N + 3), \ldots$ would be the enumeration of $P - \{\sigma(1), \ldots, \sigma(N)\}$ in increasing order. Since A is unbounded, $\{\sum_{n=N+1}^{k} a_{\sigma(n)}: k \in \mathbf{N}\}$ would also be unbounded. In particular, there would be a $K \in \mathbf{N}$ such that $\sum_{n=1}^{k} a_{\sigma(n)} > L$ whenever $k > K$. But then, according to (a) and (b), we should have $\sigma(k) \in Q$ whenever $k > K$, a contradiction. Similarly, there must be infinitely many values of k for which $\sigma(k) \in P$. It follows that σ maps $\mathbf{N}$ onto $\mathbf{N}$.

Finally, if $\epsilon > 0$, there is an $N \in \mathbf{N}$ such that $|a_n| < \epsilon$ when $n > N$, because $\sum a_n$ converges. There is also a $K \in \mathbf{N}$ such that $\sigma(k) > N$ when $k > K$, and there is an $M > K$ such that $\sigma(M) \in P$ and $\sigma(M + 1) \in Q$. Let us show that if $k \geq M$, $|\sum_{i=1}^{k} a_{\sigma(i)} - L| < \epsilon$. Note that $\sum_{i=1}^{M-1} a_{\sigma(i)} \leq L$ (by (a)) $< \sum_{i=1}^{M} a_{\sigma(i)}$ (because $\sigma(M + 1) \in Q$) $= \sum_{i=1}^{M-1} a_{\sigma(i)} + a_{\sigma(M)} < \sum_{i=1}^{M-1} a_{\sigma(i)} + \epsilon \Rightarrow |\sum_{i=1}^{M-1} a_{\sigma(i)} - L| < \epsilon$. If the assertion we are trying to prove were false, there would be a smallest $k \geq M$ such that $|\sum_{i=1}^{k} a_{\sigma(i)} - L| \geq \epsilon$. Then $|\sum_{i=1}^{k-1} a_{\sigma(i)} - L| < \epsilon$. Suppose $k \in Q$. Then by (a) and (b), $L + \epsilon > \sum_{i=1}^{k-1} a_{\sigma(i)} > L$ and $\sum_{i=1}^{k-1} a_{\sigma(i)} > \sum_{i=1}^{k} a_{\sigma(i)} = \sum_{i=1}^{k-1} a_{\sigma(i)} + a_{\sigma(k)} > \sum_{i=1}^{k-1} a_{\sigma(i)} - \epsilon$. It follows that $|\sum_{i=1}^{k} a_{\sigma(i)} - L| < \epsilon$, a contradiction. There is a similar contradiction if $k \in P$.

To make $\sum_{k=1}^{\infty} a_{\sigma(k)}$ diverge to $+\infty$, modify the inductive definition of σ as follows.

(a') If $\sum_{k=1}^{n-1} a_{\sigma(k)} \leq \frac{1}{2} \sum_{k=1}^{n-1} |a_{\sigma(k)}|$, let $\sigma(n)$ be the smallest element of $P - \{\sigma(1), \ldots, \sigma(n-1)\}$.

(b') If $\sum_{k=1}^{n-1} a_{\sigma(k)} > \frac{1}{2} \sum_{k=1}^{n-1} |a_{\sigma(k)}|$, let $\sigma(n)$ be the smallest element of $Q - \{\sigma(1), \ldots, \sigma(n-1)\}$.

The only new problem is to show that σ maps **N** onto **N**. But if $\sigma(k) \in Q$ for all but finitely many values of k, B would be bounded below. If $\sigma(k)$ were in P for all but finitely many values of k (say for $k > K$), we should have $a_{\sigma(k)} = |a_{\sigma(k)}|$ for $k > K$. Then

$$\frac{\displaystyle\sum_{k=1}^{n} a_{\sigma(k)}}{\displaystyle\sum_{k=1}^{n} |a_{\sigma(k)}|} = \frac{\displaystyle\sum_{k=1}^{n} |a_{\sigma(k)}| + \sum_{k=1}^{K} (a_{\sigma(k)} - |a_{\sigma(k)}|))}{\displaystyle\sum_{k=1}^{n} |a_{\sigma(k)}|} \to 1$$

as $n \to \infty$. This would contradict (a') and (b').

The technique for making $\sum_{k=1}^{\infty} a_{\sigma(k)}$ diverge to $-\infty$ is similar.

To get oscillatory divergence, first sum up enough positive terms to make the sum >1, then enough negative terms to make it <-1, then enough positive terms to make it >2, then enough negative terms to make it <-2, and so on.

Exercises in Section VI C
(a) $R = +\infty$.
(b) $R = 1$.
(c) $R = 1$.

Exercises in Section VI E
Part I

1. Starting with the initial approximation $g_0(x) = 1$, we get $g_1(x) = 1 + x$, $g_2(x) = 1 + x + x^2/2$, $g_3(x) = 1 + x + x^2/2 + x^3/(2 \cdot 3)$, $g_4(x) = 1 + x + x^2/2 + x^3/(2 \cdot 3) + x^4/(2 \cdot 3 \cdot 4)$.

2. Starting with the initial approximation $g_0(x) = (0, 1)$, we get $g_1(x) = (x, 1)$, $g_2(x) = (x, 1 - x^2/2)$, $g_3(x) = (x - x^3/(2 \cdot 3), 1 - x^2/2)$, $g_4(x) = (x - x^3/(2 \cdot 3), 1 - x^2/2 + x^4/(2 \cdot 3 \cdot 4))$, $g_5(x) = (x - x^3/(2 \cdot 3) + x^5/(2 \cdot 3 \cdot 4 \cdot 5), 1 - x^2/2 + x^4/(2 \cdot 3 \cdot 4))$. The approximations are partial sums of the power series expansion for $(\sin x, \cos x)$ about 0.

5. $f(x) = \exp (x^2/2)$. $(\exp (y) = e^y.)$

6. *Further hint:* Show that $|Tf(x) - Tg(x)| \leq \frac{1}{2} |f - g|_{\infty}$, so that $|Tf - Tg|_{\infty} \leq \frac{1}{2} |f - g|_{\infty}$, and use the Contraction Mapping Theorem. $(0 \leq x \leq 1 \Rightarrow |x \cos (xy)| \leq 1.)$

Part II

1. (a) $f(x) = f(a)e^{x-a}$.
   (b) $f(x) = f(a) \exp (x^2 - a^2)$. $(\exp (y) = e^y.)$
   (c) $f(x) = \sin (x - a + \arcsin f(a))$. $(|f(a)| \leq 1.)$
   (d) $f(x) = (2 \log (x/a) + (f(a))^2)^{1/2}$.

3. (a) $y(x) = -x \pm (2x^2 - K)^{1/2}$, where $K = 2a^2 - (y(a) + a)^2$, and the $\pm$ sign is chosen so that $-a \pm (2a^2 - K)^{1/2} = y(a)$.
   (b) $y(x) = -x \log (K - \log |x|)$, where $K = \log |a| + \exp (-y(a)/a)$.

4. By the Chain Rule for Partial Derivatives (see Theorem G 7)

$$(d/dx)E(x, f(x)) = \frac{\partial E}{\partial x}(x, f(x)) + \frac{\partial E}{\partial y}(x, f(x))\, f'(x).$$

$E(x, f(x))$ is a constant function on $[x_0, x_1]$ if, and only if, its derivative is 0.

5. If $P = \partial E/\partial x$ and $Q = \partial E/\partial y$, then $\partial P/\partial y = \partial^2 E/\partial y\partial x = \partial^2 E/\partial x\partial y = \partial Q/\partial x$, by the Theorem on Interchange of Partial Differentiation Operators (which is proved in Section IX C, using much more powerful techniques than are actually required to prove it).

Conversely, suppose $\partial P/\partial y = \partial Q/\partial x$. Put $E_1(x, y) = \int_{x_0}^{x} P(t, y_0)\, dt +$
$\int_{y_0}^{y} Q(x, s)\, ds$ and $E_2(x, y) = \int_{y_0}^{y} Q(x_0, s)\, ds + \int_{x_0}^{x} P(t, y)\, dt.$ Then

$$E_1(x, y) - E_2(x, y) = -\int_{x_0}^{x} [P(t, y) - P(t, y_0)]\, dt + \int_{y_0}^{y} [Q(x, s) - Q(x_0, s)]\, ds$$

$$= -\int_{x_0}^{x} \left[\int_{y_0}^{y} \frac{\partial P}{\partial y}(t, s)\, ds\right] dt + \int_{y_0}^{y} \left[\int_{x_0}^{x} \frac{\partial Q}{\partial x}(t, s)\, dt\right] ds$$

$$= \int_{x_0}^{x} \left[\int_{y_0}^{y} \frac{\partial Q}{\partial x}(t, s)\, ds\right] dt - \int_{x_0}^{x} \left[\int_{y_0}^{y} \frac{\partial P}{\partial y}(t, s)\, ds\right] dt$$

$$= \int_{x_0}^{x} \left[\int_{y_0}^{y} \left[\frac{\partial Q}{\partial x}(t, s) - \frac{\partial P}{\partial y}(t, s)\right] ds\right] dt$$

$$= \int_{x_0}^{x} \left[\int_{y_0}^{y} 0\, ds\right] dt = 0,$$

by the Theorem on Interchange of the Order of Repeated Integrals (a strong form of this theorem, Fubini's Theorem, is proved in Section IX C). Thus E_1 and E_2 are the same function, so we may define $E(x, y)$ to be $E_1(x, y)$ and also to be $E_2(x, y)$. By the Fundamental Theorem of Calculus (also in Section IX C), $\partial E/\partial x = \partial E_2/\partial x = (\partial/\partial x) \int_{x_0}^{x} P(t, y)\, dt = P(x, y),$ and $\partial E/\partial y = \partial E_1/\partial y = (\partial/\partial y) \int_{y_0}^{y} Q(x, s)\, ds = Q(x, y).$

6. (a) $y(x) = K/x$, where $K = ay(a)$.
   (b) $y(x) = K/x - x$, where $K = a(y(a) + a)$.
   (c) $y(x) = -1/x \pm (1/x^2 - K/x)^{1/2}$, where $K = a(1/a^2 - (y(a) + 1/a)^2)$, and $\pm$ is chosen so that $y(a) = -1/a \pm (1/a^2 - K/a)^{1/2}.$

7. (a) $f(x) = f(a) \exp((a^2 - x^2)/2).$
   (b) $f(x) = -1 \pm \sqrt{1 - K/x}$, where $K = a(1 - (f(a) + 1)^2)$, and $\pm$ is chosen so that $f(a) = -1 \pm \sqrt{1 - K/a}.$

Exercises in Section VI F

1. To solve the DE $f' = \Phi f + \Psi$, proceed as follows.
   Step 1. Introduce the related homogeneous DE $f' = \Phi f$. According to Exercise VI E II 2, its solution is $f(x) = f(x_0) \exp\left(\int_{x_0}^{x} \Phi(t)\, dt\right).$

Step 2. To find the Green's function, observe that the solution to the equation $y = g_1 \exp \left(\int_{x_0}^s \Phi(t)\, dt \right)$ is $G_1(s) \cdot y = g_1 = y \exp \left(- \int_{x_0}^s \Phi(t)\, dt \right)$.

Step 3. According to Theorem 7, a particular solution to the DE $f' = \Phi f + \Psi$ which takes the value 0 at x_0 is

$$f(x) = \left[\int_{x_0}^x \Psi(s) \exp \left(- \int_{x_0}^s \Phi(t)\, dt \right) ds \right] \exp \left(\int_{x_0}^x \Phi(t)\, dt \right)$$

$$= \int_{x_0}^x \Psi(s) \exp \left(\int_{x_0}^x \Phi(t)\, dt - \int_{x_0}^s \Phi(t)\, dt \right) ds$$

$$= \int_{x_0}^x \Psi(s) \exp \left(\int_s^x \Phi(t)\, dt \right) ds.$$

Step 4. The general solution to the DE $f' = \Phi f + \Psi$ is

$$f(x) = f(x_0) \exp \left(\int_{x_0}^x \Phi(t)\, dt \right) + \int_{x_0}^x \Psi(s) \exp \left(\int_s^x \Phi(t)\, dt \right) ds.$$

2. The solution to the DE $f''(x) + f(x) = \sin x$ with initial conditions $f(0) = C$ and $f'(0) = D$ is $f(x) = (\sin x)/2 - (x \cos x)/2 + C \cos x + D \sin x$.

3. (c) Since the DE $(D - \lambda I)^n f = 0$ is of order n, its space of solutions is of dimension n. In part (b), the n solutions $e^{\lambda x}, xe^{\lambda x}, \ldots, x^{n-1}e^{\lambda x}$ have been found.

$$c_0 e^{\lambda x} + c_1 x e^{\lambda x} + \cdots + c_{n-1} x^{n-1} e^{\lambda x} = 0 \Leftrightarrow (c_0 + c_1 x + \cdots + c_{n-1} x^{n-1})e^{\lambda x}$$

$$= 0 \Leftrightarrow c_0 + c_1 x + \cdots + c_{n-1} x^{n-1} = 0 \Leftrightarrow c_0 = c_1 = \cdots = c_{n-1} = 0,$$

because a polynomial vanishes identically only if all its coefficients are 0. Thus these n solutions are linearly independent, so they span the space of all solutions. The general solution has the form $P(x)e^{\lambda x}$, where P is a polynomial of degree $< n$.

6. The solution to the DE $f'' - 3f' + 2f = 1$ with initial conditions $f(0) = A$ and $f'(0) = B$ is $f(x) = e^{2x}/2 - e^x + 1/2 + (B - A)e^{2x} + (2A - B)e^x$.

Exercises in Section VI G

2. Suppose $f = (f^1, \ldots, f^m)$ is differentiable at v. For each $k = 1, \ldots, m$, let $L_k : (x_1, \ldots, x_m) \mapsto x_k$ be the linear functional on $\mathbf{R}^m$ that picks out the kth coordinate. Then $| L_k | = 1$, and $f^k = L_k \circ f$. Since $\|f^k(v + x) - f^k(v) - L_k \circ df_v(x)\|_\infty = \|L_k[f(v + x) - f(v) - df_v(x)]\|_\infty \leq | L_k | \|f(v + x) - f(x) - df_v(x)\|_\infty$, it follows that f^k is differentiable at v and that $df_v^k = L_k \circ df_v$. That is, $df_v(x) = (df_v^1(x), \ldots, df_v^m(x))$. Conversely, if $f^1, \ldots, f^m$ are differentiable at v, let D: $x \mapsto (df_v^1(x), \ldots, df_v^m(x))$. Then for each $k = 1, \ldots, m$, $L_k(f(v + x) - f(v) - Dx) = f^k(v + x) - f^k(v) - df_v^k(x)$, so that $\|f(v + x) - f(v) - Dx\|/\|x\|_\infty = \max_{1 \leq x \leq m} |f_k(v + x) - f_k(v) - df_v^k(x)|/\|x\|_\infty \to 0$ as $x \to 0$. Thus f is differentiable at v and $df_v = D$.

3. Following the hint, we observe that

$$f(x) - f(v) = f(x_1, \ldots, x_n) - f(v_1, x_2, \ldots, x_n) + f(v_1, x_2, \ldots, x_n)$$
$$- f(v_1, v_2, x_3, \ldots, x_n) + \cdots + f(v_1, \ldots, v_{n-1}, x_n) - f(v_1, \ldots, v_n),$$

and that

$$f(v_1, \ldots, v_{j-1}, x_j, \ldots, x_n) - f(v_1, \ldots, v_j, x_{j+1}, \ldots, x_n)$$

$$= \frac{\partial f}{\partial x_j} (v_1, \ldots, v_{j-1}, \eta_j, x_{j+1}, \ldots, x_n) (v_j - x_j),$$

where η_j lies between x_j and v_j. If we put $\nabla f(v) = \left(\dfrac{\partial f}{\partial x_1}(v), \ldots, \dfrac{\partial f}{\partial x_n}(v) \right)$, then

$$f(x) - f(v) - [x - v, \nabla f(v)]$$

$$= \sum_{j=1}^{n} (x_j - v_j) \left[\frac{\partial f}{\partial x_j}(v_1, \ldots, v_{j-1}, \eta_j, x_{j+1}, \ldots, x_n) - \frac{\partial f}{\partial x_j}(v) \right]$$

$|x_j - v_j|/\|x - v\|_\infty \le 1$, while the bracketed terms approach 0 as $x \to v$ because $\partial f/\partial x_j$ is continuous at v for $j = 1, \ldots, n$. This shows that $\lim\limits_{x \to v} |f(x) - f(v) - [x - v, \nabla f(v)]|/\|x - v\|_\infty = 0$, so f is differentiable at v and $df_v(h) = [h, \nabla f(v)]$. Note that we actually need only assume that $\partial f/\partial x_j$ exists in a neighborhood of v and is continuous at v.

Exercises in Section VII A

4. If $\mathfrak{F}$ is a countable family of open balls in a metric space $\langle M, d \rangle$ with the property set forth in Definition 14, then the countable set of centers of these balls is a countable dense subset of M, so M is separable (see Corollary 13). Conversely, if D is a countable dense subset of M, then $\mathfrak{F} = \{B_{1/n}(d): n \in \mathbf{N}, d \in D\}$ is a countable family of open balls. If $x \in M$ and U is a neighborhood of x, there is an $\epsilon > 0$ such that $B_\epsilon(x) \subseteq U$. Choose $n \in \mathbf{N}$ such that $1/n < \epsilon/2$. Since D is dense in M, there is a $d \in D$ such that $d \in B_{1/n}(x) \Rightarrow d(x, d) < 1/n \Rightarrow x \in B_{1/n}(d) \subseteq B_{2/n}(x) \subseteq B_\epsilon(x) \subseteq U$. Thus $\mathfrak{F}$ has the property set forth in Definition 14.

5. $\mathbf{Z}$ is an infinite subset of $\mathbf{R}$ that has no point of accumulation, so $\mathbf{R}$ is not compact. $\mathbf{Q}$ is a countable dense subset of $\mathbf{R}$, so $\mathbf{R}$ is separable (see Corollary II C 3).

Exercises in Section VII B

3. If $\langle M, d \rangle$ is a compact metric space, then Corollaries A 10 and A 11 show that a subset S of M is closed if, and only if, S is compact. Now if $f: M \to N$ is one-to-one and onto, and $g = f^{-1}$, then $g^{-1}(S) = f(S)$ is compact whenever S is compact, by Theorem 1. That is, $g^{-1}(S)$ is compact whenever S is closed. By Corollary A 10, $g^{-1}(S)$ is closed whenever S is closed, so by Theorem IV D 3, $g = f^{-1}$ is continuous.

Exercises in Section VIII A

2. Since B_n is the image of the compact interval $[1/n, 1]$ by the continuous function $x \mapsto (x, \sin(1/x))$, B_n is compact and connected. Let $k \mapsto (x_k, y_k)$ be a sequence in S. If $x_k \ge 1/n$ for all but finitely many $k \in \mathbf{N}$, then all but finitely many terms of this sequence lie in B_n. Since B_n is compact, there is a convergent subsequence. Otherwise, there is a subsequence $j \mapsto (x_{k(j)}, y_{k(j)})$ such that $\lim\limits_{j \to \infty} x_{k(j)} = 0$. Since $y_k \in [-1, 1]$ for every k, and since $[-1, 1]$ is compact, there is a subsequence $i \mapsto (x_{k(j(i))}, y_{k(j(i))})$ such that $i \mapsto y_{k(j(i))}$ converges to a limit $L \in [-1, 1]$. Then $\lim\limits_{i \to \infty} (x_{k(j(i))}, y_{k(j(i))}) = (0, L) \in A$. In either case, the sequence $k \mapsto (x_k, y_k)$ has a convergent subsequence, so S is compact.

If U and V are disjoint open subsets of $\mathbf{R}^2$ that cover S, and if $(1, \sin 1) \in U$, then $B_n \subseteq U$ for every $n \in \mathbf{N}$. In fact, B_n is connected, U and V cover B_n, and

$(1, \sin 1) \in U \cap B_n$. Since $S \cap U$ is open and closed in S, $B^c = (\bigcup B_n)^c \subseteq (S \cap U)^c = S \cap U$. But $B^c = S$, for if $(0, y) \in A$, $0 < x \le 2\pi$, and $\sin x = y$, then the sequence $n \mapsto ((x + 2\pi n)^{-1}, \sin (x + 2\pi n))$ is a sequence in B that converges to $(0, y)$. Thus $S \subseteq U$. This shows that S is connected.

Finally, let $f: [0, 1] \to S$, and suppose that $f(1) = (1, \sin 1)$ and $f^{-1}(A) \ne \emptyset$. We shall reach a contradiction from the assumption that f is continuous. It will follow that no arc in S can join $(1, \sin 1)$ to a point in A, so that S is not arcwise connected. Suppose f were continuous. Let the components of $f(y)$ be $(f_1(y), f_2(y))$. Since A is closed in S, B is open in S. Let $x = \operatorname{lub} f^{-1}(A)$. $f^{-1}(A)$ is closed, so $x \in f^{-1}(A)$. $f^{-1}(B)$ is an open neighborhood of 1 in $[0, 1]$, so $x \notin f^{-1}(B) \Rightarrow x < 1$. $x \in]x, 1]^c \Rightarrow 0 = f_1(x) \in (f_1(]x, 1]))^c$. Since $]x, 1]$ is connected, $f_1(]x, 1]) \subseteq [0, 1]$ is an interval that contains 1 and does not contain 0, but $(f_1(]x, 1]))^c$ does contain 0. Thus $f_1(]x, 1]) =]0, 1]$. For each n, $I_n = [x + (1 - x)/n, 1]$ is compact. Thus $f_1(I_n)$ is a compact subset of $[0, 1]$ that does not contain 0, so there is an $\epsilon_n > 0$ such that $[0, \epsilon_n] \cap f(I_n) = \emptyset$. Choose $y_n = 2/(k_n \pi)$ in $]0, \epsilon_n[$ such that $\sin (1/y_n) = (-1)^n$. Since $f(]x, 1]) =]0, 1]$, there is an $x_n \in]x, x + (1 - x)/n[$ such that $f_1(x_n) = y_n$. Then the sequence $n \mapsto f(x_n) = (y_n, \sin (1/y_n)) = (y_n, (-1)^n)$ does not converge, although $\lim_{n \to \infty} x_n = x$, a contradiction.

6. (a) If $x \in M$, let $C_x = \bigcup \{S \subseteq M : x \in S \text{ and } S \text{ is connected}\}$. According to Theorem 8, C_x is connected. Since $\{x\}$ is connected, $\{x\} \subseteq C_x$, so $x \in C_x$. If $x \in S$ and S is connected, then $S \subseteq C_x$, by the definition of C_x. If also $C_x \subseteq S$, then $S = C_x$. This shows that C_x is a connected component of M which contains S.

   (b) If C and D are connected components of M and $C \cap D \ne \emptyset$, then $C \cup D$ is connected by Theorem 8. It follows that $C = C \cup D = D$, since $C \subseteq C \cup D$ and $D \subseteq C \cup D$.

   (c) Let C be a connected component of the open subset M of V. If $x \in C$, there is a ball $B_\epsilon(x)$ that is contained in M, since M is open. But $B_\epsilon(x)$ is connected (see the proof of Theorem 14). Since C is connected, and $x \in C \cap B_\epsilon(x)$, $C \cup B_\epsilon(x)$ is connected. Since $C \subseteq C \cup B_\epsilon(x)$ and C is a component, $C = C \cup B_\epsilon(x) \Rightarrow B_\epsilon(x) \subseteq C$. This shows that C is open.

   (d) Let U be an open subset of $\mathbf{R}$. $\mathbf{Q} \cap U$ is countable (Definition I B 3 and Corollary I B 9). According to (a) and (b), each $x \in \mathbf{Q} \cap U$ belongs to a unique component C_x of U. According to (c), C_x is open and connected, so C_x is an open interval by Theorem 4. $\{C_x : x \in \mathbf{Q} \cap U\}$ is countable by Theorem I B 10, and $U = \bigcup \{C_x : x \in \mathbf{Q} \cap U\}$ because $\mathbf{Q}$ is dense in U, so that every component of U contains an element of $\mathbf{Q}$.

Exercises in Section IX A

2. Clearly $\int \|f\| \ge 0$ for each $f \in$ Step (n, V). Also $\int \|cf\| = \int |c| \|f\| = |c| \int \|f\|$. If $f, g \in$ Step (n, V), $(\|f\| + \|g\|)(x) \ge \|f + g\|(x)$ for each $x \in \mathbf{R}^n$. Therefore $\int \|f\| + \int \|g\| = \int (\|f\| + \|g\|) \ge \int \|f + g\|$ by Theorem 9(d). That is, $f \mapsto \int \|f\|$ has all the properties of a norm except possibly that $\int \|f\|$ may be 0 when $f \ne \mathbf{0}$. To see that this can indeed happen, note that if $\mathcal{G}$ is a grid on $\mathbf{R}^n$ and $f \in$ Step (n, V) vanishes on $(\mathbf{R}^n - (R(\mathcal{G})^c) \cup \bigcup C(\mathcal{G}))$, then $\mathcal{G}$ is a presentation of $\|f\|$ and $\int \|f\| = 0$. For example, if $v \in V$, then we could have $f(x) = v$ for every $x \in \bigcup \{\partial R : R \in C(\mathcal{G})\}$.

Exercises in Section IX B

3. The following Definitions, Theorems, and Remarks in Chapters IV, VII, and VIII apply to seminormed spaces and pseudometric spaces without significant changes. All others fail to be valid, or at least must be restated or proved in some other way.

Chapter IV, Section A, 2, 3 (The examples are all of normed spaces, with the exception of 3(f). If $\langle V, \| \ \|_v \rangle$ and $\langle W, \| \ \|_w \rangle$ are seminormed spaces, $B(V, W)$ is a seminormed space.)

Chapter IV, Section B, 2, 3, 4, 5 (with the condition $z \neq 0$ in the proof replaced by $\|z\| \neq 0$).

Chapter IV, Section C, 1, 2(b), (c), (d) (2(a) may be *false*), 3, 4, 5, 6, 7, 8, 9, 10, 11, 12, 13, 14, 15, 16, 17.

Chapter IV, Section D, 1, 2, 3, 4, 5, 6, 7, 8.

Chapter IV, Section E, 1, 2, 4, 5, 6.

Chapter VII, Section A: 1, 2, 3, 4, 5, 6, 7, 8, 9, 11, 12, 13, 14. (10 is *false* for pseudometric spaces.)

Chapter VII, Section B: 1, 2, 3, 4, 5. (6 and 7 are *false* for pseudonormed spaces.)

Chapter VII, Section C: 1, 2, 3, 4, 5, 6, 7 (with the provision that $\langle N, e \rangle$ be a *metric* space, although $\langle M, d \rangle$ need only be pseudometric), 8, 9.

Chapter VIII, Section A: 1, 2, 3, 4, 5, 6, 7, 8, 9, 10, 11, 12, 13, 14.

Remarks. (a) Throughout this book, we have tacitly used the following fact.

PROPOSITION: If $n \mapsto s_n$ is a sequence in the metric space $\langle M, d \rangle$ that converges to L and to L', then $L = L'$.

Proof. $d(L, L') = \lim_{n \to \infty} d(s_n, s_n) = 0$, since d is continuous and $L = \lim_{n \to \infty} s_n = L'$.

COROLLARY. If S is a complete subset of the metric space $\langle M, d \rangle$, then S is closed.

Proof. If $n \mapsto s_n$ is a sequence in S that converges to an element L of M, then $n \mapsto s_n$ is a Cauchy sequence. Accordingly, it has a limit $L' \in S$. By the Proposition, $L = L' \in S$. ●

Corollary VII A 10 follows from this Corollary (see Theorem IV C 10). Since the Proposition is obviously false for pseudometric spaces, so are its Corollary and Corollary VII A 10.

(b) Some of the results in Chapter VI fail for pseudometric and seminormed spaces. An outstanding example is the Contraction Mapping Theorem, which is *false* for pseudometric spaces. However, the results of Sections VI A and VI B are valid for pseudometric and seminormed spaces.

4. (a) Suppose $\{f\}^c = \{f'\}^c$, $\{g\}^c = \{g'\}^c$, $f + g = h$, $f' + g' = h'$. Then $|h - h'|_1 \leq |f - f'|_1 + |g - g'|_1 = 0 + 0 = 0$. If $|p - h|_1 = 0$, then $|p - h'|_1 \leq |p - h|_1 + |h - h'|_1 = 0$. That is, $\{h\}^c \subseteq \{h'\}^c$. Similarly, $\{h'\}^c \subseteq \{h\}^c$, so $\{h\}^c = \{h'\}^c$. This shows that there is a unique $\{h\}^c$ that bears the relation $+$ to $\langle \{f\}^c, \{g\}^c \rangle$, so $+$ is a function. (That is, the choice of the set $\{h\}^c$ is independent of which representatives of the sets $\{f\}^c$ and $\{g\}^c$ we pick.) $|rf - rf'|_1 = |r| \, |f - f'|_1 = 0$, so it follows in the same way that $\cdot$ is a function. Finally, $| \, |f|_1 - |f'|_1| \leq |f - f'|_1 = 0$, so $|\ \ |_1$ is a function as well.

(b) If $\{f\}^c$, $\{g\}^c$, $\{h\}^c \in L^1(n, V)$, then $(\{f\}^c + \{g\}^c) + \{h\}^c = \{f + g\}^c + \{h\}^c = \{f + g + h\}^c = \{f\}^c + \{g + h\}^c = \{f\}^c + (\{g\}^c + \{h\}^c)$. Thus addition in $L^1(n, V)$ obeys the associative law. In a similar way, we can check that $\langle L^1(n, V), +, \cdot \rangle$ satisfies the other axioms of Definition III A 1, so that it is a vector space over **R**. $| \{f\}^c |_1 = 0 \Leftrightarrow | f |_1 = 0 \Leftrightarrow \{f\}^c = \{0\}^c$ and, in a similar way, we can verify that $| \ |_1$ satisfies the other axioms in Definition IV A 1, so that it is a norm. It remains to show that the normed space $\langle \langle L^1(n, V), +, \cdot \rangle, | \ |_1 \rangle$ is complete. Let $k \mapsto \{f_k\}$ be a Cauchy sequence in $L^1(n, V)$. Then $| f_k - f_j |_1 = | \{f_k\}^c - \{f_j\}^c |_1$, so $k \mapsto f_k$ is a Cauchy sequence in $\mathcal{L}^1(n, V)$. By Theorem 13, there is an $f \in \mathcal{L}^1(n, V)$ such that $0 = \lim_{k \to \infty} | f - f_k |_1 = \lim_{k \to \infty} | \{f\} - \{f_k\}^c |_1$.

9. (a) $\chi_{\varnothing} = 0$. Thus $\lambda^*(\varnothing) = | \chi_{\varnothing} |_1 = 0$.

(b) $S \subseteq T \Rightarrow \chi_S(x) \leq \chi_T(x)$ for every $x \in \mathbf{R}^n$. Therefore, $\lambda^*(S) = | \chi_S |_1 \leq | \chi_T |_1 = \lambda^*(T)$ by Lemma 8(c).

(c) If $\sum_{k=1}^{\infty} \lambda^*(S_k) = +\infty$, there is nothing to prove, since by convention $+\infty > x$ for every $x \in \mathbf{R}$. Otherwise, let $\epsilon > 0$, and for each $k \in \mathbf{N}$ let $j \mapsto s_{kj}$ be a bounding sequence for χ_{S_k} such that $\lim_{j \to \infty} \int s_{kj} < \lambda^*(S_k) + 2^{-k}\epsilon$. Put $s_j' = \sum_{k=1}^{j} s_{kj}$. Then $\int s_j' = \sum_{k=1}^{j} \int s_{kj} < \sum_{k=1}^{j} (\lambda^*(S_k) + 2^{-k}\epsilon) < \epsilon + \sum_{k=1}^{\infty} \lambda^*(S_k)$. It follows that $j \mapsto s_j'$ is a bounding sequence for χ_U, where $U = \bigcup_{k=1}^{\infty} S_k$, and that $\lambda^*(U) = | \chi_U |_1 < \epsilon + \sum_{k=1}^{\infty} \lambda^*(S_k)$. Since ϵ can be arbitrarily small, $\lambda^*(\bigcup_{k=1}^{\infty} S_k) \leq \sum_{k=1}^{\infty} \lambda^*(S_k)$.

(d) Let $S_1, S_2, S_3, \ldots$ be mutually disjoint measurable sets. Suppose first that $\sum_{k=1}^{\infty} \lambda(S_k) < +\infty$. Let $U = \bigcup_{k=1}^{\infty} S_k$. Then because $S_1, S_2, S_3, \ldots$ are mutually disjoint, $\chi_U = \sum_{k=1}^{\infty} \chi_{S_k}$. By applying Beppo Levi's Theorem to the sequence of partial sums of this series, we conclude that $\chi_U \in \mathcal{L}^1(n, \mathbf{R})$ and that $\lambda(\chi_U) = | \chi_U |_1 = \int \chi_U = \sum_{k=1}^{\infty} \int \chi_{S_k} = \sum_{k=1}^{\infty} \lambda(S_k)$. Theorem 9(d) shows that if $\chi_U \in \mathcal{L}^1(n, \mathbf{R})$, then U is measurable.

If $\sum_{k=1}^{\infty} \lambda(S_k) = +\infty$, we proceed as follows. Let R be an open rectangle in $\mathbf{R}^n$. Then $\chi_R \wedge \chi_{S_k} \in \mathcal{L}^1(n, \mathbf{R})$, $\sum_{k=1}^{\infty} \int \chi_R \wedge \chi_{S_k} \leq \lambda(R)$, and $\sum_{k=1}^{\infty} \chi_R \wedge \chi_{S_k} = \chi_R \wedge \chi_U$. It follows from what we have just proved that $\chi_R \wedge \chi_U \in \mathcal{L}^1(n, \mathbf{R})$. Thus U is measurable. Now choose a sequence $R_1 \subseteq R_2 \subseteq R_3 \subseteq \cdots$ of open rectangles in $\mathbf{R}^n$ such that $\bigcup_{k=1}^{\infty} R_k = \mathbf{R}^n$. If $\lambda(S_k) = +\infty$ for some k, then $\lambda(U) \geq \lambda(S_k)$, so $\lambda(U) = +\infty$. Otherwise, we can apply Beppo Levi's Theorem to conclude that $\lambda(U) \geq \lim_{i \to \infty} \sum_{k=1}^{j} \int \chi_{R_i} \wedge \chi_{S_k} = \sum_{k=1}^{j} \int \chi_{S_k}$ for every $j \in \mathbf{N}$, since $\lim_{i \to \infty} \chi_{R_i} \equiv 1$. Thus, letting $j \to \infty$, we see that $\lambda(U) \geq \sum_{k=1}^{\infty} \int \chi_{S_k} = +\infty$. Once again, $\lambda(U) = +\infty$.

10. If f is a nonnegative function in $\mathcal{L}^1(1, \mathbf{R})$, let χf be the characteristic function of $S = \{(x, y) \in \mathbf{R}^2 : 0 \leq y \leq f(x)\}$ (see Figure 38, A). First, suppose $f \in \text{Step}(1, \mathbf{R})$. Let $G_1 = \{x_1, x_2, \ldots, x_n\}$ be a grid that presents f. Put $G_2 = \{f(R) : R \in C(G_1)\} \cup \{0, 1, f(x_1), \ldots, f(x_n)\}$. (We put 1 in G_2 to insure that G_2 contains at least two points.) Then (G_1, G_2) is a grid on $\mathbf{R}^2$ that presents χf (see Figure 38, B). We may suppose that $x_1 < x_2 < \cdots < x_n$. For each $i = 1, \ldots, n-1$, let $y_i = f((x_i + x_{i+1})/2)$, and let $f_i : \mathbf{R} \to \mathbf{R}$ be defined as follows: if $x_i < x < x_{i+1}, f_i(x) = y_i$. If $x \leq x_i$ or $x \geq x_{i+1}, f_i(x) = 0$. Then $G_i = (\{x_i, x_{i+1}\}, \{0, y_i\})$ is a presentation of χf_i, and $\int \chi f_i = \sum \{\lambda(R)\chi f_i(R) : R \in C(G_i)\} = (x_{i+1} - x_i)(y_i - 0) \cdot 1 = \int f_i$. Since $G = (G_1, G_2)$ is a refinement of G_i, $\int \chi f_i = \sum \{\lambda(R)\chi f_i(R) : R \in C(G)\}$. Thus $\int \chi f = \sum \{\lambda(R)\chi f(R) : R \in C(G)\} = \sum \{\lambda(R) \sum_{i=1}^{n-1} \chi f_i(R) : R \in C(G)\} =

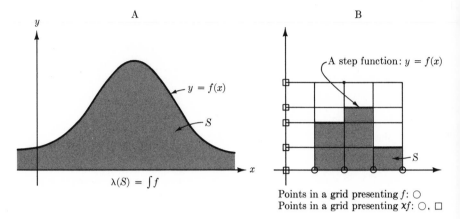

Figure 38

$\sum_{i=1}^{n-1} \sum \{\lambda(R)\chi f_i(R) : R \in C(\mathcal{G})\} = \sum_{i=1}^{n-1} \int \chi f_i = \sum_{i=1}^{n-1} \int f_i = \int f$. This proves the assertion $\lambda(S) = \int \chi f = \int \int f$ in case $f \in$ Step $(1, \mathbf{R})$.

Next, if $f \in \mathfrak{N}(1, \mathbf{R})$, let $k \mapsto s_k$ be a bounding sequence for f. Then $k \mapsto \chi s_k$ is a bounding sequence for χf and $|\chi f|_1 \le \lim_{k \to \infty} \int \chi s_k = \lim_{k \to \infty} \int s_k$. It follows that $|\chi f|_1 \le |f|_1$. In particular, $\chi f \in \mathfrak{N}(2, \mathbf{R})$. Next, suppose $f \in \mathcal{L}^1(1, \mathbf{R})$, and let $k \mapsto t_k$ be a sequence in Step $(1, \mathbf{R})$ such that $\lim_{k \to \infty} |f - t_k|_1 = 0$. Put $t'_k = t_k \vee 0$. Since $f(x) \ge 0$ for all $x \in \mathbf{R}$, $|f(x) - t'_k(x)| \le |f(x) - t_k(x)|$ for all $x \in \mathbf{R}$. Thus $\lim_{k \to \infty} |f - t'_k|_1 = 0$ as well. If we can show that $\lim_{k \to \infty} |\chi f - \chi t'_k|_1 = 0$, we will be finished, for this will show that $\chi f \in \mathcal{L}^1(1, \mathbf{R})$, and that $\lambda(S) = \int \chi f = \lim_{k \to \infty} \int \chi t'_k = \lim_{k \to \infty} \int t'_k = \int f$. Thus it will be enough to show that $|\chi f - \chi t'_k|_1 \le 2 |f - t'_k|_1$, for

Let $f \in \mathcal{L}_1(1, \mathbf{R})$ be nonnegative and let t be any nonnegative step function. If $k \mapsto s_k$ is a bounding sequence for $f - t$, then $\lim_{k \to \infty} \chi(t + s_k)(x, y) = 1$ if $0 \le y \le f \vee t(x)$, and $\lim_{k \to \infty} \chi(0 \vee (t - s_k))(x, y) = 0$ if $y > f \wedge t(x)$. It follows that $k \mapsto \chi(t + s_k) - \chi(0 \vee (t - s_k))$ is a bounding sequence for $\chi f - \chi t$, since $(\chi f - \chi t)(x, y) \ne 0 \Leftrightarrow f \wedge t(x) < y \le f \vee t(x)$, and in any case $|(\chi f - \chi t)(x, y)| \le 1$. But $\int [\chi(t + s_k) - \chi(0 \vee (t - s_k))] = \int (t + s_k) - \int 0 \vee (t - s_k) = \int [(t + s_k) - (0 \vee (t - s_k))] \le \int [(t + s_k) - (t - s_k)] = 2 \int s_k$. It follows that $|\chi f - \chi t|_1 \le \lim_{k \to \infty} \int [\chi(t + s_k) - \chi(0 \vee (t - s_k))] \le \lim_{k \to \infty} 2 \int s_k$. Taking the glb with respect to all bounding sequences for $f - t$, we conclude that $|\chi f - \chi t|_1 \le 2 |f - t|_1$.

11. The definitions of R-normable function, $\mathfrak{N}_R(n, V)$, and $|\ \ |_R$ were already given in Remark 24. The theory of Riemann integration then proceeds as follows.

2R. PROPOSITION. $\mathfrak{N}_R(n, V)$ is a linear subspace of $\mathfrak{F}(\mathbf{R}^n, V)$, and $|\ \ |_R$ is a seminorm on $\mathfrak{N}_R(n, V)$.

Proof. The proof is identical to that of Proposition 2, with the simplification that in this case all bounding sequences are constant sequences.

3R. PROPOSITION. If $f \in$ Step (n, V), then $f \in \mathfrak{N}_R(n, V)$ and $\mathbf{I} f \mathbf{I}_R = \int \|f\|$.

Proof. The proof that $\mathbf{I} f \mathbf{I}_R \leq \int \|f\|$ is the same as the corresponding part of the proof of Proposition 2. The converse is easier. If s is a bounding function for f, then $s(x) \geq \|f(x)\|$ for every $x \in \mathbf{R}^n$. Therefore, by Theorem A 9(d), $\int s \geq \int \|f\|$. Taking the glb over all such bounding functions, we conclude that $\mathbf{I} f \mathbf{I}_R \geq \int \|f\|$.

Remarks 4 remain equally valid for the Riemann integral.

5R. PROPOSITION. Let $\langle V, \| \; \| \rangle$ be a Banach space and let $k \mapsto g_k$ be a sequence in Step (n, V) that converges to an element $f \in \mathfrak{N}_R(n, V)$ in the sense that $\lim\limits_{k \to \infty} \mathbf{I} f - g_k \mathbf{I}_R = 0$. Then the sequence $k \mapsto \int g_k$ has a limit in V. Moreover, if $k \mapsto h_k$ is another sequence in Step (n, V) that converges to f, then $\lim\limits_{k \to \infty} \int h_k = \lim\limits_{k \to \infty} \int g_k$.

Proof. The proof is just like that of Proposition 5.

6R. DEFINITION. Let $\langle V, \| \; \| \rangle$ be a Banach space. $R(n, V)$, the space of **Riemann-integrable functions** from $\mathbf{R}^n$ to V, is the closure of Step (n, V) in $\langle \mathfrak{N}_R(n, V), \mathbf{I} \; \mathbf{I}_R \rangle$. If $f \in R(n, V)$, then $\int f$ is the common value assumed by all the limits $\lim\limits_{k \to \infty} \int s_k$, where $\lim\limits_{k \to \infty} \mathbf{I} f - s_k \mathbf{I}_R = 0$, and $k \mapsto s_k$ is a sequence in Step (n, V).

7R. THEOREM RELATING RIEMANN AND LEBESQUE INTEGRATION.
$\mathfrak{N}_R(n, V) \subseteq \mathfrak{N}(n, V)$, $R(n, V) \subseteq \mathfrak{L}^1(n, V)$. If $f \in \mathfrak{N}_R(n, V)$, then $\mathbf{I} f \mathbf{I}_1 \leq \mathbf{I} f \mathbf{I}_R$. If $f \in R(n, V)$, then the Riemann and Lebesgue integrals of f are equal.

Proof. Remark 24 shows that $\mathfrak{N}_R(n, V) \subseteq \mathfrak{N}(n, V)$ and that $\mathbf{I} f \mathbf{I}_1 \leq \mathbf{I} f \mathbf{I}_R$. Suppose $f \in R(n, V)$. Then there is a sequence $k \mapsto s_k$ in Step (n, V) such that $\lim\limits_{k \to \infty} \mathbf{I} f - s_k \mathbf{I}_R = 0$. Since $\mathbf{I} f - s_k \mathbf{I}_1 \leq \mathbf{I} f - s_k \mathbf{I}_R$, it follows at once that $\lim\limits_{k \to \infty} \mathbf{I} f - s_k \mathbf{I}_1 = 0$. Thus $f \in \mathfrak{L}^1(n, V)$, and the Lebesgue and Riemann integrals of f both equal $\lim\limits_{k \to \infty} \int s_k$. This justifies using the same notation for the Lebesgue and Riemann integrals.

8R. LEMMA. Let $\langle V, \| \; \| \rangle$ be a normed space, and let $f, g \in \mathfrak{F}(n, V)$.
(a) $g \in \mathfrak{N}_R(n, V) \Leftrightarrow \|g\| \in \mathfrak{N}_R(n, \mathbf{R})$.
(b) If $g \in \mathfrak{N}_R(n, V)$, then $\mathbf{I} g \mathbf{I}_R = \mathbf{I} \|g\| \mathbf{I}_R$.
(c) If $\|f(x)\| \leq \|g(x)\|$ for every $x \in \mathbf{R}^n$ and $g \in \mathfrak{N}_R(n, V)$, then $f \in \mathfrak{N}_R(n, V)$ and $\mathbf{I} f \mathbf{I}_R \leq \mathbf{I} g \mathbf{I}_R$.

Proof. Just like the proof of Lemma 8.

9R. THEOREM. Let $\langle V, \| \; \| \rangle$ be a Banach space.
(a) $R(n, V)$ is a linear subspace of $\mathfrak{N}_R(n, V)$.
(b) If $f \in R(n, V)$, then $\|f\| \in R(n, V)$, and $\|\int f\| \leq \int \|f\| = \mathbf{I} f \mathbf{I}_R$.
(c) If $\langle W, \|\| \; \|\| \rangle$ is another Banach space and $L: V \to W$ is a bounded linear map, then for each $f \in R(n, V)$, $L \circ f \in R(n, W)$ and $L \int f = \int (L \circ f)$.
(d) If $f, g \in R(n, \mathbf{R})$, then $f \vee g$ and $f \wedge g$ are in $R(n, \mathbf{R})$.
(e) If $f, g \in R(n, \mathbf{R})$ and $f \leq g$, then $\int f \leq \int g$.

Proof. The proof is just like the proof of Theorem 9, with the simplification that all bounding sequences are constant sequences.

10R. ARZELA'S DOMINATED CONVERGENCE THEOREM. Let $\langle V, \|\ \|\rangle$ be a Banach space, and let $k \mapsto f_k$ be a sequence in $R(n, V)$ with the following properties:

(a) there is a function $f \in R(n, V)$ such that $\lim_{k \to \infty} f_k(x) = f(x)$ for every $x \in \mathbf{R}^n$, and

(b) there is a function $b \in \text{Step } (n, \mathbf{R})$ such that $\|f_k(x)\| \le b(x)$ for every $x \in \mathbf{R}^n$. Then $\int f = \lim_{k \to \infty} \int f_k$.

Proof. Since the hypotheses of this theorem are stronger than the hypotheses of Theorem 10, we can apply Theorem 10 and Theorem 7R to prove Theorem 10R.

Remarks. (a) The hypothesis that the limit f is Riemann-integrable cannot be dropped, nor can we conclude that $f \in R(n, \mathbf{R})$ from the other hypotheses of Theorem 10R, as shown by the example that concludes Remark 24.

(b) The proof of Theorem 10R from Theorem 10 shows that the hypotheses of the Theorem can be weakened to this extent: we need only assume that $b \in \mathfrak{N}(n, \mathbf{R})$, and that $\lim_{k \to \infty} f_k(x) = f(x)$ λ-almost everywhere. However, this strengthened form of the Theorem cannot even be stated without developing part of the Lebesgue theory, and if we are going to develop the Lebesgue theory, we might as well go all the way.

(c) It seems unsatisfactory for the same reason to prove Theorem 10R by going outside the Riemann theory of integration. In fact, every proof I know uses some of the Lebesgue theory at least implicitly. I know two well-known texts on advanced calculus that rather dishonestly avoid the Lebesgue theory theory by the following glib remark. Obviously, if for each $k \in \mathbf{N}$, S_k is a finite union of open intervals contained in $[0, 1]$, and if $\lambda(S_k) \ge \epsilon > 0$ for all k, then there is a point $x \in [0, 1]$ that belongs to infinitely many of the sets S_k. Their assertion is glib for the following reason: so far as I know, the only proofs of this fact use the Lebesgue theory at least implicitly. Here is one: let $\psi_j = \lim_{k \to \infty} \chi_{S_{j+1}} \vee \chi_{S_{j+2}} \vee \cdots \vee \chi_{S_{j+k}} = \chi_{U_j}$, where $U_j = \bigcup_{k=j}^{\infty} S_j$. Let $\chi = \lim_{j \to \infty} \psi_j$. Then it is easy to see that $\chi(x) \ne 0 \Leftrightarrow x \in S_k$ for infinitely many values of k. By Lebesgue's Dominated Convergence Theorem, $\int \psi_j \ge \int \chi_{S_{j+1}} \ge \epsilon$ for every j, and $\int \chi = \lim_{j \to \infty} \int \psi_j \ge \epsilon$. In particular, $\chi \ne 0$.

12. Suppose $f \in R(n, \mathbf{R})$. Given any $\epsilon > 0$, there is an $s \in \text{Step } (n, \mathbf{R})$ such that $|f - s|_R < \epsilon/2$, according to Definition 6R (see the answer to Exercise 11). By definition, this means that there is a $t \in \text{Step } (n, \mathbf{R})$ such that $t(x) \ge |f(x) - s(x)|$ for every $x \in \mathbf{R}^n$ and that $\int t < \epsilon/2$. Then for every $x \in \mathbf{R}^n$, $h(x) = s(x) + t(x) \ge f(x) \ge s(x) - t(x) = g(x)$, and $\int (h - g) = \int 2t < \epsilon$. Conversely, suppose that $f \in \mathfrak{F}(\mathbf{R}^n, \mathbf{R})$, and that for every $\epsilon > 0$, there are g, $h \in \text{Step } (n, \mathbf{R})$ such that $g \le f \le h$ and $\int (h - g) < \epsilon$. Then, in particular, for each n we can choose g_n and h_n such that $g_n \le f \le h_n$ and $\int (h_n - g_n) < 1/n$. $|f(x)| \le |g_1(x)| + |h_1(x)|$ for every $x \in \mathbf{R}^n$, so that $f \in \mathfrak{N}_R(n, \mathbf{R})$. Also, $|f(x) - g_n(x)| \le |h_n(x) - g_n(x)|$ for every $x \in \mathbf{R}^n$, so $|f - g_n|_R \le |h_n - g_n|_R < 1/n$, by Proposition 8R and Theorem 9R(b). Thus $\lim_{n \to \infty} |f - g_n|_R = 0$, so $f \in R(n, \mathbf{R})$.

Exercises in Section IX C

1. Suppose $f \in \mathcal{L}^1(I, \mathbf{R})$. Since $0 \le f_*^*(x) \le f^*(x)$ for every $x \in \mathbf{R}$, $0 \le \int_{a_k}^{b_k} f =$

$\int f_k^* \leq \int f^*$, so that $\left\{ \int_{a_k}^{b_k} f : k \in \mathbf{N} \right\}$ is bounded. Conversely, suppose this set is bounded. $f_k^* \in \mathcal{L}^1(1, \mathbf{R})$ for each $k \in \mathbf{N}$ by Proposition B20. Since $0 \leq f_1^* \leq f_2^* \leq f_3^* \leq \cdots$ and since $\lim_{k \to \infty} f_k^*(x) = f^*(x)$ for every $x \in \mathbf{R}$, it follows from Beppo Levi's Theorem that $f^* \in \mathcal{L}^1(1, \mathbf{R})$. Thus $f \in \mathcal{L}^1(I, \mathbf{R})$.

Additional problems

This appendix contains additional problems, mostly of a more routine nature than the exercises in the body of the text. The latter should be considered a part of the exposition, and are mentioned throughout the text. These problems are supplementary material, to be used as needed.

Problems for Section I B

1. Show that if A and B are sets, $A \cap A = A \cup A = A$, $A \cap (B - A) = \emptyset$, and that $A \subseteq B \Leftrightarrow A \cup B = B \Leftrightarrow A \cap B = A$.
2. Simplify the following expression, using Problem 1 and Proposition IB17: $[B \cap ((B \cup B) - (A \cap B))] \cup [A \cap (B - (A \cup A))] \cup [A \cap (A \cup B)]$.
3. Which of the following relations are functions?
   (a) $\{\langle x, y \rangle \in \mathbf{N} \times \mathbf{N} : x = y^2 + y + 3\}$.
   (b) $\{\langle y, x \rangle \in \mathbf{N} \times \mathbf{N} : x = y^2 + y + 3\}$.
   (c) $\{\langle x, y \rangle \in \mathbf{N} \times \mathbf{N} : x = y + 3\}$.
   (d) $\{\langle y, x \rangle \in \mathbf{N} \times \mathbf{N} : x = y + 3\}$.
   (e) $\{\langle x, y \rangle \in \text{Frenchmen} \times \text{Frenchmen} : y \text{ is the father of } x\}$.
   (f) $\{\langle x, y \rangle \in \text{Frenchmen} \times \text{Frenchmen} : y \text{ is a brother of } x\}$.
4. If A contains n elements, B contains m elements, and $A \cap B$ contains k elements, how many elements are in $A \cup B$?
5. If A contains n elements, how many subsets does A have?
6. How many real numbers have more than one decimal expansion? For example, $1 = 1.0000 \cdots = .9999 \cdots$.
7. Prove that the Euclidean plane cannot be the union of countably many straight lines, using a counting argument. *Hint:* Consider the intersections of countably many lines with a fixed circle.

Problems for Section I C

1. Find the error, unwarranted assumption, or glib statement in each of the following places.

(a) R. G. Bartle, *The Elements of Real Analysis*, Wiley, New York, 1964, bottom of page 288.

(b) J. B. Roberts, *The Real Number System in an Algebraic Setting*, Freeman, San Francisco, 1962, middle of page 114, statement of theorem on page 104.

(c) D. Kleppner and N. Ramsey, *Quick Calculus*, Wiley, New York, 1965, Appendix A7, page 256.

(d) This book.

2. What is wrong with the following argument?

$$\frac{1}{-1} = \frac{-1}{1},$$

$$\sqrt{\frac{1}{-1}} = \sqrt{\frac{-1}{1}},$$

$$\frac{\sqrt{1}}{\sqrt{-1}} = \frac{\sqrt{-1}}{\sqrt{1}},$$

$$\sqrt{1} \times \sqrt{1} = \sqrt{-1} \times \sqrt{-1},$$

$$1 = -1.$$

3. What is wrong with the following argument (see Chapter V)?
$$e^x = e^{2\pi i(x/2\pi i)} = \left(e^{2\pi i}\right)^{x/2\pi i} = 1^{x/2\pi i} = 1.$$

4. What is wrong with the following argument?

THEOREM. If m and n are integers that have no common factors except 1 and -1, then there are integers p and q such that $pm + qn = 1$.

Proof. If m and n have a common factor f different from 1 and -1 (say $m = rf$, $n = sf$), then every expression of the form $pm + qn = (pr + qs)f$ is a multiple of f. Since 1 is not divisible by f, 1 cannot be of the form $pm + qn$.

5. What is wrong with the following argument?

THEOREM. For every $n \in \mathbf{N}$, either n is even or $n + 1$ is even.

Proof. It is enough to show that $(n - 1)(n + 1) + 2(n + 1)$ is a perfect square. But this is evident because $(n - 1)(n + 1) + 2(n + 1) = (n + 1)^2$.

Problems for Section II A

1. Show how the field of complex numbers can be distinguished by its properties from each of the three fields $\mathbf{Z}_2$, $\mathbf{Q}$, $\mathbf{R}$.

2. Verify that the construction of the isomorphism ϕ described in the text actually does yield a function with the stated properties. (This is a hard problem because it requires a long, tedious argument and many of the results of Section IIC.)

3. Exhibit a very simple model for the system of axioms obtained by omitting axiom IIA1(d) from the list IIA1, 2, 3; thereby show that this abbreviated system of axioms is consistent.

Problems for Section II B

1. Prove that the following principles are true in any (not necessarily commutative) group $\langle G, \circ \rangle$:

(a) *Left cancellation:* $x \circ y = x \circ z \Rightarrow y = z$. *Hint:* Multiply both sides on the left by x^{-1} (the inverse of x).

(b) *The left neutral element is also a right neutral element:* if $e \circ x = x$ for every $x \in G$, then $x \circ e = x$ for every $x \in G$. *Hint:* Show that $x^{-1} \circ (x \circ e) = x^{-1} \circ x$ and use (a).

(c) *There is only one identity:* if e and e' are identities, then $e = e'$.

(d) *A left inverse is also a right inverse:* $x^{-1} \circ x = e \Rightarrow x \circ x^{-1} = e$. *Hint:* Show that $x^{-1} \circ (x \circ x^{-1}) = x^{-1} \circ e$ and use (a).

(e) *There is only one inverse for each element:* if $y \circ x = e$ and $z \circ x = e$, then $y = z$.

2. Suppose that $\langle G, \circ \rangle$ satisfies the axioms for a group except that the axiom asserting the existence of a *left* neutral element is replaced by an axiom asserting the existence of a *right* neutral element. Show that $\langle G, \circ \rangle$ need not be a group. *Hint:* In the simplest counterexample, G has just *two* elements.

3. Prove that the following principles are true in any field $\langle F, +, \times \rangle$.

(a) $-A = (-1) \times A$ for every $A \in F$.

(b) 0 has no multiplicative inverse.

(c) If $A, B, C \in F$, $A \neq 0$, there are *at most two* solutions X in F to the quadratic equation $AX^2 + BX + C = 0$.

(d) If $A, B, C, D, P, Q \in F$, then for there to be *exactly one* pair $\langle X, Y \rangle \in F \times F$ such that $AX + BY = P$ and $CX + DY = Q$, it is necessary and sufficient that $AD - BC \neq 0$. *Hint:* Show that $AD - BC \neq 0$ implies the existence of a unique solution, but $AD - BC = 0$ implies that if there is any solution $\langle X, Y \rangle$, then $\langle X - B, Y + A \rangle$ and $\langle X - D, Y + C \rangle$ are also solutions. Give separate attention to the case that $A = B = C = D = 0$.

4. Show that in any ordered field, $x > 0$, $y > 0 \Rightarrow x/y > 0$.

Problems for Section II C

1. Prove that if $0 < s < 1$, $\lim_{n \to \infty} n^2 s^n = 0$.

2. For each of the following sets of real numbers, state whether it is bounded above or below. If it is bounded above, find its lub; if it is bounded below, find its glb. (There is one tricky case in which you cannot carry out these instructions. Why?)

(a) $\{x \in \mathbf{R} : x^4 - x^2 < 0\}$.

(b) $\{x \in \mathbf{R} : x^4 - x^2 \leq 0\}$.

(c) $\{x \in \mathbf{R} : x^3 - x^2 < 0\}$.

(d) $\{x \in \mathbf{R} : x^4 - x^2 + 1 < 0\}$.

(e) $\{(-2)^{-n} : n \in \mathbf{N}\}$.

3. Prove that the Completeness Axiom IIA3 is not satisfied by the ordered field $\langle \mathbf{Q}, \mathbf{Q}^+, +, \times \rangle$.

4. Formulate and prove the analogue of Theorem IIC1 for greatest lower bounds.

Problems for Section III A

1. Prove that the systems described in Examples IIIA2 do satisfy the axioms for a vector space.

2. Prove that in any vector space, $(-1) \cdot v = -v$.

3. In each of the following cases, four vectors in $\mathbf{Q}^3$ are given. They must be dependent. Express one as a linear combination of the others.

(a) $(1, 3, 2)$, $(4, 5, 6)$, $(1, 1, 0)$, $(1, 0, 3)$.

(b) $(1, 0, 0)$, $(0, 1, 0)$, $(0, 0, 1)$, $(7, 9, 12)$.

(c) $(1, 1, 1)$, $(1, 1, 0)$, $(1, 0, 1)$, $(0, 1, 1)$.

4. Which of the following sets of vectors form a basis for $\mathbf{Q}^3$?
   (a) $(1, 0, 0), (1, 0, 2)$.
   (b) $(1, 0, 0), (0, 1, 0), (0, 1, 3), (1, 0, 3)$.
   (c) $(1, 1, 0), (1, 0, 1), (0, 1, 1)$.
   (d) $(1, 2, 2), (1, 3, 3), (2, 5, 5)$.
5. $(1, 1, 0), (0, 1, 1), (1, 1, 1), (1, 0, 1)$ are four vectors in $(\mathbf{Z}_2)^3$. Express one as a linear combination of the others with coefficients in $\mathbf{Z}_2 = \{0, 1\}$.
6. How many subspaces does $(\mathbf{Z}^2)^2$ have?
7. How many one-dimensional subspaces does $(\mathbf{Z}_2)^3$ have?

Problems for Section III B
1. Find *all* solutions to each of the following systems of linear equations.
   (a) $x + 2y + 2z = 1$
        $x + 2y + 3z = 0$
        $x + 3y + z = 0.$
   (b) $x + 2y + 2z = 1$
        $x + 2y + 3z = 0$
        $x + 2y + z = 0.$
   (c) $x + 2y + 2z = 1$
        $x + 2y + 3z = 0$
        $x + 2y + z = 2.$

Problems for Section III C
1. Fill in the details of the proof of Theorem IIIC7.
2. Prove that in an inner product space the equations $\|x\| = \|y\| = 1, \|x + y\| = 2$ imply that $x = y$.
3. Find an equation for the plane in $\mathbf{R}^3$ through the point $(1, 2, 3)$ that is perpendicular to the vector $(5, 7, 2)$.
4. Find an equation for the plane in $\mathbf{R}^3$ that is tangent to the sphere $x^2 + y^2 + z^2 = 14$ at the point $(1, 2, 3)$.
5. Orthonormalize the vectors $(1, 2, 2), (1, 1, 1)$ with respect to the usual inner product on $\mathbf{R}^3$.

Problems for Section III D
1. Let $\mathcal{C}^\infty$ be the space of functions from $\mathbf{R}$ to $\mathbf{R}$ that have derivatives of all orders, and let $D: \mathcal{C}^\infty \to \mathcal{C}^\infty$ be the differentiation map $D: f \to f'$. What is the kernel of D?
2. Let L be the linear map from $\mathbf{R}^2$ to $\mathbf{R}^2$ whose matrix with respect to the basis $(1, 0), (0, 1)$ for $\mathbf{R}^2$ is $\begin{pmatrix} 5 & 1 \\ 4 & 2 \end{pmatrix}$. That is, $L(1, 0) = 5(1, 0) + 4(0, 1)$ and $L(0, 1) = (1, 0) + 2(0, 1)$. What is the matrix for L with respect to the basis $(1, 2), (4, 2)$ for $\mathbf{R}^2$?
3. Calculate $\dim \ker L$ and $\dim \operatorname{range} L$ for the case where $L: \mathbf{R}^3 \to \mathbf{R}^3$ is given by the matrix
$$\begin{pmatrix} 1 & 0 & 0 \\ 0 & 1 & 0 \\ 1 & 1 & 0 \end{pmatrix}$$
with respect to the basis $(1, 0, 0), (0, 1, 0), (0, 0, 1)$ for $\mathbf{R}^3$.
4. What is the kernel of the linear map L described in problem 3?

Problems for Section III E

1. Let $L: \mathbf{R}^2 \to \mathbf{R}^2$ be the linear map whose matrix with respect to the usual basis is $\begin{pmatrix} A & B \\ C & D \end{pmatrix}$. Show that L has an inverse if, and only if, $AD - BC \neq 0$. *Hint*: See Problem 3(d) for Section IIB.

2. Calculate the matrix of L^{-1} where L is the linear map described in Problem 1, in case $AD - BC \neq 0$.

3. Calculate the inverse of the matrix
$$\begin{pmatrix} 1 & 1 & 2 \\ 0 & 2 & 5 \\ 0 & 0 & 3 \end{pmatrix}.$$

   (The **inverse matrix** corresponds to the inverse linear map.)

4. Calculate the matrix product
$$\begin{pmatrix} 1 & 2 & 3 \\ 4 & 5 & 6 \\ 7 & 8 & 9 \end{pmatrix} \times \begin{pmatrix} 1 & 1 & 1 \\ 1 & 0 & 1 \\ 0 & 1 & 2 \end{pmatrix}.$$

5. Fill in the details of the proof of IIIE1.

Problems for Section III F

1. Show that if σ is a permutation and $\tau_1, \ldots, \tau_k$ are transpositions such that $\tau_1 \circ \tau_2 \circ \cdots \circ \tau_k = \sigma$, then sgn σ is 1 or -1 according to whether k is even or odd.

2. Can the permutation $1 \mapsto 2$, $2 \mapsto 3$, $3 \mapsto 4$, $4 \mapsto 1$ be expressed as the product of an even number of transpositions?

3. If $A = (a, b)$ and $B = (c, d)$ are vectors from the origin as shown in Figure 12, compute the area of the parallelogram which they span by techniques of analytic geometry. *Hint:* You may want to observe that the vector $A' = (-b, a)$ is perpendicular to A, and to use the formula $[A', B] = \|A'\| \|B\| \cos \phi = \|A\| \|B\| \sin (90° - \phi)$, where ϕ is the angle between the vectors A' and B.

4. Show that the determinant of the matrix
$$\begin{pmatrix} A & B & C \\ 0 & D & E \\ 0 & 0 & F \end{pmatrix}$$

   is ADF.

5. Compute the determinant of the matrix
$$\begin{pmatrix} 1 & 2 & 3 \\ 1 & 2 & 4 \\ 1 & 0 & 1 \end{pmatrix}.$$

Problems for Section III G

1. Show that the linear map L from $\mathbf{R}^n$ to $\mathbf{R}^n$ has an inverse if, and only if, 0 is *not* an eigenvalue of L.

2. Compute the characteristic polynomial and the eigenvalues of the matrix $\begin{pmatrix} A & B \\ C & D \end{pmatrix}$ in terms of A, B, C, D.

3. Do the same for the matrix $\begin{pmatrix} A & C \\ B & D \end{pmatrix}$.

4. Find all eigenvalues of the matrix

$$\begin{pmatrix} A & B & C \\ 0 & D & E \\ 0 & 0 & F \end{pmatrix}.$$

Hint: See Problem IIIF4.

5. Find all eigenvalues and eigenvectors of the matrix

$$\begin{pmatrix} 0 & 1 & 0 \\ 0 & 0 & 1 \\ 1 & 0 & 0 \end{pmatrix}.$$

6. If v and w are eigenvectors for the linear map L corresponding to *distinct* eigenvalues λ and μ, prove that v and w are linearly independent. *Hint:* Apply the linear operators $L - \lambda I$ and $L - \mu I$ to the linear combination $rv + sw$.

Problems for Section IV A

1. Prove that Examples IVA3(d) and (e) do satisfy the axioms for a normed space.
2. Compute an explicit expression for the norm of a linear functional in $B(\langle \mathbf{R}^n, \| \ \|_\infty \rangle, \langle \mathbf{R}, | \ | \rangle)$. See the remark following Problem 4.
3. Do the same for $B(\langle \mathbf{R}^n, \| \ \|_1 \rangle, \langle \mathbf{R}, | \ | \rangle)$. See the remark following Problem 4.
4. Do the same for $B(\langle \mathbf{R}^n, \| \ \|_2 \rangle, \langle \mathbf{R}, | \ | \rangle)$. *Hint:* Use the Cauchy-Schwartz inequality.

   In each of the Problems 2, 3, and 4, use the fact that a linear functional L on $\mathbf{R}^n$ has the form $L: (x_1, \ldots, x_n) \to a_1 x_1 + \cdots + a_n x_n$ (see Exercise IIIE4). For example, the answer to Problem 4 is $\| L \| = (a_1^2 + \cdots + a_n^2)^{1/2}$.

Problems for Section IV B

1. Verify that the Examples IV B2 do satisfy the axioms for a metric space.
2. Prove that the sequence $x_0 = 1$, $x_{n+1} = (1/2)(x_n + 2/x_n)$ is a Cauchy sequence, and that the square of its limit is 2. *Hint:*

$$x_{n+1} - x_n = \frac{x_n - x_{n-1}}{2} \left(1 - \frac{2}{x_n x_{n-1}} \right).$$

Show that $x_n \geq 1$ for all n, and conclude that $|x_{n+1} - x_n| < 2^{-n}$ for all n. If $\lim_{n \to \infty} x_n = L$, then $L = \frac{1}{2}(L + 2/L)$.

3. Let d be the discrete metric on M. Show that if $n \mapsto x_n$ is a Cauchy sequence in $\langle M, d \rangle$, then there is an N such that $x_n = x_N$ for all $n \geq N$.

Problems for Section IV C

1. Show by example that the intersection of countably many open subsets of $\mathbf{R}$ can be open, closed, or neither. Do the same for the union of countably many closed sets.
2. Show that a sequence in a metric space can have at most one limit.
3. Show that a complete subset of a metric space is closed.
4. If S is a nonempty subset of a metric space $\langle M, d \rangle$ and $x \in M$, put $d(x, S) = $ glb $\{d(x, y): y \in S\}$. Show that for any $\epsilon > 0$, $\{x \in M: d(x, S) < \epsilon\}$ is open.
5. Show that $\{x \in M: d(x, S) = 0\} = S^c$.
6. Show that every closed subset of a metric space is the intersection of countably many open sets. *Hint:* Look at $\{x \in M: d(x, S) < 1/n\}$.

7. Consider the set S described in Example IV C 16(b), and give S the metric d arising from the inner product on $\mathbf{R}^2$. Let $E = \{(x, y) \in S : y \leq 0\}$. Is E open, closed, both, or neither in $\langle S, d \rangle$?

Problems for Section IV D

1. Let $\langle M, d \rangle$ be a metric space. Show that d is a continuous function from $\langle M \times M, d \times d \rangle$ to $\mathbf{R}$. *Hint:* Use the triangle inequality.
2. Let $\langle V, \| \ \| \rangle$ be a normed space. Show that $\| \ \|$ is a continuous function from V to $\mathbf{R}$. *Hint:* Use the triangle inequality.
3. Let $[\ , \ ]$ be an inner product on the vector space V over $\mathbf{R}$. Show that $[\ , \ ]$ is a continuous function from $\langle V, \| \ \| \rangle \times \langle V, \| \ \| \rangle$ to $\mathbf{R}$, where $\| \ \|$ is the norm defined by the formula $\|x\|^2 = [x, x]$. *Hint:* Use the Cauchy-Schwartz inequality.
4. Prove that $x \mapsto \sqrt{|x|}$ is a continuous function from $\mathbf{R}$ to $\mathbf{R}$.
5. Define a function $f : \mathbf{R} \to \mathbf{R}$ by putting $f(x) = x$ if x is rational, and $f(x) = 0$ if x is irrational. Show that f is continuous at 0 but nowhere else.
6. In the situation of Definition IV D 5, put $e(x, y) = d_1(x_1, y_1) + \cdots + d_n(x_n, y_n)$. Show that $d(x, y) \leq e(x, y) \leq nd(x, y)$, and conclude that a sequence in $M = M_1 \times \cdots \times M_n$ converges in $\langle M, d \rangle$ if, and only if, it converges in $\langle M, e \rangle$. This justifies the Remark following Definition IV D 5.

Problems for Chapter V

1. Show that $|z| = 1 \Rightarrow \left| \dfrac{w - z}{1 - \bar{z}w} \right| = 1$. *Hint:* Multiply $w - z$ by $\bar{z}$.

2. Show that if $|z| < 1$ and $|w| < 1$, then $\left| \dfrac{w - z}{1 - \bar{z}w} \right| < 1$. *Hint:* Show by multiplying $(w - z)/(1 - \bar{z}w)$ by its conjugate that the inequality can be put in the form $A + B < 1 + AB$, where $0 \leq A < 1$ and $0 \leq B < 1$.
3. Find the complex cube roots of 1.
4. Fill in the details in the proof of Theorem V A 2.
5. Prove the formula $\sin^2 x + \cos^2 x = 1$ using the formulas $\cos x = \operatorname{Re} e^{ix}$, $\sin x = \operatorname{Im} e^{ix}$, Theorem V A 6 and Proposition V A 8.

Problems for Section VI A

1. Give an example of a sequence of discontinuous functions converging uniformly to a continuous function.
2. Let $\langle M, d \rangle$ be a complete metric space, and let $n \mapsto f_n$ be a sequence of continuous real-valued functions converging pointwise on M to a continuous limiting function L. Using the Baire Category Theorem, prove that there is a nonempty open subset U of M on which the functions f_n are uniformly bounded: there is a $B \in \mathbf{R}$ such that $|f_n(x)| \leq B$ for each $x \in U$ and each $n \in \mathbf{N}$. *Hint:* Let $E_k = \{x \in M : |f_n(x)| \leq k$ for each $n \in \mathbf{N}\}$ and show that E_k is closed and $\bigcup_{k=1}^{\infty} E_k = M$.
3. For $x \in \mathbf{R}$, let $f(x) = d(x, \mathbf{Z})$, the distance from x to the nearest integer. Prove that f is continuous and that $g : x \mapsto \sum_{n=1}^{\infty} 10^{-n} f(10^n x)$ is continuous. (g is an example of a continuous nowhere differentiable function.)

Problems for Section VI B

1. Test the following series for convergence or divergence. In which case is the convergence absolute?

(a) $\displaystyle\sum_{n=1}^{\infty} \frac{\log n}{n^2}$.

(b) $\displaystyle\sum_{n=2}^{\infty} \frac{(-1)^n}{\log n}$.

(c) $\displaystyle\sum_{n=1}^{\infty} \frac{1}{10^{-6}n^2 + 3n + 2}$.

(d) $\displaystyle\sum_{n=1}^{\infty} \frac{1}{3n + 2}$.

(e) $\displaystyle\sum_{n=1}^{\infty} \cos n$.

(f) $\displaystyle\sum_{n=3}^{\infty} \frac{1}{n \log n \log (\log n)}$.

2. Show that $\sum_{n=1}^{\infty} a_n$ converges if, and only if, $\sum_{n=k}^{\infty} a_n$ converges.
3. Show that if the series $\sum_{n=1}^{\infty} a_n$ converges, then $\lim_{n \to \infty} a_n = 0$.
4. Show that if the series $\sum_{n=1}^{\infty} a_n$ converges, then the series $\sum_{n=1}^{\infty} (a_{2n-1} + a_{2n})$ converges to the same limit.

Problems for Section VI C

1. Determine the radii of convergence of the following series.

(a) $\displaystyle\sum_{n=0}^{\infty} n! x^n$.

(b) $\displaystyle\sum_{n=0}^{\infty} e^n x^{n^2}$.

(c) $\displaystyle\sum_{n=0}^{\infty} x^{n!}$.

(d) $\displaystyle\sum_{n=0}^{\infty} (x - 2)^n$.

2. Show by examples that a power series can converge at all, some, or none of the points on the boundary of its circle of convergence.
3. Find an explicit formula for the sum of the series $\sum_{n=1}^{\infty} n^2 x^n$. *Hint:* Look at the series $\sum_{n=1}^{\infty} n(n - 1)x^{n-2}$ and the series $\sum_{n=1}^{\infty} nx^{n-1}$.
4. Suppose $a_0 = 1$, $a_1 = 1$, and $a_n = a_{n-1} + a_{n-2}$ for $n > 1$. (The sequence $n \mapsto a_n$ is called the **Fibonacci Sequence**.)
   (a) Show that $a_n \leq 2^n$ for all n, so that the series $\sum_{n=0}^{\infty} a_n x^n$ has a positive radius of convergence R.
   (b) For $|x| < R$, let $f(x) = \sum_{n=0}^{\infty} a_n x^n$. By equating coefficients of like powers of x, verify that $(1 - x - x^2)f(x) = 1$ for $|x| < R$. Conclude that $f(x) = (1 - x - x^2)^{-1} = [(x - \alpha)^{-1} - (x - \beta)^{-1}]/(\beta - \alpha)$, where $\alpha = (-1 - \sqrt{5})/2$ and $\beta = (-1 + \sqrt{5})/2$.
   (c) Find an explicit formula for a_n in terms of α^n and β^n.
   (d) What is the exact value of R? (See Remark VI C 5(c).)

Problems for Section VI D

1. Let $f(x) = x + \pi/2 - \arctan x$.
   (a) Show that $0 \le f'(x) < 1$ for all $x \in \mathbf{R}$.
   (b) Show that f has no fixed point in $\mathbf{R}$.
   (c) Why does this not contradict Corollary VI D 4?
2. Let $f: \mathbf{R} \to \mathbf{R}$ and suppose that $|f'(x)| < 1$ for all $x \in \mathbf{R}$. Prove that f can have at most one fixed point.
3. Use the technique of Exercise IV D 2 to calculate the inverse of the matrix
   $$\begin{pmatrix} 1 & 1/2 \\ 0 & 1 \end{pmatrix}.$$
4. Let f be any continuous function from the closed unit interval $[0, 1]$ into itself. Prove **Brouwer's Fixed Point Theorem** in one dimension: f has a fixed point. *Hint:* Apply the Intermediate Value Theorem (Exercise II C 8(c)) to the function $x \mapsto x - f(x)$.

Problems for Section VI E

1. Convert the following system of DE's into a system of first-order equations:
   $$y'' + (y')^2 + yz' = \sin x,$$
   $$z''' + y'z' + z^2 = 0.$$
2. Show that if f is a continuous function from $[a, b]$ to $\mathbf{R}^n$, then $\int_a^b \|f(x)\|_2 \, dx \ge$ $\left\| \int_a^b f(x) \, dx \right\|_2$, where $\|x\|_2 = [x, x]^{1/2}$. *Hint:* Let $c = \int_a^b f(x) \, dx$. Show that $\left\| \int_a^b f(x) \, dx \right\|_2^2 = \left[c, \int_a^b f(x) \, dx \right] = \int_a^b [c, f(x)] \, dx$, and use the Cauchy-Schwartz inequality.
3. Show that the initial value problem $f(0) = 0$, $f'(x) = 3[f(x)]^{2/3}$ actually has infinitely many continuously differentiable solutions. *Hint:* Choose $a > 0$, and put $g(x) = 0$ if $x \le a$, and $g(x) = (x - a)^3$ if $x > a$.
4. What is the largest value of b for which Theorem VI E 6 guarantees the existence of a solution to the initial value problem $y(0) = 0$, $y' = 1 + y^2$ on the interval $[0, b]$?
5. Give an example of a DE for which the bound
   $$\|f_1(x) - f_2(x)\|_\infty \le \|f_1(x_0) - f_2(x_0)\|_\infty e^{L(x - x_0)}$$
   given by Theorem VI E 5 is actually attained for all x:
   $$\|f_1(x) - f_2(x)\|_\infty = \|f_1(x_0) - f_2(x_0)\|_\infty e^{L(x - x_0)}.$$
6. Apply the Picard process of successive approximations to the initial value problem $f(0) = 2$, $f'(x) = \frac{1}{2}xf(x)$ and see what you get. Do enough steps (5 or 6) to get a clear picture of what the successive approximations look like.
7. Do the same for the system of DE's $f(0) = 1$, $g(0) = 0$,
   $$f'(x) = f(x) + g(x),$$
   $$g'(x) = f(x) - g(x).$$
8. Convert the integral equation $f(x) = 1 + \int_0^x [f(t)]^2 \, dt$ into an initial value problem and solve it.

9. Define two real-valued functions on $\mathbf{R}^2 - \{0\}$ as follows: $P(x, y) = -y(x^2 + y^2)^{-1}$, $Q(x, y) = x(x^2 + y^2)^{-1}$. Show that although

$$\frac{\partial P}{\partial y}(x, y) \equiv \frac{\partial Q}{\partial x}(x, y),$$

there is no function $E: \mathbf{R}^2 - \{0\} \to \mathbf{R}$ such that $\partial E/\partial x = P$, $\partial E/\partial y = Q$. Why does this not contradict the result of Exercise VI E II 5? *Hint:* Show that if $P = \partial E/\partial x$ and $Q = \partial E/\partial y$, then $\int_a^b [P(f(t),g(t))f'(t) + Q(f(t),g(t))g'(t)]\, dt = E(f(b), g(b)) - E(f(a), g(a))$. Apply this result to the case that $a = 0$, $b = 2\pi$, $f(t) = \cos t$, $g(t) = \sin t$.

10. Solve these DE's.
   (a) $y' \cos y + 1/x = 0$.
   (b) $y' = e^x y$.
   (c) $y' = \dfrac{y + x}{y - x}$.
   (d) $e^y + xe^y y' = 0$.
   (e) $1 + xy' = 0$.
   (f) $xy + x^2 yy' = 0$.
   (g) $x + yy' = 0$.

Problems for Section VI F

1. Solve these DE's.
   (a) $y' + e^x y = e^x$.
   (b) $y' + y = e^x$.
   (c) $y' + xy = x^2$.
2. Find all solutions to the DE $y'' - 5y' + 6y = 1$.
3. Show that if v is an eigenvector of the linear operator L and λ is the corresponding eigenvalue, then $f: x \mapsto e^{\lambda x} v$ is a solution to the DE $y' = Ly$.
4. Using 3, find all solutions to the system of DE's (vector-valued DE)

$$y' = 6y - 2z,$$
$$z' = 2y + z.$$

5. Using 4, find all solutions to the system of DE's

$$y' = 6y - 2z + 1,$$
$$z' = 2y + z - 2.$$

6. Find all solutions to the DE $y'' - 4y' + 4y = 1$.
7. Just how many *T. toxica* would be present in the stagnant pond after two months if none are present at time 0 and $\psi(t) = 1$ milligram per hour?

Problems for Section VI G

1. Define a function $f: \mathbf{R} \to \mathbf{R}$ as follows: if x is rational, let $f(x) = x^2$, and if x is irrational, let $f(x) = 0$. Show that f is differentiable at 0 but discontinuous everywhere else.
2. Suppose

$$\sin (f(x, y) + g(x, y)) = x,$$
$$\sin (f(x, y) - g(x, y)) = y^2$$

for all (x, y) near $(0, 0)$, and that $f(0, 0) = g(0, 0) = 0$. Calculate

$$\frac{\partial f}{\partial x}(0, 0), \quad \frac{\partial f}{\partial y}(0, 0), \quad \frac{\partial g}{\partial x}(0, 0), \quad \frac{\partial g}{\partial y}(0, 0).$$

3. Prove that there is an $\epsilon > 0$ such that there exist differentiable functions f and g from $B_\epsilon((0, 0))$ into $\mathbf{R}$ satisfying $f(0, 0) = g(0, 0) = 0$ and the equations

$$\sin{(f(x, y) + g(x, y))} = x,$$
$$\sin{(f(x, y) - g(x, y))} = y^2.$$

4. What is an equation for the line perpendicular to the surface $z - x^2 - y^2 = 0$ at the point $(1, 1, 2)$?

5. What is an equation for the plane tangent to the surface $z - x^2 - y^2 = 0$ at the point $(1, 1, 2)$?

6. Prove that the differential equation $y + \sin{((x + 1))e^{y\prime}} = 0$ with initial condition $y(0) = 0$ has a solution in some neighborhood of 0. *Hint:* Use both the Implicit Function Theorem and Picard's Theorem.

Problems for Section VII A

1. Show that the union of a finite family of compact subsets of a metric space is compact.

2. Show that a metric space that has a dense totally bounded subset is itself totally bounded. *Hint:* To cover the metric space by balls of radius ϵ, first cover the dense subset by balls of radius $\epsilon/2$.

3. Show that a sequence $n \mapsto s_n$ in a metric space $\langle M, d \rangle$ converges to L if, and only if, every subsequence of $n \mapsto s_n$ has a subsequence which converges to L. *Hint:* If $n \mapsto s_n$ does not converge to L, then there is an $\epsilon > 0$ such that $d(s_n, L) \geq \epsilon$ for infinitely many $n \in \mathbf{N}$.

Problems for Section VII B

1. Let V be the vector space of continuous functions from $[0, 1]$ to $\mathbf{R}$. Show that the uniform norm $\| \ \|_\infty : f \mapsto \sup{\{f(x) : 0 \leq x \leq 1\}}$ and the norm $\| \ \|_1 : f \mapsto \int_0^1 |f(x)| \, dx$ are *not* equivalent. *Hint:* For any $f \in V$, $\|f\|_1 \leq \|f\|_\infty$, but Exercise VII B 5 gives an example of a sequence $n \mapsto f_n$ in V such that $\|f_n\|_\infty = 1$ for each n but $\lim_{n \to \infty} \|f_n\|_1 = 0$. Note that V is infinite-dimensional.

2. Show by example that it is not always true, when S is a closed subset of a non-compact metric space $\langle M, d \rangle$ and $f : M \to \mathbf{R}$ is continuous, that $f(S)$ is closed in $\mathbf{R}$.

3. Prove Theorem VII B 2 by showing that the product space is complete and totally bounded.

4. Is every bounded subset of $\mathbf{R}$ that contains its lub and its glb compact?

5. Let S be a compact subset of the metric space $\langle M, d \rangle$ and let T be a closed subset of M that has no point in common with S. Prove that there is an $\epsilon > 0$ such that $x \in S$, $y \in T \Rightarrow d(x, y) \geq \epsilon$. *Hint:* If there were no such ϵ, we could choose for each $n \in \mathbf{N}$ an $x_n \in S$ and a $y_n \in T$ such that $d(x_n, y_n) < 1/n$, and then could choose a subsequence of the sequence $n \mapsto x_n$ that converged to a point $L \in S$.

6. Show by example that the result of Problem 5 is false if we assume only that S is closed.

7. Is it true that if f is a one-to-one continuous map from a metric space M onto a compact metric space N, then M is compact? *Hint:* See Exercise VII B 4.

8. Prove *Dini's Theorem:* suppose $\langle M, d \rangle$ is a compact metric space and $n \mapsto f_n$ is a sequence of continuous real-valued functions on M converging *pointwise* to the continuous function g. Suppose further that $f_{n+1}(x) \leq f_n(x)$ for every $x \in M$ and every $n \in \mathbf{N}$. Then f_n converges *uniformly* to g as $n \to \infty$. *Hint:* Given $\epsilon > 0$, let $U_n = \{x \in M, f_n(x) - g(x) < \epsilon\}$. Show that $\{U_n : n \in \mathbf{N}\}$ is an open cover of M.

9. Let S and T be compact subsets of $\mathbf{R}$. Prove that $S + T = \{x + y = x \in S, y \in T\}$ is a compact subset of $\mathbf{R}$. *Hint:* Use Theorem IV D 7.

Problems for Section VII C

1. Prove that every continuous real-valued function f on the interval $[a, b]$ is Riemann-integrable in the following sense. For each $\epsilon > 0$, there is a $\delta > 0$ with the following property: if $a = x_0 < x_1 < \ldots < x_n = b$ and $\max \{x_{i+1} - x_i : i = 0, \ldots, n - 1\} < \delta$, then $\sum_{i=0}^{n-1} M_i(x_{i+1} - x_i) - \sum_{i=0}^{n-1} m_i(x_{i+1} - x_i) < \epsilon$, where $M_i = \mathrm{lub} \{f(y) : x_i \leq y \leq x_{i+1}\}$ and $m_i = \mathrm{glb} \{f(y) : x_i \leq y \leq x_{i+1}\}$. *Hint:* Using Corollary VII C 3, choose δ so that $|x - y| < \delta \Rightarrow |f(x) - f(y)| < \epsilon/(b - a)$.

2. Let $\mathcal{F} \subseteq B(V, W)$ be a family of bounded linear maps from the normed space V to the normed space W. Show that $\mathcal{F}$ is uniformly equicontinuous if, and only if, $\{|L| : L \in \mathcal{F}\}$ is bounded.

3. Show that a function from a metric space to $\mathbf{R}$ that satisfies a Lipschitz condition is automatically uniformly continuous.

4. Let d be the discrete metric on a set M. Show that every real-valued function on $\langle M, d \rangle$ is uniformly continuous, and indeed that the family of all real-valued functions on $\langle M, d \rangle$ is uniformly equicontinuous.

5. Let d be the discrete metric on $[0, 1]$, and let $\mathcal{F}$ be the family of all functions from $[0, 1]$ to $[0, 1]$. Then, by Problem 4, $\mathcal{F}$ is a uniformly equicontinuous family of functions from $\langle [0, 1], d \rangle$ to $\langle [0, 1], \text{usual metric} \rangle$ that satisfies conditions (a) and (c) of the Arzela-Ascoli Theorem. Nonetheless it is easy to see that $\mathcal{F}$ is not compact (see Exercise VII C 5). Why doesn't this example contradict the Arzela-Ascoli Theorem?

6. Construct the Cauchy polygon approximate solution to the initial value problem $y(0) = 0$, $y' = y$ on the interval $[0, 1]$ with mesh size $h = 0.1$, and use the value of $a(1)$ to approximate e.

7. Obviously the Cauchy polygon approximate solutions to the initial value problem of Exercise VII C 6 do not have any subsequence that converges to a solution. What part of the proof of Peano's Existence Theorem fails in this case?

8. Let $f: [a, b] \to \mathbf{R}$ be a continuous function. Construct the Cauchy polygon solution on the interval $[a, b]$ with mesh size $(b - a)/n$ to the initial value problem $y(a) = 0$, $y' = f$ and interpret the result as a Riemann sum.

Problems for Chapter VIII

A metric space $\langle M, d \rangle$ is called **locally connected** if for every $x \in M$ and every neighborhood U of x, there is a connected neighborhood V of x contained in U. M is **locally arcwise connected** if V can be chosen so as to be arcwise connected.

1. Prove that every open subset of a normed space is locally arcwise connected

(compare Theorem VIII A 14). Conclude that a locally arcwise connected metric space need not be connected.

2. Show that every locally arcwise connected metric space is locally connected.
3. Show that every connected, locally arcwise connected metric space is arcwise connected. *Hint:* Look at the proof of Theorem VIII A 14.
4. Show that the connected metric space described in Exercise VIII A 2 is not locally connected. *Hint:* Look at a neighborhood of $(0, 0)$.

The subsets S and T of a metric space are **separated** if $(S^c \cap T) \cup (S \cap T^c) = \varnothing$.

5. Show that if S and T are nonempty separated subsets of a metric space M, then $S \cup T$ is a disconnected subset of M.
6. Show that if S and T are separated subsets of M and $S \neq \varnothing$, then $x \mapsto d(x, S)$ is a function on M that takes the value 0 on S and positive values on T.
7. If S is any nonempty subset of the metric space $\langle M, d \rangle$, show that the function $x \mapsto d(x, S)$ satisfies the Lipschitz condition $|d(x, S) - d(y, S)| \leq d(x, y)$, so that it is a continuous function. *Hint:* The proof of this fact is the essence of the proof of the first Lemma following Remark VIII A 2(b).
8. Show that the intersection of any family of intervals is an interval.
9. Show that if $I_1 \subseteq I_2 \subseteq I_3 \subseteq \cdots$ is an increasing sequence of intervals, then $\bigcup_{n=1}^{\infty} I_n$ is an interval.
10. Show that if $f: M \to N$ is continuous and S and T are separated subsets of N, then $f^{-1}(S)$ and $f^{-1}(T)$ are separated subsets of M.
11. Is the interior of a connected subset of a metric space necessarily connected?
12. Strengthen Theorem VIII A 8 as follows.

Theorem. Let $\mathfrak{F}$ be a family of connected subsets of $\langle M, d \rangle$ such that every pair of sets in $\mathfrak{F}$ have a point in common. Then $\bigcup \mathfrak{F}$ is connected.

Problems for Section IX A

1. The two ways of realizing the polygon I in Figure 31, J as a union of rectangles lead to the same value for Area (I). Express this fact as a statement about integrals of step functions, and prove it.
2. Why must the hypothesis of Corollary IX A 4 require that $T(0, \ldots, 0) = 0$? Where is this fact used in the proof of the Corollary?
3. Show that for any $x \in \mathbf{R}$, $|x| = \max(x, -x)$.
4. Prove that $\min(a, b) = \frac{1}{2}(a + b - |a - b|)$, and $\max(a, b) = \frac{1}{2}(a + b + |a - b|)$.
5. Prove the following assertion implicit in the proof of Lemma IX A 11.

If $A \geq 0$, $\epsilon > 0$, and R is a rectangle in $\mathbf{R}^n$, there is a rectangle P in $\mathbf{R}^n$ such that $P^c \subseteq R$ and $A(\lambda(R) - \lambda(P)) < \epsilon$. *Hint:* If $A = 0$, there is nothing to prove. If $A > 0$, put $\delta = \epsilon/A$. Let $R = R(a, b)$. Choose t such that $0 < t < 1$ and $(1 - t^n)\lambda(R) < \delta$. Show how to choose c and d in $\mathbf{R}^n$ such that if $P = R(c, d)$, $P^c \subseteq R$, and $d_i - c_i = t(b_i - a_i)$ for $i = 1, \ldots, n$.
6. Do the hypotheses of Theorem IX A 12 imply that $f_k \to 0$ uniformly as $k \to \infty$? Compare this problem to Problem VII B 8.

Problems for Section IX B

1. Find bounding sequences for each of the following functions from $\mathbf{R}$ to $\mathbf{R}$, and use these sequences to obtain upper bounds for the norms of these functions.
   (a) $f(x) = \sin x$ if $0 \leq x \leq 2\pi$. $f(x) = 0$ if $x \notin [0, 2\pi]$.

(b) $g(x) = (1 + x^2)^{-1}$.

(c) $h(x) = x^{-1/4}$ if $x \in]0, 1]$, $h(x) = 0$ if $x \notin]0, 1]$.

2. Which of the functions in Problem 1 are Riemann-integrable and which are Lebesgue-integrable?

3. Work out the details of the proofs of Corollaries IX B 12, 15, and 16.

4. Construct the **Cantor Set** C as follows. If $[a, b]$ is a closed interval, let

$$M[a, b] = [a, a + \tfrac{1}{3}(b - a)] \cup [b - \tfrac{1}{3}(b - a), b]$$

be the set obtained by removing from $[a, b]$ its middle third. If $I_1 \cup \cdots \cup I_n = I$ is a union of disjoint closed intervals, let $MI = MI_1 \cup \cdots \cup MI_n$. Let $C_0 = [0, 1]$, and let $C_{n+1} = MC_n$ for $n = 0, 1, 2, 3, \cdots$. Let $C = \bigcap_{n=0}^{\infty} C_n$.

(a) Show that $\lambda(MI) = (\tfrac{2}{3})\lambda(I)$ for each disjoint union of closed intervals I.

(b) Conclude that $\lambda(C_n) = (\tfrac{2}{3})^n$, and that $\lambda(C) = 0$.

(c) Show that if x is an end point of one of the 2^n intervals composing C_n, then $x \in C$.

(d) Conclude that if $x \in C$ and $\epsilon > 0$, there is a $y \in C$ such that $y \neq x$ but $|y - x| < \epsilon$.

(e) Using (d), show that no point of C is open in the metric space $\langle C$, usual metric$\rangle$.

(f) Show that $\langle C$, usual metric$\rangle$ is complete. *Hint:* C is compact.

(g) Show that C is uncountable. *Hint:* Use Exercise IV E 2. This shows that there are uncountable sets of measure 0 (see Exercise IX B 2).

The following are a connected series of problems on measurable functions, square-summable functions, and Fourier series.

DEFINITION. A function $f \in \mathfrak{F}(n, V)$ is **measurable** if there is a sequence $k \mapsto s_k$ in Step (n, V) that converges to f λ-almost everywhere. Let $\mathfrak{M}(n, V)$ be the set of measurable functions in $\mathfrak{F}(n, V)$.

5. Show that $\mathfrak{M}(n, V)$ is a vector subspace of $\mathfrak{F}(n, V)$.

6. Show that $\mathfrak{M}(n, V) \cap \mathfrak{N}(n, V) = \mathcal{L}^1(n, V)$ as follows.

(a) Show that $\mathcal{L}^1(n, V) \subseteq \mathfrak{M}(n, V) \cap \mathfrak{N}(n, V)$.

(b) Show that if $v \in V$ and $f \in \mathcal{L}^1(n, \mathbf{R})$, then $f \cdot v : x \mapsto f(x)v$ is in $\mathcal{L}^1(n, V)$. *Hint:* Apply Theorem IX B 9(c) to the map $L : r \mapsto rv$ from $\mathbf{R}$ to V.

(c) Use (b) to prove that if $f \in$ Step (n, V) and $g \in \mathcal{L}^1(n, \mathbf{R})$, then $g \times f \in \mathcal{L}^1(n, V)$.

(d) Show that if $f \in$ Step (n, V), there are functions $f' \in$ Step (n, V) and $f'' \in$ Step $(n, \mathbf{R})$ such that $f = f'' \times f'$, $|f''(x)| = \|f(x)\|$ for each $x \in \mathbf{R}^n$, and $\|f'(x)\| \leq 1$ for each $x \in \mathbf{R}^n$. *Hint:* Let $\mathcal{G}$ be a presentation of f and put $f''(R) = \|f(R)\|$ for each $R \in C(\mathcal{G})$.

(e) Show that if $f \in \mathfrak{N}(n, V)$ then there is an $h \in \mathcal{L}^1(n, \mathbf{R})$ such that $h(x) \geq \|f(x)\|$ for each $x \in \mathbf{R}^n$. *Hint:* Use the definition of $\mathfrak{N}(n, V)$ and Beppo Levi's Theorem.

(f) Show that $\mathfrak{M}(n, V) \cap \mathfrak{N}(n, V) \subseteq \mathcal{L}^1(n, V)$ as follows. If $f \in \mathfrak{M}(n, V) \cap \mathfrak{N}(n, V)$, choose a sequence $k \mapsto s_k$ in Step (n, V) that converges to f λ-almost everywhere and an $h \in \mathcal{L}^1(n, \mathbf{R})$ such that $h(x) \geq \|f(x)\|$ for each $x \in \mathbf{R}^n$. For each k, express s_k as a product $s_k'' \times s_k'$ as in (d). Let $t_k = (h \wedge s_k'') \times s_k'$. Using (c), show that $t_k \in \mathcal{L}^1(n, V)$. Show that $\|t_k(x)\| \leq h(x)$ for all x and that $\lim_{k\to\infty} t_k = f$ λ-almost everywhere, and apply Lebesgue's Dominated Convergence Theorem.

7. Let $F: V^k \to W$ be continuous. Suppose $g_1, \ldots, g_k \in \mathfrak{M}(n, V)$. Show that $F(g_1, \ldots, g_k) \in \mathfrak{M}(n, W)$. *Hint:* First look at $F - F(0, \ldots, 0) = H$, which has the additional property that $H(0, \ldots, 0) = 0$. Show that $H(g_1, \ldots, g_k) \in \mathfrak{M}(n, W)$ using Corollary IX A 4, and that the constant function $x \mapsto F(0, \ldots, 0)$ belongs to $\mathfrak{M}(n, W)$.

8. Deduce from Problem 7 that if f and g are in $\mathfrak{M}(n, \mathbf{R})$, so are $f \times g$, $f + g$, and $f \wedge g$.

9. Show that $f \in \mathfrak{M}(n, V)$ if, and only if, $\chi_R \times f \in \mathfrak{M}(n, V)$ for each rectangle R in $\mathbf{R}^n$. *Hint:* Suppose $\chi_R \times f \in \mathfrak{M}(n, V)$ for each R in $\mathbf{R}^n$. Choose a sequence $k \mapsto R(k)$ of disjoint rectangles such that $\sum_{k=1}^{\infty} \chi_{R(k)} x = 1$ λ-almost everywhere. For each k, choose a sequence $j \mapsto s_{kj}$ of step functions converging λ-almost everywhere to $\chi_{R(k)} \times f$. Then look at the sequence $j \mapsto \sum_{k=1}^{j} \chi_{R(k)} s_{kj}$.

10. Show that if $B \in \mathbf{R}$ and $k \mapsto f_k$ is a sequence in $\mathfrak{M}(n, V)$ such that
    (a) $\|f_k(x)\| \le B$ for all $x \in \mathbf{R}$ and all $k \in \mathbf{N}$,
    (b) $\lim_{k\to\infty} f_k(x) = f(x)$ λ-almost everywhere in $\mathbf{R}^n$, then $f \in \mathfrak{M}(n, V)$. *Hint:* Use Problems 6 and 9 and the Dominated Convergence Theorem.

11. Show that if $k \mapsto f_k$ is a sequence in $\mathfrak{M}(n, V)$ converging λ-almost everywhere to a function $f \in \mathfrak{F}(n, V)$, then $f \in \mathfrak{M}(n, V)$, as follows.
    (a) For $v \in V$, let $G(v) = (1 + \|v\|)^{-1}v$ and for $v \in B_1(0) \subseteq V$, let $H(v) = (1 - \|v\|)^{-1}v$. Show that $G: V \to B_1(0)$ and $H: B_1(0) \to V$ are continuous and that $H = G^{-1}$.
    (b) Using Problems 7 and 10, show that $G \circ f$ is measurable.
    (c) Conclude that $f = H \circ G \circ f$ is measurable.

12. Show that a subset S of $\mathbf{R}^n$ is measurable if, and only if, $\chi_S \in \mathfrak{M}(n, \mathbf{R})$.

13. Let $E_1 \supseteq E_2 \supseteq E_3 \supseteq E_4 \supseteq \cdots$ be a decreasing sequence of measurable subsets of $\mathbf{R}^n$, with $\lambda(E_1) < +\infty$. Show that $E = \bigcap_{k=1}^{\infty} E_k$ is measurable and that $\lim_{k\to\infty} \lambda(E_k) = \lambda(E)$. *Hint:* $E_1 = E \cup \bigcup_{k=1}^{\infty} (E_k - E_{k+1})$.

14. Show that if $f \in \mathfrak{M}(n, \mathbf{R})$ then $E = \{x \in \mathbf{R}^n : f(x) > 0\}$ is measurable. *Hint:* $\chi_E = \lim_{k\to\infty} 1 \wedge k(f \vee 0)$.

15. Show that if $f \in \mathfrak{M}(n, \mathbf{R})$, then, for each $r \in \mathbf{R}$, $E_r = \{x \in \mathbf{R}^n : f(x) > r\}$ is measurable. *Hint:* Apply Problem 14 to the function $f - r$.

16. Conversely, show that if $f \in \mathfrak{F}(n, \mathbf{R})$ and, for each $r \in \mathbf{R}$, $E_r = \{x \in \mathbf{R}^n : f(x) > r\}$ is measurable, then f is measurable. *Hint:* For each $r \in \mathbf{R}$, let χ_r be the characteristic function of E_r. Let $f_k = \sum_{i=-\infty}^{\infty} (j/k)(\chi_{j/k} - \chi_{(i+1)/k})$. Show that $f_k \in \mathfrak{M}(n, \mathbf{R})$ and that $f = \lim_{k\to\infty} f_k$.

DEFINITION. $\mathcal{L}^2(n, V) = \{f \in \mathfrak{M}(n, V) : \|f\|^2 \in \mathcal{L}^1(n, \mathbf{R})\}$.

17. Show that if f and g are in $\mathcal{L}^2(n, \mathbf{R})$, then $f \times g \in \mathcal{L}^1(n, \mathbf{R})$. *Hint:* $2|f \times g| \le f^2 + g^2$. Use Problems 6 and 8.

18. Show that the map $[\, , \,]: \mathcal{L}^2(n, \mathbf{R}) \times \mathcal{L}^2(n, \mathbf{R}) \to \mathbf{R}$ defined by $[f, g] = \int f \times g$ is a **semi-inner product,** that is, for each $f, g, h \in \mathcal{L}^2(n, \mathbf{R})$,
    (a) $[f, f] \ge 0$,
    (b) $[f, f]$ may be 0 when $f \neq 0$ ($[f, f] = 0$ if, and only if, $f = 0$ λ-almost everywhere),

(c) $[f, g] = [g, f]$,

(d) for each $t \in \mathbf{R}$, $[tf, g] = t[f, g]$,

(e) $[f + g, h] = [f, h] + [g, h]$.

DEFINITION. If $f \in \mathcal{L}^2(n, V)$, $|f|_2 = (\int \|f\|^2)^{1/2} = [\|f\|, \|f\|]^{1/2}$.

19. Show that the Cauchy-Schwarz inequality $|[f, g]| \leq |f|_2 |g|_2$, and the triangle inequality, $|f + g|_2 \leq |f|_2 + |g|_2$, are true in $\mathcal{L}^2(n, \mathbf{R})$. *Hint:* The proofs of these inequalities in Chapter III Section C work just as well without the assumption that $[f, f] = 0 \Rightarrow f = 0$, except that it is necessary to give a separate proof that $[f, g] = 0$ if $|f|_2 = |g|_2 = 0$. This follows from the Hint for Problem 17.

20. Prove that the seminormed space $\langle \mathcal{L}^2(n, \mathbf{R}), |\ |_2 \rangle$ is complete. *Hint:* As in the proof of Theorem IX B 13, it is enough to show that if $f_k \in \mathcal{L}^2(n, \mathbf{R})$ for each $k \in \mathbf{N}$ and if $\sum_{k=1}^{\infty} |f_k|_2$ converges, then there is an $f \in \mathcal{L}^2(n, \mathbf{R})$ such that $\lim_{N \to \infty} |f - \sum_{k=1}^{N} f_k|_2 = 0$. Note that $|f|_2 = |\ |f|\ |_2$ and therefore that the series

$$\sum_{k=1}^{\infty} |f_k| \text{ converges } \lambda\text{-almost everywhere to a function } g \text{ in } \mathcal{L}^2(n, \mathbf{R})$$ (apply Beppo Levi's Theorem to the sequence $N \mapsto (\sum_{k=1}^{N} |f_k|)^2$). It follows that $\sum_{k=1}^{\infty} f_k$ converges λ-almost everywhere to a function $f \in \mathfrak{M}(n, \mathbf{R})$, that $|f(x)| \leq g(x)$ for λ-almost all $x \in \mathbf{R}^n$, and hence that $f \in \mathcal{L}^2(n, \mathbf{R})$. Finally, $|f(x) - \sum_{k=1}^{N} f_k(x)| \leq 2g(x)$ for λ-almost all $x \in \mathbf{R}^n$, so the Dominated Convergence Theorem can be used to show that $\lim_{N \to \infty} \int |f - \sum_{k=1}^{N} f_k|^2 = 0$.

21. Prove the *Riemann-Lebesgue Lemma*. If $f \in \mathcal{L}^1(1, \mathbf{R})$, then

$$\lim_{\lambda \to \infty} \int_{-\infty}^{\infty} f(x) \cos \lambda x\, dx = \lim_{\lambda \to \infty} \int_{-\infty}^{\infty} f(x) \sin \lambda x\, dx = 0.$$

Hint. Let $\epsilon > 0$, be given. If $f \in \mathcal{L}^1(1, \mathbf{R})$, there is a $g \in \text{Step}\,(1, \mathbf{R})$ such that $|f - g|_1 < \epsilon/2$. Since $|\sin \lambda x| < 1$ for all $x \in \mathbf{R}$, it follows that $|f \sin \lambda x - g \sin \lambda x|_1 < \epsilon/2$, so that $|\int f \sin \lambda x - \int g \sin \lambda x| < \epsilon/2$, for all $\lambda \in \mathbf{R}$. It remains to show that $\lim_{\lambda \to \infty} g \sin \lambda x = 0$. Since g is a sum of characteristic functions of intervals, it is enough to consider the case that $g = \chi_{[a,b]}$. But the proof that $\lim_{\lambda \to \infty} \int_a^b \sin \lambda x\, dx = 0$ results from a trivial computation.

22. Prove that if $f \in \mathcal{L}^1(1, \mathbf{R})$, $\lim_{\lambda \to \infty} \int_{-\infty}^{\infty} f(x) \sin^2 \lambda x\, dx = \lim_{\lambda \to \infty} \int_{-\infty}^{\infty} f(x) \cos^2 \lambda x\, dx = \frac{1}{2} \int f$. *Hint:* Show, using the Riemann-Lebesgue Lemma, that

$$\lim_{\lambda \to \infty} \int_{-\infty}^{\infty} f(x)(\cos^2 \lambda x - \sin^2 \lambda x)\, dx = 0.$$

DEFINITIONS. Define a sequence $n \mapsto b_n$ of functions from $[0, 2\pi]$ to $\mathbf{R}$ as follows: $b_0: x \mapsto 1/\sqrt{2\pi}$, $b_{2n}: x \mapsto (1/\sqrt{\pi}) \cos nx$, $b_{2n-1}: x \mapsto (1/\sqrt{\pi}) \sin nx$ for $n \in \mathbf{N}$. For f, g in $\mathcal{L}^2([0, 2\pi], \mathbf{R}) = \{h \in \mathcal{F}([0, 2\pi], \mathbf{R}): h^* \in \mathcal{L}^2(1, \mathbf{R})\}$, put $[f, g] = \int f \times g = \int f^* \times g^* = [f^*, g^*]$, and $|f|_2 = |f^*|_2$. The **nth Fourier coefficient** $c_n(f)$ of a function $f \in \mathcal{L}^1([0, 2\pi], \mathbf{R})$ is $c_n(f) = [f, b_n] = \int f b_n$. The **Fourier series** for f, whether or not it converges, is $\sum_{n=0}^{\infty} c_n(f) b_n$. The following problems show that in appropriate circumstances the Fourier series for f converges to f.

23. Show that if $f \in \mathcal{L}^2([0, 2\pi], \mathbf{R})$ then $f \in \mathcal{L}^1([0, 2\pi], \mathbf{R})$. *Hint:* Apply Problem 17 with $g = \chi_{[0,2\pi]}$.

24. Show that if $n \neq m$, $[b_n, b_m] = 0$ but $[b_n, b_n] = 1$. Conclude that $\{b_n\}$ is an orthonormal set of vectors in $\mathcal{L}^2([0, 2\pi], \mathbf{R})$. (See Exercise III C 2.)

25. Suppose the series $\sum_{n=0}^{\infty} a_n b_n$ $(a_n \in \mathbf{R})$ converges in $\langle \mathcal{L}^2([0, 2\pi], \mathbf{R}), |\ \ |_2 \rangle$ to a function f. Show that $a_n = c_n(f)$ for each n. *Hint:* Show that $[f, b_n] = \sum_{m=0}^{\infty} a_m[b_m, b_n]$ and use Problem 24.

26. Suppose $f \in \mathcal{L}^1([0, 2\pi], \mathbf{R})$ and that f is *differentiable* at $y \in [0, 2\pi]$. Show that $\sum_{n=0}^{\infty} c_n(f)b_n(y) = f(y)$ as follows.

   (a) Show that

$$\sum_{n=0}^{2K} c_n(f)b_n(y) = \frac{1}{\pi}\int_0^{2\pi} f(x)\,(\tfrac{1}{2} + \sum_{n=1}^{K} \cos n(x - y))\,dx$$

$$= \frac{1}{\pi}\int_0^{2\pi} f(x)[\sin((K + \tfrac{1}{2})(x - y))/(2\sin \tfrac{1}{2}(x - y))]\,dx.$$

   Hint: Use Exercise V A 3.

   (b) Prove the theorem in case f is a constant function.

   (c) Show that the function $x \mapsto (f(x) - f(y))/2\sin \tfrac{1}{2}(x - y)$ is an element of $\mathcal{L}^1([0, 2\pi], \mathbf{R})$. *Hint:* By L'Hôpital's rule this function is bounded in a neighborhood of y.

   (d) Using (c) and the Riemann-Lebesgue Lemma (Problem 21), show that $\sum_{n=0}^{\infty} c_n(f - f(y))b_n(y) = 0$.

   (e) Using (d) and (b), complete the proof.

27. Prove that for each $f \in \mathcal{L}^2([0, 2\pi], \mathbf{R})$ and every $\epsilon > 0$, there is a step function s on $[0, 2\pi]$ such that $|f - s|_2 < \epsilon$, as follows.

   (a) Show that f is the limit in $\langle \mathcal{L}^2([0, 2\pi], \mathbf{R}), |\ \ |_2 \rangle$ of a sequence of bounded measurable functions. *Hint:* Look at the sequence $n \mapsto n \wedge (f \vee (-n))$ and use the Dominated Convergence Theorem.

   (b) Show that if g is a bounded measurable function on $[0, 2\pi]$, there is a sequence $n \mapsto s_n$ of step functions on $[0, 2\pi]$ that converges to g in $\langle \mathcal{L}^2([0, 2\pi], \mathbf{R}), |\ \ |_2 \rangle$. *Hint:* Because g is measurable, there is a sequence $n \mapsto t_n$ of step functions converging λ-almost everywhere to g^*. Let χ be the characteristic function of $[0, 2\pi]$ and let B be a bound for g. Let $s_n = B \wedge (\chi t_n \vee (-B))$ and use the Dominated Convergence Theorem.

   (c) Use (a) and (b) to complete the proof.

28. Prove *Bessel's Inequality.* If $f \in \mathcal{L}^2([0, 2\pi], \mathbf{R})$, then for each $N \in \mathbf{N}$, $\sum_{n=0}^{N} [c_n(f)]^2 \le |f|_2$. *Hint:* Let $g = \sum_{n=0}^{N} c_n(f)b_n$. Using Problem 24, show that $(|g|_2)^2 = \sum_{n=0}^{N} [c_n(f)]^2$, that $[g, f - g] = 0$, and that $(|g|_2)^2 + (|f - g|_2)^2 = (|f|_2)^2$.

29. Prove that if $f \in \mathcal{L}^2([0, 2\pi], \mathbf{R})$, the Fourier series for f converges to f in the seminormed space $\langle \mathcal{L}^2([0, 2], \mathbf{R}), |\ \ |_2 \rangle$, as follows.

   (a) Using Problems 20 and 28, show that the Fourier series for f converges to some element g in $\mathcal{L}^2([0, 2\pi], \mathbf{R})$. *Hint:* Note that $(|\sum_{n=M}^{N} c_n(f)b_n|_2)^2 = \sum_{n=M}^{N} [c_n(f)]^2$.

   (b) Show that if f is a step function, then $g = f$ λ-almost everywhere. *Hint:* Use Problem 26.

   (c) Show that in any case, $[f - g, b_n] = 0$ for each n, thus reducing the problem to that of showing that if $h \in \mathcal{L}^2([0, 2\pi], \mathbf{R})$ and $[h, b_n] = 0$ for every n, then $|h|_2 = 0$.

   (d) Complete the proof. *Hint:* Suppose $|h|_2 > 0$ but $[h, b_n] = 0$ for every n. Using Problem 27, choose a step function s such that $|h - s|_2 < \tfrac{1}{2}|h|_2$.

Note that $c_n(s) = c_n(s - h)$, so that $\sum_{n=0}^{\infty} [c_n(s)]^2 < \frac{1}{4}[| h |_2]^2 < [| s |_2]^2$. Show that this would contradict (b).

Remark. This problem shows that the functions b_n behave like an orthonormal basis for the infinite-dimensional semi-inner product space $\mathcal{L}^2([0, 2\pi], \mathbf{R})$, with convergent infinite series replacing the finite sums of Exercise III C 2.

30. Let f be a differentiable function from $\mathbf{R}$ to $\mathbf{R}$. Show that f' is measurable. *Hint:* Use Problem 11.

Problems for Section IX C

1. Suppose u and v are continuously differentiable on $[a, b]$. Prove the formula for **integration by parts:** $u(b)\, v(b) - u(a)\, v(a) - \int_a^b vu' = \int_a^b uv'$. *Hint:* Use Corollary IX C 5.

2. Suppose f is continuous on $[a, b]$ and g maps $[c, d]$ differentiably into $[a, b]$. Show that $(d/dx) \int_a^{g(x)} f = f(g(x))g'(x)$ for each $x \in [c, d]$. *Hint:* Use the Chain Rule.

3. Suppose g is continuously differentiable on $[a, b]$ and f is continuously differentiable on $g([a, b])$. Prove the formula for **change of variable in a definite integral:** $\int_a^b (f \circ g)g' = \int_{g(a)}^{g(b)} f$. *Hint:* Use Problem 2.

4. Calculate the following.

(a) $\displaystyle\int_{-\infty}^{\infty} \frac{dx}{1 + x^2}$

(b) $\displaystyle\int_{-\infty}^{\infty} 2|x|e^{-x^2}\, dx.$

5. (a) Show that $(d/dx) \displaystyle\int_0^{\infty} (\sin y/y)e^{-xy}\, dy = -\int_0^{\infty} e^{-xy} \sin y\, dy$

$$= -\operatorname{Im} \int_0^{\infty} e^{-xy}e^{iy}\, dy = \frac{-1}{1 + x^2}.$$

(b) Use (a) to show that $\displaystyle\int_0^{\infty} (\sin y/y)e^{-xy}\, dy = \pi/2 - \text{arc}\tan x.$

6. Show that $f(y) = \displaystyle\int_{-\infty}^{\infty} e^{-x^2/2}e^{ixy}\, dx$ satisfies the differential equation $f'(y) = -yf(y)$. *Hint:* After differentiating under the integral sign, integrate by parts.

7. Let $f(x) = \sum_{n=0}^{\infty} a_n x^n$ be a power series with radius of convergence $R > 0$. Show that if $|x| < R$, $f'(x) = \sum_{n=1}^{\infty} na_n x^{n-1}$.

8. Let $f\colon (x, y) \mapsto |xy|e^{-x^2-y^2}$. Calculate $\int f$.

9. Prove *Tonelli's Theorem:* Let f be a measurable function from $\mathbf{R}^{n+m}$ to $\mathbf{R}$ that assumes only nonnegative values. Suppose that for λ-almost all $x \in \mathbf{R}^n$, $f(x, \cdot) \in \mathcal{L}^1(\mathbf{R}^m, \mathbf{R})$, and that $g\colon x \mapsto \int f(x, \cdot)$ is an element of $\mathcal{L}^1(\mathbf{R}^n, \mathbf{R})$. Then $f \in \mathcal{L}^1(\mathbf{R}^{n+m}, \mathbf{R})$. *Hint:* First show that Tonelli's Theorem is true if f is *bounded* and there is a rectangle R in $\mathbf{R}^{n+m}$ such that f vanishes on $\mathbf{R}^{n+m} - R$. Then by using Beppo Levi's Theorem, pass to the case where f is merely bounded. Finally, noting that $f = \lim_{n \to \infty} n \wedge f$, pass to the general case. *Remark:* Fubini's Theorem assumes that $f \in \mathcal{L}^1(\mathbf{R}^{n+m}, \mathbf{R})$ and then proves that $f(x, \cdot) \in \mathcal{L}^1(\mathbf{R}^m, \mathbf{R})$ and that $x \mapsto \int f(x, \cdot)$ is a function in $\mathcal{L}^1(\mathbf{R}^n, \mathbf{R})$. Tonelli's Theorem states the *converse*, on the additional assumptions that f is *nonnegative* and *measurable*. See Problem 6 for Section IX B.

Problems for Section IX D

1. Fill in the details of the proofs of Propositions IX D 1 and 3.

2. Calculate $\int_0^{2\pi} [\exp (e^{i\theta}) e^{i\theta} / (e^{i\theta} - \frac{1}{2})] \, d\theta$ ($\exp (z) = e^z$).

3. Suppose f is analytic on all of $\mathbf{C}$ except that f has a pole at each of the points $n \pm i$, $n \in \mathbf{N}$. What is the largest possible radius for a disk centered at a point on the real axis on which f has a convergent power series expansion?

4. An analytic function f has a **zero of order n** at the point w if f and its first $n - 1$ derivatives vanish at w, but $f^{(n)}(w) \neq 0$.
   (a) Show that if f has a zero of order n at w, then in some neighborhood U of w, $f(z) = (z - w)^n g(z)$, where g is a nonvanishing analytic function on U.
   (b) Show that if f has a zero of order n at w, then in some neighborhood U of w, $(f'(z)/f(z)) = (n/(z - w)) + h(z)$, where h is analytic on U.

5. Suppose f is analytic on a neighborhood of $B_R^c(a)$ and that f does not vanish on $\partial B_R(a)$. Show that there are just finitely many points $w_1, \ldots, w_n$ in $B_R(a)$ at which f vanishes, and that if f has a zero of order K_j at w_j for $j = 1, \ldots, n$, then

$$K_1 + K_2 + \cdots + K_n = \frac{1}{2\pi} \int_0^{2\pi} \frac{f'(a + Re^{i\theta}) \, Re^{i\theta} \, d\theta}{f(a + Re^{i\theta})}.$$

   Hint: Make n applications of Problem 4(b). This is called the *Argument Principle*.

6. Suppose f is analytic and nonconstant on $B_R(z_0)$ and that $f(z_0) = 0$. Show that there is an $\epsilon > 0$ such that if $w < \epsilon$, then there is a $z \in B_R(z_0)$ such that $f(z) - |w| = 0$. *Hint:* Use Problem 4 to find a $\delta < R$ such that f never vanishes on $B_\delta^c(z_0) - \{z_0\}$. Then observe that the function

$$w \mapsto \frac{1}{2\pi} \int_0^{2\pi} \frac{f'(z_0 + \delta e^{i\theta}) \delta e^{i\theta} \, d\theta}{f(z_0 + \delta e^{i\theta}) - w}$$

   is continuous, and apply Problem 5.

7. Prove the *Open Mapping Principle*. Let U be a connected open subset of $\mathbf{C}$ and $f: U \to \mathbf{C}$ be a nonconstant analytic function. If V is an open subset of U, then $f(V)$ is also open. *Hint:* Apply Problem 6.

8. Carry through as much of Chapter IX Section C as you can for functions that take their values in a complex Banach space. In particular, prove Theorems IX C 6, 7, 8, 9, 10, and 11 in this setting.

9. Prove the *Maximum Principle*. If U is a connected open subset of $\mathbf{C}$, $f: U \to \mathbf{C}$ is an analytic function, and $|f(z)|$ attains its maximum at some point z_0 in U, then f is a constant function. *Hint:* Use the Open Mapping Principle.

10. Let $f: \mathbf{C} \to \mathbf{C}$ be an analytic function and suppose that $|f(z)| \to \infty$ as $|z| \to \infty$. Prove that for some $z \in \mathbf{C}$, $f(z) = 0$.

11. Prove that no analytic function can map $\mathbf{C}$ onto $\{z \in \mathbf{C}: |z| < 1\}$.

12. Suppose that $0 < R < \pi$. Prove the integral formula

$$f(z) = \frac{e^z}{2\pi} \int_0^{2\pi} \frac{f(a + Re^{i\theta}) \, Re^{i\theta} \, d\theta}{\exp (a + Re^{i\theta}) - e^z}$$

   if $|z - a| < R$. *Hint:* The function $w \mapsto f(w)(w - z)/(e^w - e^z)$ has a removable singularity at z.

Symbols and notation

**Symbols
and
Notation**   **Explanation**   **Page**

$E_k(\mathcal{G})$ kth set of end points of the grid $\mathcal{G}$ 181

f^* extension of f 195

$\mathcal{F}(S, V)$ space of all V-valued functions on S 181

$f(\cdot, y), f(x, \cdot)$ partial functions   202

i imaginary unit, square root of -1 101, 102

I identity map   64

$\mathcal{L}^1(n, V)$ space of integrable functions from
                                    $\mathbf{R}^n$ to V 189

$\mathcal{L}^2(n, V)$ space of square-summable func-
                                    tions from $\mathbf{R}^n$ to V 258–261

$L^1(n, V)$ space of equivalence classes of
                                    λ-almost equal summable func-
                                    tions from $\mathbf{R}^n$ to V 198

$\mathcal{L}^1(S, V)$ space of integrable functions from
                                    S to V $(S \subseteq \mathbf{R}^n)$ 195

$\mathcal{L}(V, W)$ space of linear maps from V to W 61

$\mathfrak{M}(n, V)$ space of measurable functions from
                                    $\mathbf{R}^n$ to V 257–261

$\mathbf{N}$ set of positive integers   13

$\mathfrak{N}_R(n, V)$ space of R-normable function from
                                    $\mathbf{R}^n$ to V 197, 240–242

$\mathfrak{N}(n, V)$ space of normable functions from
                                    $\mathbf{R}^n$ to V 188

$\mathbf{Q}$ set of rational numbers   15

$\mathbf{R}$ set of real numbers   28

$\mathbf{R}^+$ set of positive real numbers   28, 29

$\mathbf{R}^0$ set of nonzero real numbers   33

$R(a, b)$ open rectangle determined by a
                                    and b 181

$R(\mathcal{G})$ rectangle spanned by the grid $\mathcal{G}$ 181

$R(n, V)$ space of Riemann-integrable func-
                                    tions from $\mathbf{R}^n$ to V 198, 241–242

$\mathrm{Step}(n, V)$ space of step functions from $\mathbf{R}^n$ to V 182

$\mathbf{Z}$ set of integers   7

$\mathbf{Z}_2$ set of integers modulo 2   30

$\in, \notin$ member of, not a member of   5

**Symbols
and
Notation** | **Explanation** | **Page**

Symbol	Explanation	Page
λ	volume function on $\mathbf{R}^n$	181
λ^*	Lebesgue outer measure	197
$\pi(S)$	set of permutations of S	34, 70, 248
$\sum$	summation, $\sum_{x \in E} f(x) = \sum \{f(x) : x \in E\}$ (see pages 110 and 182)	109–113
∇	gradient	142
$\bullet$	stop sign	vi, 22
$\int$	integral	184, 188–189
$\int_a^b$	integral from a to b	199
$\wedge$	minimum	183
$\vee$	maximum	183
$\bigcup, \cup, \bigcup_{n=1}^{\infty}$	union	6, 7
$\bigcap, \cap, \bigcap_{n=1}^{\infty}$	intersection	6, 7
$\subseteq$	inclusion	6
☛	material of special importance	22
$\Rightarrow$	if . . . , then	22
$\Leftrightarrow$	if, and only if	17, 22
$\rightarrow, \mapsto$	function	11, 12
$^{-1}$	inverse	12, 29
$\times$	product	14, 28, 93, 102
$\oplus, \otimes$	operations modulo 2	30
$\circ$	composition of functions	12
$\varnothing$	empty set	6
$\langle \ , \ \rangle$	ordered pair	8
$\| \ \|$	absolute value	82
$[\ , \ ]$	inner product	54, 258
$[a, b], [a, b[,]a, b[$, etc.	intervals	174
$\{ \ \}, \{ \ : \ \}$	set	6, 7
$\| \ \|$	norm	54, 82
$\|\| \ \|\|$	norm	190
$\| \ \|_1$	ℓ_1-norm	83

Symbols and Notation	Explanation	Page
$\| \ \|_2$	ℓ_2-norm, inner product norm or Euclidean norm	83
$\| \ \|_\infty$	maximum norm	83
$\| \ \|$	norm	161
$\| \ \|_1$	$\mathcal{L}^1$-norm	83, 188
$\| \ \|_2$	$\mathcal{L}^2$-norm	259
$\| \ \|_\infty$	uniform norm	83
$\| \ \|_R$	Riemann norm	197

Index